In the United States District Court,

NORTHERN DISTRICT OF CALIFORNIA.

THE UNITED STATES, *vs.* ANDRES CASTILLERO. — ON CROSS APPEAL.

Claim for the Mine and Lands of New Almaden

ARGUMENT

OF

HON. J. P. BENJAMIN

DELIVERED ON THE

24th, 25th and 26th October, and 5th November, 1860,

IN REPLY TO THE GOVERNMENT'S SPECIAL COUNSEL.

REPORTED BY SUMNER & CUTTER.

SAN FRANCISCO:
COMMERCIAL STEAM BOOK AND JOB PRINTING ESTABLISHMENT.
1860.

In the United States District Court,

NORTHERN DISTRICT OF CALIFORNIA.

THE UNITED STATES,
vs.
ANDRES CASTILLERO.
ON CROSS APPEAL.

Claim for the Mine and Lands of New Almaden

ARGUMENT

OF

HON. J. P. BENJAMIN

DELIVERED ON THE

24th, 25th and 26th October, and 5th November, 1860,

IN REPLY TO THE GOVERNMENT'S SPECIAL COUNSEL.

REPORTED BY SUMNER & CUTTER.

SAN FRANCISCO:
COMMERCIAL STEAM BOOK AND JOB PRINTING ESTABLISHMENT.
1860.

NOTE.

The following Argument is printed from the notes of Messrs. Sumner & Cutter, who reported it. It is due to Mr. Benjamin to say that he had no opportunity to revise it before leaving California.

ARGUMENT.

WEDNESDAY, October 24th, 1860.

MR. BENJAMIN said:—

May it please your Honors: It seems to me convenient in opening this case, to ask your Honors to recur in memory to the condition of the subject-matter of this litigation at the date of the breaking out of hostilities between the United States and Mexico in the year 1846. I desire first to call the attention of the Court to so much of the testimony contained in the Transcript of this case, as will satisfy the Court in relation to the knowledge that had been acquired by the officers of the Government of the United States in relation to the fact of the possession and ownership of this mine prior to the breaking out of the war.

On the 13th of May, 1846, the Congress of the United States declared by resolution, and the President of the United States announced by proclamation to the people of the United States, that war existed between the United States and Mexico, by virtue of the commencement of actual hostilities on the part of the Republic of Mexico.

On the 7th of July, 1846, according to the established jurisprudence of the Supreme Court of the United States, recognized also by Congress, the conquest of California was effected by proclamation made of that conquest by the forces of the United States.

On the 2d of February, 1848, the war was closed by the Treaty of Peace; and by that treaty California, previously a conquered province, became a ceded territory—part of the domain of the United States.

Prior to the breaking out of the war, the Government of the United States had public officers in California, whose duty it was to report to the Departments at Washington the condition of things as they existed in this (then) Mexican Province. I propose to call the attention of the Court to what was known by the Gov.ernment of the United States, through its officers, in relation to the possession and ownership of this mine early in the year 1846, and prior to the breaking out of any hostilities whatever.

On page 2678 of this record, we have the testimony of a man of the highest distinction, whose name is known not only from one end of our own Republic to the other, but to every civilized nation on earth. I refer to John C. Fremont. He had crossed the continent under the orders of his Government. He came hither for the purpose of ascertaining the most direct route across the continent. He came for the purpose of examining the country to which the jurisdiction of the United States then extended, and of ascertaining its adaptation to the purposes of habitation, occupation, cultivation, and its general value to the country.

In January, 1846, in pursuit of information to be communicated to his Government in relation to everything that could interest his country, Capt. Fremont visited the New Almaden Mine; and there he found it in the possession of Andres Castillero, as owner. He found Castillero working the mine. He saw the piles of ore. He became impressed with the great value of the mine. He, and a gentleman who accompanied him, each spoke to Castillero in relation to the purchase of the mine, and the propositions of both were by Castillero coldly received. I state this merely *en passant*, in answer to a remark which fell from the lips of the special counsel for the Government the other day: that Andres Castillero had made a false statement to the Government of Mexico, when he returned and said to the Junta that foreign houses wanted an interest in this mine, but that he had reserved all its benefits for his own country.

Fremont in his examination says:

I think no one went with me to the mine but Hinckley; arrived there, I should think, about noon; at the mine Capt. Hinckley introduced me to Mr. Castillero, the owner of the mine, who showed me about; showed me the excavation from whence he had the ore, and showed me two or three heaps of ore, and gave me some specimens, some of which I brought away. Before visiting the mine, Captain Leidesdorff and myself had had some conversation together with regard to purchasing the mine. When there, I spoke slightly with Castillero on the subject, and Mr. Hinckley also said something to him, at greater length, tending to the same end; but Castillero was not at all disposed to converse about selling. At this time, I think Castillero was engaged in building a house below in the valley, to be used for the occupation of himself or workmen.

He also went through the process roughly of extracting the quicksilver from the ore, by putting some on red-hot iron and collecting the fumes in a cup. We remained there perhaps some two hours.

This is not the whole of the statement of the visit of Capt. Fremont.

Captain Fremont on that visit acquired information which he subsequently put to practical purposes. He had a conversation with Castillero about the means by which mines were acquired. Castillero told him that he had acquired his mine by denouncement. He told him what a denouncement was; how it was made; and gave Fremont the information upon which that American officer, years later, denounced the mines upon his own property in Mariposa.

I know not whether this testimony is to be contradicted or not. I have listened to the arguments upon the other side of this case until sometimes I was tempted to doubt my own existence by reason of the persistent denial of facts that appear to me plain as the noon-day sun. We have not been favored by any comment by brother Randolph upon this testimony of Captain Fremont; and I am not aware whether his testimony is to be attacked or his character impeached—I hope neither.

Mr. Randolph, (interrupting.) No, sir. Col. Fremont is my personal friend.

Mr. Benjamin.—He is your personal friend. Very well, then. May it please the Court, we have got some common starting point now. We have got one witness, whom we can all believe. Captain Fremont is to be believed! That was the condition of the mine then.

But these things rest not upon the frail tenure of human memory. Captain Fremont was not the only officer of the United States that knew of the existence of this mine. There was another active, intelligent, competent public officer, whose duty it was, under instructions of the Government, to examine into and report upon these things.

Thomas O. Larkin, United States Consul, did examine, did make a report; and that report is spread upon this Transcript, as taken from the archives of the Government of the United States in the City of Washington.

Your Honors will find on page 2657, extracts from the correspondence in the Department of State, certified by Lewis Cass. This certificate bears date 28th November, 1859. The documents there presented are mere extracts from letters which your Honors will find at length on page 2669, and following. I read from a letter addressed by Thomas O. Larkin to James Buchanan, Secretary of State:

CONSULATE OF THE UNITED STATES OF AMERICA,
Monterey, California, May 4th, 1846.

SIR: The undersigned has the honor to forward to the Department, the following information respecting the mines of California, most of them discovered within six or nine months; for many years previous to this the inhabitants have supposed the places in question contained metal of some kind. Ninety miles (by sea) south of San Diego, there are some very extensive copper mines, belonging to Don Juan Bandini; the undersigned is informed by D. José Rafael Gonzales, that on his rancho, sixty or eighty miles south of Monterey, there are coal mines; at San Pablo, in the Bay of San Francisco, there are others. At the Mission of San Juan, twenty-five miles north of Monterey, there are sulphur beds, or mines; fifty to eighty miles north of Monterey, there is said to be several silver mines. There are several places throughout California where the people obtain a bituminous pitch to cover the roofs of their houses; some make a floor of it by mixing earth with it. At these places, rabbits, squirrels and birds often get half-buried in the pitch, and some die; even horses and horned cattle are lost there. A few miles north of Santa Barbara, the sea for several miles upon the coast is colored by the pitch oozing from the banks. Five or six miles from the town of San José, and near the Mission of Santa Clara, there are mountains with veins of quicksilver ore, discovered by Don Andres Castillero, of

Mexicò, in 1845, which the undersigned has twice seen produce twenty per cent. of pure quicksilver, by simply putting the pounded rock in an old gun-barrel, one end placed in the fire, the other end in a pot of water for the vapor to fall into, which immediately becomes condensed; the metal was then strained through a silk handkerchief. The red ore produces far better than the yellow. There appears no end to the production of the metal from these mountains. Working of quicksilver is but now commenced under great disadvantages from not having any of the materials generally used in extracting that metal.

So, then, prior to the declaration of war this mine was not only possessed but it was *owned*, as the Government of the United States was informed, by a Mexican. And it was not only so owned, but it was worked, although the working was imperfect. The Consul had twice seen the work. He describes it. The accuracy of the description is unchallenged. The fact is not controverted. It comes from under the very eye of the Attorney General (Black), who hoped that it would be impossible for us to get our proofs in this case.

This is not all that is contained in this letter. The letter tells your Honors what the laws and customs of Mexico were understood to be in California at that time, in relation to the acquiring of mines. Mr. Larkin goes on to say:

By the laws and customs of Mexico, respecting mining, every person or company, foreign or native, can present themselves to the nearest authorities and denounce any unworked mine; the authorities will then, after the proper formalities, put the denouncer in possession of a certain part of it, or all, which is, I believe, according to its extent; the possessor must hereafter occupy and work his mine, or some other may denounce against him; in all cases the Government claims a certain proportion of the products; up to the present time there are few or no persons in California with sufficient energy and capital to carry on mining, although a Mexican officer of the army, a padre and a native of New York are, on a very small scale, extracting quicksilver from the San Josó mine.

Now, who was this "Mexican officer"? Who was this "Padre?" Who was this "native of New York"? The Mexican officer was Andres Castillero. The padre was Padre Real. The native of New York was Chard. And we make our acknowledgments on this, as we shall have occasion to do in several other instances, to the astute and persevering special counsel for the Government, who brought out the fact from Chard on his cross-examination, that he was a native of New York; thus making his testimony correspond precisely with the archives of the United States Consul at Monterey, which at that time we did not have in our power.

Then, may it please your Honors, without reference to another syllable of testimony in this case, I have secured through our Government officers and Government archives, conclusive proof of the fact that so far as the Government of the United States was concerned, it had not merely the constructive notice which the law would imply, but it had actual notice, brought home to it prior to the declaration of war, that this mine was private property, claimed, possessed and worked by private owners.

But this is not all. From all parts of the world the evidence of the facts sought to be controverted in this case rises up before us. From the distant Sandwich Islands comes back the echo, at that date, of this discovery and ownership, and of the means by which the ownership was acquired.

At page 2223, your Honors have an extract from *The Polynesian*, printed in the Sandwich Islands on the 25th of July, 1846.

And what does that contain? A letter written by this same Consul, under date of 24th of June of that year, addressed to the Minister of Finance of that Government, giving him an account of this discovery—of the denouncement and ownership of the mine.

First, from this *Polynesian* is extracted a letter of G. P. Judd, the Minister of Finance, to the editor of *The Polynesian*, inclosing Thomas O. Larkin's letter.

Now, here is Mr. Larkin's letter, dated "24th June, 1846;" it is addressed to G. P. Judd, Esq.:

CONSULATE OF THE UNITED STATES OF AMERICA,
Port of Monterey, California, July 2, 1846.

SIR:—I have the pleasure of forwarding to you a specimen of California quicksilver ore, from a mine seventy miles north of Monterey, and ten miles from the Pueblo of San José, discovered in 1845 to have quicksilver in it. The place was known for eighteen years, and supposed by the Californians to be a silver mine; they, in 1828, having with some foreign quicksilver extracted the other metal. In 1845, a Mexican being in the vicinity, heard that the mountains contained rock different from any other, went to examine it, and immediately denounced the place before the nearest Alcalde, and then made known what it contained.

The owner, with a priest, in a small and imperfect manner has commenced extracting the metal. The mine is in the top of a steep mountain, a mile or more from the plain, to which it is brought down on a mule, piled up with a whaler's pot covered over it, well cemented with clay, some six or eight cords of fire-wood placed over and fired; in fourteen or sixteen hours the quicksilver is found below in a small wooden tank of water, though much of the rock is thrown away afterwards that has not been well heated. They obtain about fifteen per cent. of the metal.

The receipt of that letter in the Sandwich Islands is known by the party who edited and published this paper (*The Polynesian*). His testimony is unimpeached. The fact of that announcement appearing on that day in that paper is proved by a gentleman of San Francisco, who read it, and had his attention particularly attracted to it. The fact of the receipt of the letter by Mr. Judd is apparent on the face of this Transcript, by the production made by young Larkin from his father's papers, of Judd's reply—having Judd's sign-manual affixed. And that reply is proven.

Mr. Randolph—Of what date is that?

Mr. Benjamin—This is dated on the 20th of July, 1846. It is produced by young Larkin from among the papers found after his father's death, and the signature of Mr. Judd is proven by those intimate with him—men who had lived with him.

This letter gives the name of the vessel which brought Mr. Larkin's letter:

Treasury Office, Honolulu,
Oahu, 20th July, 1846.

Sir: I have the honor to acknowledge receipt of your letter per bark "Angola," of the 24th ulto., together with a specimen of California quicksilver ore.

I thank you sincerely for your kind attention in sending me the specimen, as well as the very interesting particulars relative to the mine.

I am happy in thinking that this is but the beginning of discoveries which will tend to make California a valuable country.

My onerous duties as His Majesty's Minister of Finance will not permit me to assay the ore in order to test its quality, as I might do if still in practice as a medical man.

I have sent your letter to the editor of the "Polynesian" for insertion, and have no doubt it will be found as interesting to others as it has been to me.

I beg to assure you that I shall always be happy to receive information on any subject of interest connected with California, whenever it may suit your convenience to favor me.

I am, sir, your obedient servant,

G. P. JUDD.

Thomas O. Larkin, Esq.,
U. S. Consul, &c., &c., &c., Monterey.

[Endorsed.]
G. P. Judd, July 20th, 1846, S. Islands.

Now, may it please the Court, in the year 1860 we are told that this mine never was denounced; that all our proofs to the effect that the mine ever was denounced are falsehoods; that the papers of the denouncement are forgeries; that the witnesses who prove the denouncement and the papers are perjurers.

And this comes from the Government that holds in its own archives the proof to the contrary. And this comes from the Attorney General of the United States who, when implored, besought in every possible manner by the claimants to send and satisfy himself by inspection of the records in this country and Mexico—to take his own commissioners and send them out at our expense, and satisfy himself directly of the justice and equity of our claims—utterly refused to do so, and contented himself with denouncing them as forgeries.

He answers to all our appeals of this nature: "I will not satisfy myself; I will not look; I will look at nothing. I say it is forgery, and I will not assist in procuring any evidence which may tend to prove the contrary."

Fool! Fool!

"Truth crushed to earth will rise again."

All the efforts of men cannot strangle her. Strangle her in San Francisco, she reappears in the Sandwich Islands. Choke her there, and, lo! she is under your very eyes in the archives at Washington. She permeates the air; she is in the atmosphere you breathe. As well might man endeavor to destroy the Almighty as to destroy his essence—the truth. And yet this is what the United States is trying to do; seeking, as the assistant of a band of unscrupulous speculators, to acquire a valuable property developed by the capital and labor of others; undertaking to say that that does not belong to us which they were thoroughly and officially informed did belong to us before the United States Government had an interest in one foot of ground in California.

Now, may it please your Honors, having shown you that this mine was private property to the knowledge of our Government before the declaration of war, I proceed to the examination of some of the testimony in this case that puts that fact beyond the possibility of human controversy. I do not mean to say that it cannot be denied. I do not mean to say that no human being can assert the contrary; because, with all due respect to my brother Randolph, I must say to him, that his power of assertion exceeds anything that I ever imagined could exist in a human being; and having now seen and confessed that the contrary can be *asserted*, I venture to say that it cannot be *proved* to the satisfaction of any intelligent, disinterested mind.

I will proceed now to take up what is disclosed in this record in relation to the action of Andres Castillero, from the time he found his mine to the date of his departure from California—to which he has never returned.

He finds his mine in November, 1845; leaves California in April, 1846, and has never returned.

Now, the first point to which I wish to call your Honor's attention is this: Years and years after the title to this property claimed by us had been presented to the inspection of the United States Land Commissioners, with such evidence of title as was sufficient to satisfy them *prima facie* of our right to the mine, not one syllable was breathed against or in question of this denouncement, or of the truth of these documents. It was long subsequent to that period that a controversy arose on the allegation made that these papers were forged, antedated and fraudulent. Serious charges, requiring prompt repulse.

For this purpose, search was directed to be made in the archives of Mexico, and there was found what we had never seen before; that is to say, that prior to the date of the grant made by the Mexican Government to us, there had been a short correspondence between the President of the Republic and the Governor of California—a thing never dreamed of before. That correspondence was brought here. And now, if your Honors will refer to page 1805, you will see what it was.

On the 13th of February, 1846, the Governor of California communicated to the President of Mexico, through the Minister of Relations, the fact of the existence of this mine, of its denouncement, and of the joy experienced by the public officers in California by reason of so happy a discovery. And the announcement was made by inclosing to the President of Mexico the letter which the Governor of California had received from Andres Castillero, making the announcement to him. This letter of Andres Castillero, which I will first read to the Court, is dated on the 10th of December, 1845, and will be found at the foot of the page which I have just given [1805]:

To His Excellency Governor Pio Pico:

MISSION OF SANTA CLARA,
10th December, 1845.

My Most Esteemed Friend:—I send you a sample of the quicksilver I have taken out in the presence of many witnesses. The mine has been denounced by me, and between a few we have formed a company. I am sure that yourself and the Departmental Junta will appreciate a discovery which will form the riches of the country, and we wish that the vacant lands near to our works be conceded to us, to cut wood, with the order of possession.

There is such an abundance of quicksilver, that eight arrobas of ore gave one of metal; and, in my belief, there is much, and a great extraction can be made.

I will also thank you to order that possession be given to me of the Island of Santa Cruz, which was granted to me by the Supreme, at the same time as to the Messrs. Carrillos; the cattle which is to be shipped, is already bought, and Don Anastacio Carrillo can give me the possession.

May you continue in good health, and order your wishes to your friend, who esteems you, and regards your orders.

ANDRES CASTILLERO.

P. S.—My kind regards to my friends Covarrubias and Valle; and to Mr. Hijar, why he does not reply to the three letters I have sent him, giving him also my remembrances.

Now, this letter may be said to constitute the first action of Castillero, after the denouncement of this mine. It is barely a week afterwards. What does he propose then? He says: "I have now denounced the mine; I want a grant of land close to it, on which to cut wood; I want that from this Department; I want it from you, the Governor of California. I want a title here. I also send to you my title to the Island of Santa Cruz, and beg that you will order me to be put in possession.

Now, this letter was transmitted by Governor Pico to the President of Mexico, through the Minister of Relations, on the 13th of February, 1846, in these words:

OFFICE OF THE DEPARTMENT OF CALIFORNIAS.

April 6th, 1846. Received and noted with satisfaction, and with respect to the other matters contained in the letter, let him inform attentively what he may think fit. [Rubric.]

Excellent Sir:—By the accompanying letter of Señor Don Andres Castillero, which I respectfully transmit to your Excellency, original, you will be informed of the important discovery made in this Department of a quicksilver mine. In consequence thereof, I avail myself of the good opportunity to send to your Excellency, by the Commissioner of this Government, Don José Maria Covarrubias, the quicksilver, which, as a sample, was sent to me by Señor Castillero, and to which he refers in the above mentioned letter.

With such a motive, I beg your Excellency will be pleased to put this in the superior knowledge of his Excellency the President, showing him the quicksilver which said mine produces, so that his Excellency may be made aware and satisfied at so happy a discovery.

I repeat anew to your Excellency the assurances of my consideration and respect.

God and Liberty. Angeles, February 13th, 1846.

PIO PICO.

To His Excellency the Minister of Exterior Relations.

Upon the margin of this letter is the order of the Minister of Relations in reference to a reply. This will be found on page 1805. I will call your Honor's attention to this marginal reference, particularly, because it is an answer to the whole tirade about the "Hannah."

This is the answering order of the President of Mexico, to be sent to the Governor of California. The answer is accordingly sent on the same day. See the next page of the Transcript, page 1806:

Excellent Sir:—His Excellency, the President *ad interim*, has seen with satisfaction by the letter from Señor Castillero, which your Excellency sent me with your official communication of the 13th of February last, the important discovery which has been made of a mine of quicksilver in that Department. His Excellency having seen the sample of that ingredient cited in said letter, and which your Excellency sent me by Don José Maria Covarrubias, I have the honor to say this to you by Supreme order in reply to the said communication, and that with respect to the other matters referred to in Señor Castillero's letter, that Government will please report attentively what it may deem convenient.

D., April 6th, 1846.

To His Excellency the Governor of the Department of Californias, Port of Los Angeles.

Now, a very remarkable fact occurred. When this paper came back from Mexico—and this was not until January, 1859—the question arose: Where is the original of this correspondence, that emanates from California, goes to Mexico, and comes back here from Mexico, in 1860? Nobody had ever heard of any such correspondence here. The parties claimant employed Mr. Hopkins, the keeper of the archives, to examine the early California records and find the original of these communications. Did they exist in California? No man knew of it. Mr. Hopkins' testimony is here on file. He swears that at the request of the parties claimant, he searched and found the original. I will read his testimony.

Now, if your Honor will look at page 3068, you will find what Mr. Hopkins tells us about the California archives, and how this paper came to be found here. They were not found until last year, and not until our attention had been directed to them by the copy sent here from the City of Mexico:

I hereby certify, that I am, and have been for the last five years, in charge of the Spanish archives in the office of the United States Surveyor-General for California; that during this time, I have, from time to time, been employed by Messrs. Halleck, Peachy & Billings, to search said archives for any document in relation to the discovery and development, by one Andres Castillero, of the quicksilver mine of New Almaden, and generally to search for any evidence tending directly or indirectly to sustain the title of said Castillero to said quicksilver mine; that in the prosecution of said searches I have found—

First. A letter from Andres Castillero to Governor Pio Pico, dated Monterey, December 15th, 1845, saying: That at the distance of thirty leagues from Monterey, he had discovered a mineral of azogue of the best quality, etc., etc. This letter was first seen by me about the month of May, 1856.

Second. The *borrador* of a letter from Governor Pio Pico to the Minister of Exterior Relations in Mexico, dated February 13th, 1846 (Angeles), in relation to the discovery by Andres Castillero of a quicksilver mine, forwarding at the same time a letter of Castillero in relation to the matter, together with a sample of the ore, which he asks may be presented to the President; the *borrador* is in the handwriting of Agustin Olvera, and was first discovered by me in the early part of 1859.

Now we are on the track. Olvera is brought in. I need not say who he is. I do not know him personally. I know him by reputation. I have been told by those familiarly acquainted with the early native California population, that he is a gentleman of the highest respectability; that he was admired as well as respected by his fellow-citizens. I have been told that he was selected by them as one of the Presidential electors at the last general election; and I judge, that from the granite of his character, the attacks of the special counsel for the Government would have recoiled, defeated. I think he ventured to put to him no cross-questions.

Mr. Randolph.—You will observe that I pursued the same course in regard to other witnesses of the same class. There was no need of cross-examining such witnesses.

Mr. Benjamin.—Ah! now I understand. The special counsel for the Government never cross-examines unless he thinks it will be for his benefit. And this is all proper and right; for if you cannot gain anything by cross-examining never attempt it.

Olvera's testimony is found at page 2609. These letters are presented to him. He is asked if he wrote them; or if he knows in whose handwriting they are.

At page 2610, he says—speaking first of this rough draft:

Look at this document, produced from the office of the United States Surveyor-General of California, purporting to be a *borrador*, or rough draft of a communication, addressed on the 13th February, 1846, to the Minister of Exterior Relations of Mexico, in which is made known to him the discovery of a quicksilver mine in California, a copy of which *borrador* is on file in this case, marked "Exhibit Pio Pico No. 1, W. H. C." and state in whose handwriting is said document?

[The counsel for the United States admits that said document is found on file among the archives in the office of the United States Surveyor-General.]

A. It is in my own handwriting.
Q. When was it written?
A. I don't know; but I suppose I wrote it on the day it bears date.

Now, observe that this testimony was taken in 1859, in relation to a letter written thirteen years before, found in the archives in this city, by the keeper, who is an officer of the United States, and identified as in the handwriting of Olvera. He does not remember the date on which it was written, but he supposes that it was written on the day on which it bears date.

Q. How came you to write that *borrador?*
A. Because I was Secretary of the Assembly at that time; and it was also my duty to assist in the office of the Governor's Secretary, when called upon, and I suppose that the Governor directed me to write it.

He says that it is a copy of the *borrador* written by himself. The letter of Castillero was sent by Pico.

So that here we have this irrepressible TRUTH rising up again; rising up in despite of all opposing efforts of the Government through the Attorney General; rising up thirteen years afterwards to confound the men engaged in the conspiracy formed against the owners of this property.

But this was not all that was found here in the archives. What else did Mr. Hopkins find?

He found, also, a letter from Manuel Castro, page 3068:

Third. A letter from Manuel Castro, Prefect of the Second District, to the Secretary of State, dated Monterey, December 31st, 1845; in relation to the denouncement and working by Andres Castillero of a quicksilver mine in the jurisdiction of San José, and saying that he has made a petition for two square leagues of land in the immediate neighborhood of the same. This document was also discovered by me about the beginning of the year 1859.

Now, let us look at these letters; or, in other words, let us look for a moment at the California General Archives referred to in my index, at page 9; from which it will appear that they are to be found at page 2546 of the Transcript.

We find at page 2550, Manuel Castro's letter, directed to the Governor. This is dated on the 31st of December, 1845:

Prefecture of the 2d District:—Don Andres Castillero has denounced and is now working a quicksilver mine which was found in the jurisdiction of the town of San José Guadalupe, on private property; and this prefecture which interests itself in the encouragement of all branches of industry in the department, felicitates your Excellency, and through you the most Excellent Señor Governor of the Department, upon so beneficent a discovery; inclosing also a petition which the said Señor Castillero makes, soliciting a piece of land of two square leagues which is adjacent to the said mine, so that your Excellency, if you think fit, order what may be proper, so that this prefecture may be able to make the necessary reports.

Please accept my esteem and distinguished consideration.

God and Liberty.

MANUEL CASTRO.

Monterey, December 31, 1845.

For Secretary of Departmental Government.

Now, what is the letter of Andres Castillero, inclosed in this letter of the Prefect to the Government? Your Honors will find it at page 2552:

Excellent Señor Governor, Don Pio Pico.

MONTEREY, December 15, 1845.

Esteemed Friend whom I value:—I have the grateful satisfaction of informing you, if you have not received my other letter through the Prefecture, that at thirty leagues from here, in the jurisdiction of the town of San José Guadalupe, I have discovered a mine of quicksilver of the best quality, eight arrobas of ore having yielded one arroba of liquid metal.

Señor Don Pablo Noriega, the bearer, will present to you a petition from me, which is based upon an order (disposicion) of the Supreme Government, asking that you order that possession be given me of the Island of Santa Cruz, the Messrs. Carrillo having already chosen that of Santa Rosa, which impediment prevented my occupying it, but now I have already purchased the cattle to occupy it, and I will be obliged by your sending me by said gentleman the title and order of possession.

May you continue in good health. Salute Messrs. Valle and Covarrubias for me, and order your affectionate friend who kisses your hand.

ANDRES CASTILLERO.

What can be answered to all this? The archives are searched and nothing was found till 1859. Why? Because what Mr. Hopkins was looking for were formal petitions to the Governor, and acts of the Departmental Assembly in relation to the petition for two leagues. He was not looking out for correspondence with the President of Mexico not known to exist, and accordingly did not find it.

But when, in 1859, we return from the City of Mexico with a copy of the espediente in the Department there, showing that such a correspondence as this had occurred, his attention was directed to what had taken place in Mexico; and at once he finds these originals in the archives here.

Did we put them there? Did he put them there? How came they there? Is the Surveyor-General—is anybody in that office—engaged in fabricating titles, and putting them there while Mr. Hopkins is searching for them, and in order that Mr. Hopkins may find them? How came they there? All these papers, we are told, are forgeries and frauds. This we hear, not as proven by anything in the record, but by repeated assertions.

And here allow me to say, that one would be almost inclined to believe that by some rule of this Court causes are permitted to be argued on the assertions of counsel, and not by reference to the proof in the record. Then, to the assertion, that not a solitary paper presented here, by us, is true and genuine, we oppose our denial. And if your Honors establish such a rule, how are you to determine whose assertions are true? I take for granted that you have established such a rule, for we have heard statement after statement—that men were perjurers, that documents were forged—all made with perfect conviction by my brother Randolph, but without a solitary word in the entire Transcript to back these assertions.

I heard brother Randolph, the other day, accuse a poor man who came into Court, compelled by the process of your Honors, to testify that he knew certain books—because it was his misfortune years ago to be a clerk in the Consul's office—of perjury and forgery. I say that man was denounced as a perjurer and forger because he stated that he knew certain books. Yet he had written in those books, and knew them to be genuine. Not a syllable in the record, not a word in the Transcript, to impeach the character of this gentleman; and for aught the Court knows, he stands as high in point of reputation for personal integrity as the humble individual who addresses you, or even the special counsel for the Government. There is nothing in the record to the contrary; yet that man is denounced in the name of the Government by my brother Randolph, over and over again, in a volume of words; and by their repetition in distant countries, his family, his relations, his friends, will shrink back with a blush of shame whenever his name is pronounced, because they could not believe that he would be so denounced by an officer of the Government when there was not a syllable of testimony on which to lay the foundation for such charges. This is cruel, cruel indeed!

Now, may it please your Honors, Andres Castillero is still in California. Let us see if there is any other archive in California, establishing beyond contradiction the existence of this denouncement and registry—independent, I mean, of the papers themselves produced to the Court.

Let me refer the Court to some of the archives in San José. Look at page 807 of the Transcript. By the Mexican law the outgoing officer takes a receipt from the incoming officer of all the former delivers up to the latter.

On the first day of January, 1846, or the second day rather, Antonio Maria Pico ceased to be the officer properly charged with the keeping of the archives of San José, and Pedro Chabolla took his place; and on that day, in compliance with law, an inventory was made of everything delivered up by the outgoing to the incoming officer, and the latter gave his receipt therefor. That inventory is produced at page 807 of the Transcript. It is there in Spanish. It comprehends everything in the line of "papers" existing in the Alcalde's office. Nothing is omitted, every thing is enumerated, even to the old tin candlestick, to the piece of board used in the office for publishing orders, to an old padlock, of which the key was lost—everything delivered up by the old Alcalde is put down in the inventory and receipted for by the new officer. Now, here is the inventory of what was delivered up on the 2d of January, 1846; and if your Honors will look towards the bottom of page 807, you will find in Spanish, though you will readily understand the meaning of the words:

1. *Posecion de la Mina de Sta. Clara á D. Andres Castillero.*

The old Alcalde, on the 2d January, 1846, delivered to the new one "the record of the possession given to Don Andres Castillero of the mine of Santa Clara."

The possession had been given but two or three days before. So that you have the Mexican archives in the City of Mexico; you have the California general archives; you have the archives in the Alcalde's office in San Jose; you have the papers of contemporaneous dates coming from the Sandwich Islands; you have the records of Consular communications describing this mine in Castillero's possession and working, from the Department of State in the City of Washington—all of which stand in direct proof of the origin of our just and equitable claim to this property. And yet your Honors will be treated with the same assertions as heretofore: that not one word of this is true; that these omnipresent forgers in all parts of the world are forging newspapers, records, letters, archives, in every possible place in which it could be conceived that a forgery would be beneficial to their interests. In this the special counsel for the Government is but carrying out the attempt of the Attorney General of the United States—to prevent your Honors seeing any evidence upon our side at all. The settled, persistent, continuous efforts to head off these claimants by every possible means that human ingenuity could suggest, are shown in this record. If they can only strangle all the evidence, then they can take our property. If they cannot suppress all of the evidence, they will conceal as much as possible, and then coolly deny, without contradictory proof, the truth of whatever evidence is introduced. And all this is done in violation of the plighted faith of the nation, given before the civilized world.

In 1848, when this country was acquired, our Government said to Mexico, in the face of Christendom we will respect the property titles of your people. They shall stay with us and become our citizens if they choose. If they do not so choose, they shall go home in peace. Their property rights shall be respected. The plighted faith of the United States of America is given to you, that your citizens shall not be despoiled of what was their own before you ceded the territory to us. We promise this; we pledge the honor of the Nation to its fulfillment, and this promise has been fulfilled to all, except to those who hold title to property in California under Andres Castillero.

Now, may it please your Honors, from amongst the papers produced in evidence in this case, testimony is rising up every hour to confound the designs of those who are seeking to rob us of our property.

After eight or ten years of painful uncertainty and search, the original documents of our title are found amongst the papers of a deceased individual to whom we once intrusted them as our agent, and who basely betrayed his trust, and endeavored to pervert his possession of them into an instrument for despoiling us—they had passed from his hands to those of my brother

Hall McAllister. They are found by accident. They are produced in Court, and they throw a flood of light on this whole proceeding.

Now, your Honors will find in this paper what seems to me the most incontrovertible evidence, proof conclusive that these papers were made early in the year 1846. And that proof comes from the certificate of Mr. Hopkins. How does this occur?

Most fortunately for us, the Alcalde Chavolla, in making a copy of the act of possession to hand over to Castillero for his title, omitted a line. When he handed the paper to Castillero, the latter saw the deficiency; and upon the face of the paper appears the correction in his own handwriting. Now, we have already seen that he left California early in the year 1846, and has never since returned. Will it be pretended that this paper was sent to Mexico to Castillero, in order that he might put in an interlineation? How did the handwriting of Castillero get upon this certified copy, unless it was made in '46, before his departure from California? There are but two possible solutions to these questions. Either this interlineation was made by Andres Castillero in '46, before he left California, or this paper has been sent back to Castillero, in Mexico, for him to interline these words. Certain it is that these words are here, interlined. Certain it is that Andres Castillero has never been back to California since his departure in 1846, in the month of April of that year. Tell me now how these words got there in his handwriting? Hopkins swears that they are in Castillero's handwriting, and most fortunately does it so occur that the interlined words contain his signature. There is the date and signature.

It occurred in this wise: The Notary or Alcalde in making out a copy of this testimonio, in regard to giving Castillero possession of this mine, copied out the different denouncements; and in copying the first denouncement, neglected to copy into it the date and signature, and these are filled up in the handwriting of Andres Castillero, as proven by Mr. Hopkins' certificate. Again, I say, how did it get there?

Now, this little circumstance was not known until about two months ago; so we find that nothing transpires in relation to these papers, here or elsewhere, which does not point to their truth and genuineness. Nothing occurs in any part of the civilized world in connection with this case which can serve to raise or create a doubt as to the authenticity of the documents. All concur in forcing conviction of the truth of our claim on the most incredulous human mind. They force conviction in spite of all opposition and prejudice against us—compelling belief, so that no reasoning man can escape it.

Mr. Hopkins' testimony is on file in relation to this matter, and none of it is disputed.

Now, may it please your Honors, we have got certain facts thus far established; established by what may be considered as the history of the day, of the age.

We have the fact that the mine was discovered, was taken possession of by an individual as its owner; that this possession was not clandestine, as has been frequently urged against it, but was open, notorious, and communicated to all the public authorities of California and Mexico, and that so far from disapproving that possession, those authorities expressed gratification and delight, and that all this occurred before the United States acquired one foot of ground in California. We have the fact, that as soon as Andres Castillero discovered this mine, he told everybody that he had denounced it. His letters abound in that assertion. We have the fact that the United States Consul informed the public through the newspapers that Castillero had denounced this mine. We have the fact that the United States Consul informed the Government of the United States that Castillero was in possession and working this mine as owner. We have the fact that Castillero declared to Fremont that he had denounced this mine. We have the fact that the archives contain what purports to be a record of denouncement and possession. We have the schedule of papers in the Alcalde's office in 1846, in San José, in which these Castillero documents of denunciation and possession are enumerated. We have copies from the Mexican archives. We have the archives in the keeping of Mr. Hopkins, substantiating our claims, and we have about forty witnesses who all swear to the facts demonstrative of the genuineness of our title. Not one man swears to the contrary. And yet, I am here in the presence of an intelligent Court, arguing to your Honors, that these papers are true, and are not forgeries.

And all this is proven, not merely by our own witnesses, but by the evidence introduced by the Government. We might, I think, safely agree that your Honors should take up this record and decide it by the evidence introduced by the Government. For, we say again, that *that* evidence itself gives a direct contradiction to the assertions made by the counsel of the United States. I will not say the *Government.* Of course the Government of the United States makes no such averment. They are made in its name, but do not emanate from it.

Now, I say I put this point—of the existence of this record of possession—upon the testimony of the Government witnesses alone; and I agree that you shall forget everything else, —take up everything else in this record, and walk to the windows and throw it out, as the special counsel invites you to do.

MR. RANDOLPH—Your certificates?

MR. BENJAMIN—Exactly; take them and tear them all up, and throw them out of the window. I have no fear of the result, because the Government has taken the pains to prove our papers for us. I say again, you may take all our evidence and throw it out of the window, and decide this case on the evidence produced by the Government.

And yet, my brother Randolph has a curious way of treating even his own witnesses. When they say anything that he don't want them to say, he calls them "perjurers."

There is one witness who, I believe, brother Randolph does not call a perjurer, although he identifies these papers. I refer to Mr. Weekes. He brings Weekes up into Court. Now, he informs us he is going to prove that our papers are forgeries. Now I have found you out! Come up here, Mr. Weekes! Did you give a copy of certain documents to Alexander Forbes in January, 1848?

I did. Were the original in your office? They were. They were in a little paper book which I saw as far back as when Burton was Alcalde in 1846—and again in 1848, when I was Alcalde.

Weekes is then shown the original record of registry and

possession from the County Recorder's Office, and he identifies it.—*Trans.* 306.

Now, brother Randolph dismisses this witness for the Government, who has proven contrary to his expectation, that the papers are genuine, with the exclamation that he is nothing but a "drunken Alcalde."

Mr. Randolph—Do you vouch for his sobriety?

Mr. Benjamin—I am not a witness in this case. There then we have the testimony of one Government witness. Being a "drunken Alcalde," the Government turns him out of Court.

Then the Government introduced A. M. Pico, and examines him very closely. Unfortunately, Pico persists in stating that these documents are genuine. What says the Government in such a predicament? "Your Honors know what credit to attach to California witnesses."

And so with all the witnesses that come up at the call of the Government, and tell brother Randolph: "you are mistaken in regard to these papers; they are genuine, and we know them to be such." All are disposed of, one after another, with "California witnesses—your Honors know what *they* are;" drunken Alcaldes!—what credit can you attach to their oath?"

Well, how are we ever to prove this thing? We can never prove anything according to the theories of the Government. Our own witnesses are all to be denounced as perjurers, and the Government's own witnesses are dismissed as false and drunken. All on the other side consists in bare assertions. Now, my brother Randolph can out-assert us all day long, and if your Honors have established that rule of ascertaining truth we give up our case at once. But if you are going to decide this case upon the proof contained in the Transcript, then it does not admit of discussion so far as the question of the genuineness of this record of possession is concerned. In this particular, there is no case before you.

I was much struck with the contradictions in my brother's argument. At one moment he charges that we commenced forging in 1850: in a few moments he says that the forgeries began in 1847; and still again, that the record of possession was forged in 1848. In this assertion he ran foul of several pieces of stubborn record evidence, and amongst others he encountered the testimony of his main witness.

James Alex. Forbes, as Vice Consul of Her Britanic Majesty, wrote an official letter, which was found in the archives of San José in 1847. And in that letter, brought from the archives and filed in this case, Her Britannic Majesty's Consul complains to the Alcalde that persons have intruded on the "*juridical possession of the mine.*" By looking at page 810, your Honors will see what the Consul says to John Burton, Alcalde:

British Vice Consulate for California,
Santa Clara, 14th August, 1847.

Sir: I have received information from the person in charge of the Quicksilver mine of Sta Clara that two persons have commenced digging a pit by the direction of Mr. G. Cook, *within the limits of the juridical possession of the said mine*, and upon remonstrating with them, they have refused to discontinue their operations.

Permit me to refer you to the documents which exist in your office upon which was founded your conviction of the justice of your decision in relation to the claim of Mr. Cook in March last, and to request that you will be pleased to adopt such measures for protecting the rights of the owners of the said mine and of those who are legally interested in the same as you may deem most conducive to that end.

I have the honor to be sir,

Your most obedient servant,

J. Alex. Forbes,
Vice Consul.

John Burton, Esq.,
Justice of Peace, Pueblo de San José.

Thus, may it please your Honors, we have gone back from the pretended forgery of 1848, to James Alex. Forbes' letter in 1847, and the existence of the documents of possession in the Alcalde office—the previous month of March, 1847.

Here my brother Randolph—and we must renew our acknowledgments to him—produces the record of the previous suit in the month of March. See page 815 of the first volume of the Transcript:

G. C. COOK
vs.
JAMES ALEX. FORBES.

On complaint of plaintiff that Andres Castillero, James Alex. Forbes, José Maria del Real, José Castro, Secundino Robles, and Chato, *alias* Theodore Robles, were to work on his land contrary to law, and prayed that they be removed.

Here we have got these miners to work on the land claimed by Cook; and this in March, 1847.

A summons was issued bearing date March 17th, 1847, returnable on Friday next, which was served and returned according to law. The parties having appeared the case is continued, until the mine can be surveyed.

Just here: What was to be surveyed? What was to be surveyed?

MR. RANDOLPH—That's the question.

MR. BENJAMIN—If there was no title to the land, if there was nothing which the Alcalde could order to be surveyed under this title, how could a survey have been ordered at all? How could the surveyor do anything without any title with which to make a survey. Something then in the shape of a definite title paper did then exist, by which the surveyor was enabled to execute his duty of perfecting a survey. What was it? James Alex. Forbes says, "*The juridical possession of the mine, according to the documents which exist in your* (the Alcalde's) *office.*"

Two men were sent to the mines to measure and report thereon, who went and returned their report.

What did they measure? Did they go there and measure any quantity of land *ad libitum?* Or did they find something definite to measure? They go to the mouth of the mine. How far do they measure around it? There must have been some paper declaring the limits which they were to measure. They go and make the measurement, and return and report. What is the result? It appears in a letter of James Alex. Forbes that the Justice decided in favor of the mine owners, on the survey of their mining possession. What was that mining possession? On the 5th of May, 1847, you have James Alex. Forbes' letter, so often referred to, in which is a description of the mining possession as being a *mining possession of* 3,000 *varas on all sides.* That is what James A. Forbes certifies was surveyed. That was what existed in the archives, and was called a "juridical possession." That was the direction the surveyors received from the Justice.

Thus, may it please your Honors, we have Andres Castillero claiming the mine in California before his departure. We have him proved to possess a full knowledge of the mode by which mines are acquired, and the mode by which lands are acquired. His mining possession he asks from the Alcalde. His lands he asks from the Governor. He had previously obtained a grant of land from the Supreme Government. It was an island. And that island was immediately given to him in accordance with the order of the Supreme Government, when he presented his claim. Being now absent from Mexico, and desirous of obtaining a two-league grant in California, he was obliged to make application to the local authorities.

But, before action is had on this petition for two leagues, Andres Castillero leaves California; and when he gets to the City of Mexico, remembering his success in relation to the former grant of the Island of Santa Cruz, he determined to make his petition for a two-league grant directly to the Government; and he abandoned his proceeding before the local authorities here. Accordingly, the moment he reaches the City of Mexico, we find him carrying out precisely what had been designed in California.

He goes before the Mining Junta. The first thing he does is to apply for aid in carrying on his mining operations. He says to the Junta, I want my *mining possession* confirmed by the Supreme Government. In all his plans there stated, he followed out precisely what he had commenced in California. There was no variation, no interruption, no change of purpose. His plan was the straightforward one of a man who had one object in view, and who seeks and adopts the best measures to secure its prompt execution.

He had previously sent a special messenger with his specimens of ore and his letters, and now that the Governor required to send a special commissioner to Mexico, he attends to his interests in person.

But it is said that he sent no special messenger. It is said that Lazaro Piña went with him, and did not go previously on the "Hannah."

The statement on the other side is, as I suppose, that the "Hannah" did not leave the port of Monterey at the time indicated in evidence. This is one of the most puzzling subjects in this whole case for me to treat. I do not understand the pretext of the argument to the contrary. I cannot make it out. I can clearly show to your Honors by direct testimony that the "Hannah" did leave the port of Monterey at the time we specify. How is the contrary made out? I listened attentively to the gentleman for two days, and could not get at what he meant. I could not make out by what species of ratiocination he hoped to satisfy the Court that all the documents proving that fact did not exist; that they were all forged.

What is the evidence?

At page 2605, we have two letters of Andres Castillero, written to General Vallejo in February or March, 1846. Now, unless General Vallejo had forged these letters, the proof is there complete—that Lazaro Piña did leave the port of Monterey, and that he was sent from that port by Castillero early in March. What is the proof? Andres Castillero writes to General Vallejo. Now, let your Honors read those letters; look at them, and you will see on the face of the papers themselves, the intrinsic, absolute proof of their verity. You will see from the statement in these letters that they cannot but be true. They refer to matters shown to have connected both the parties in strict intimacy. General Vallejo explains that a short time before Andres Castillero acted as godfather at the christening of one of his children. Hence he is called "*compadre*," a sacred tie amongst the Spanish population.

Now what does he write to his "*compadre*"? He writes on the 21st February, 1846:

MISSION OF SANTA CLARA, Feb. 21, 1846.

Colonel Commander of the Line of the North,
Don Mariano Guadalupe Vallejo:

My Dear Compadre:—I have the pleasure to transmit to you

a letter from His Excellency the President, in which he informs me of the reasons why the expedition should not have come here, but in an official communication received by Don José Castro, he is advised by the Ministry of War that Colonel Don Tanacio Triesta had set out.

By the brigantine schooner which brought these communications, we have received information that the Republic is in perfect peace. This vessel sails shortly, and will carry communications of what has occurred lately.

Myself or Piña will leave in it, or both together. I am only detained waiting the arrival of the division which may touch here in a day or two.

Tell me what commands you have for the Capital, and I shall execute them faithfully in person, or order Piña to do so, as the circumstances may be.

Nothing has been done regarding the uniting of the Government to the Commandancy-General, and this matter will not be decided until I have an interview with the Supreme Government, or communicate with it.

Do not fail to reply to this soon, and write to me with the frankness which is congenial to you.

Respects to my comadre, and a kiss to my godson, and order your affectionate compadre and obedient servant,

ANDRES CASTILLERO.

On the 11th March he writes to General Vallejo:

MISSION OF SANTA CLARA, March 11, 1846.

Colonel Don Mariano Guadalupe Vallejo:

My Dear Compadre:—In the mountain range of the Gabilan, over against the Natividad, Captain J. C. Fremont has fortified himself, and he has been joined by some foreigners; but the Commandant-General is upon them, and I set out immediately to accompany them. This I advise you of for your information.

Piña embarked on the fourth of this month in Monterey, and was dispatched in perfect order. He will travel post to Mexico.

Respects to my comadre and family, and order your affectionate compadre and obedient servant,

ANDRES CASTILLERO.

P. S.—Expressions to Don Salvador and Don Victor.

Were these letters forged in Mexico by Castillero, and sent

up here for General Vallejo to swear to? So the Government's attorney would intimate.

The statement of General Vallejo is direct and positive—that he received these letters in due course from the day of their date. He presents the originals. Do they contain the truth? How do we prove that they contain the truth? We produce the archives of the United States Consul, and in the letters of the Consulate the fact is found that the "Hannah" did leave at the time we specify; that Larkin stopped her two or three days in order to send important letters by her; that he sent a courier to meet her at Santa Barbara to take later dates; and that letters by the "Hannah" of that date were received in Mazatlan.

Now, may it please your Honors, if Andres Castillero forged these letters, he must have known what was in the Consulate archives. Because it does so occur that this vessel, the "Hannah," was intended to leave on the 4th. She was "up" to leave on that day, and therefore it is, that Castillero, from Santa Clara, writes to General Vallejo that Piña embarked on the 4th, and was dispatched in perfect order.

But the American Consul finds Fremont in a dangerous position; and discovering that Fremont wanted to write home, he detains the "Hannah" three days. When the Consul's memorandum book is brought into Court, which states that the "Hannah" left for Mazatlan on the 7th of March, 1846, no human being could read these several papers and give any other construction to them—always excepting the Government counsel.

There are numerous other letters reading the same way. The books of the Consul are in Court.

[Mr. Randolph objected to introducing as evidence any portion of the Consul's books, except such parts as were directly proven at the time the books were introduced and sworn to before the Commissioner taking the testimony.]

Mr. Benjamin—The book was introduced before the Commissioner. Here it is.

On page 2660, I read from Mr. Swasey's testimony—

Ques. 7.—Where did you next go, and what was your occupation?

Ans.—I went to Monterey, where I arrived about the last of February or first of March, 1846, and immediately engaged with Mr. Thomas O. Larkin as his Consular clerk.

Ques. 8—What were your duties as such?

Ans.—Writing and copying official and other dispatches. I also kept some of his private books.

Ques. 9.—Did he keep copies of all his official correspondence; if yea, in what books were they kept?

Ans.—He did. They were kept in one book, entitled on the back: "Correspondence with the Department of State. Thos. O. Larkin, Monterey." Also, in another book, entitled: "Copies of Official Letters. Thos. O. Larkin, Monterey." They are large record books, strongly bound in calf. These books are now here before me.

Now I say, as a matter of law, that the Commissioner represents the Court in receiving the testimony, and these are the same books which were before him. The Consular archives are brought into Court—the books were placed before the Commissioner first, and now they are here for inspection. The witness swears that he was the clerk who wrote in them, and that these are genuine Consular archives. In opposition to these we have—what? The denial of the Government. Passages are read from official documents of the year 1846, with a view to have your Honors draw the conclusion that these are forged papers. And then when the books themselves are handed in for inspection, the special counsel for the Government gets up and tells your Honors that the case is closed, and you must not look at the books. That is it. Shut your eyes. If they can only keep your eyes shut they can gain this case. Don't look at anything. This is perfectly consistent with his entire management of this case. If the Judges will only shut their eyes, the Government may possibly gain this case.

But we mean to open this case to full examination. We intend that the Court shall see all. We never yet feared the closest scrutiny. We never raised an objection to the taking of any testimony. If in the whole range of facts existing upon earth in the shape of evidence, there is one which in the judgment of the counsel militates against our title, we say let him bring it in now. We never shut the doors. You have got the universe before you; range it. Find something. Find, if you can, *some one* fact inconsistent with the genuineness of these papers, and we give up our case. That is not all. We produce volumes of Mexican official newspapers. We first bring them before competent witnesses. We prove them to be genuine newspapers, published in Mexico. We offer these volumes in evidence. They are Mexican daily newspapers for the year 1846. This is the "Diario Oficial." The other volume is a bound copy of "El Monitor Republicano," for the year 1846.

We point out to the witnesses in these papers certain extracts having reference to this mine. We have them copied into the record for our use.

MR. RANDOLPH—That is in evidence; the rest is not.

MR. BENJAMIN—Now, brother Randolph has spent two days before the Court, arguing about what? That it is not likely, it is not to be believed, it is against all human probability, that the "Hannah" carried the dispatches which this gentleman, Piña, was to carry. Why? Because the Mexican newspapers would have spoken of Fremont's affairs; whereas they only speak of the mine. Now these newspapers were before my brother's face; and he examined them with reference, of course, to other passages than those we offered in evidence, as well as in reference to what we did offer—he himself offering extracts from them in evidence. [See *Trans.* pp. 2361 to 2365.] We could not copy the whole newspaper into this record. But it so happened that the very pages from which we copied our statements about Andres Castillero, *do* contain the statements about Fremont.

[MR. RANDOLPH objected to the introduction before the Court of any portion of the Mexican papers which were not displayed or referred to in the Transcript.]

MR. BENJAMIN—That is it. Shut the eyes of your Honors, and the Government may gain this case.

MR. RANDOLPH—I do not know of any evidence but what

is in the record. If the claimants have any evidence outside of the record, they should have introduced it in the proper place and time. I had no opportunity to examine these papers after the examination of them before the Commissioner. I had to make my argument on the basis of what was contained in the record.

Justice Hoffman—What would you estimate the effect of your argument on this point to be, if after these assertions of Mr. Benjamin's you close the books? How is the Court to know that these papers do *not* contain a notice of the Fremont affair, if we cannot look into them?

Mr. Randolph—If your Honors please, I pointed to the printed Transcript, and showed the evidence put into Court on that subject. It is to that evidence that your Honors are confined in the trial of this cause; when there is no rule of evidence by which I am called upon to answer the matter introduced to our notice for the first time at the present moment.

Mr. Benjamin—If your Honors please, I will read the statements in this paper concerning Captain Fremont, as a part of my speech. I anticipated this objection. It is in perfect accordance with everything which has been done by the special counsel for the Government here, and the *Attorney General* at Washington. Keep out the evidence, they say, and we can gain the case.

Mr. Randolph—What do we know about what has been placed in those newspapers?

Mr. Benjamin—Well, I am now going to read some sentences which we have *newly forged!*—since we heard brother Randolph's argument—and placed in this pretended Mexican newspaper.

Mr. Randolph—I have little doubt about its being a forgery.

Mr. Benjamin—Very well. I will, with your Honors' permission, proceed to read this "*forgery*" as a part of my speech. Your Honors will have the paper to examine hereafter; and you can place such reliance upon this "*forgery*" as you think proper, while you are weighing my brother Randolph's argument upon this particular omission which we have within the last two days taken particular pains to supply.

Now, in the "Diario Oficial" of the 5th of April, 1846, is the statement of this arrival from California, and of the Fremont affair:

The Supreme Government has also received news from Upper California, down to the 5th of last month, and nothing particular had occurred in that Department.

Commandants Castro and Vallejo express themselves full of enthusiasm to defend that important Territory, and are resolved to sustain it even if the enemy should pass over their dead bodies. Our hearts swell at noticing the noble enthusiasm of the Mexicans who, at the places of greatest risk, bravely sustain the sacred rights of their country.

Now what did my brother Randolph say? He said, "your 'Republicano' newspaper, published in the City of Mexico, contains nothing but the forged story about your quicksilver affairs. All Mexico is represented as mad with astonishment and joy about the quicksilver discoveries in California." Can the Court believe, urges my learned brother, that they received intelligence of nothing but the quicksilver mine, when Fremont was here with his flag flying on the Gavilan? At great trouble and expense we procured these files of old Mexican newspapers, placing such extracts in the record as had reference to this mine, which we supposed was all that was in litigation. We had no idea that Fremont's expedition was in litigation before your Honors. But as soon as it was said by my brother, that this omission proves that the files are not genuine, we "*forged*" these volumes afresh, and here they are in Court, or, at least, here a small portion of one of them is in my speech.

The "Monitor Republicano" of the 12th of April, contains this extract from the Tepic newspaper, of which we quote in the record only the first paragraph, relating to the discovery of the mine, that being all that concerned us.

We have now forged this additional portion, so that your Honors may be prepared to answer that portion of brother Randolph's argument relating to the supposed omission to mention

Fremont's expedition. This forgery appears in my argument, not in the Transcript.

After stating all about the quicksilver mine, this Mexican newspaper goes on to say:—

Recently, the greatest alarm has been produced by the entry into the country of a small force of Americans under the command of Capt. Fremont, of the Engineers. The military authorities of the department were preparing themselves to fulfill their duties by driving these invaders out of the limits of the Republic.

The copy of the document which we transcribe (which has been given to us by a respectable merchant of this place to whom it was sent) will inform our readers of this remarkable event.

Then it copies a letter of José Castro, directed to the Minister of War and Marine, stating these facts:

In my communication of the 5th current, I announced to you the arrival of a captain at the head of fifty men, who came, as he said, by order of the Government of the United States to survey the limits of Oregon. This person presented himself at my headquarters some days ago, accompanied by two individuals, with the object of asking permission to procure provisions for his men that he had left in the mountains, which was given to him. But two days ago I was much surprised at being informed that this person was only two days' journey from this place; in consequence, I immediately sent him a communication, ordering him, that on the instant of its receipt he should put himself on the march and leave the department, but I have yet received no answer; but in order to make him obey, in case of resistance, I sent out a force to observe their operations, and to-day I march in person to join it and to see that the object is attained. The hurry with which I undertake my march, does not permit me to be more diffuse, and I beg that you will inform his Excellency the President, assuring him that not only shall the national integrity of this department be defended with the enthusiasm of good Mexicans, but those who attempt to violate it will find an impregnable barrier in the valor and patriotism of every one of the Californians.

Receive the assurances of my respect, etc.,

God and Liberty. Monterey, 6th March.

JOSE CASTRO.

To the Minister of War and Marine.

Now, what becomes of the argument about the "Hannah's" not leaving the port of Monterey, and not reaching Mazatlan at the time we name? It appears that we have not only forged in these papers an account of the arrival by the "Hannah" of important quicksilver intelligence, but we have also forged an account of news from Fremont's expedition, of the latest dates. Of course this is all forged and fabricated. Everything connected with this trip of the "Hannah" is forged; and I should not be surprised to hear that the "Hannah" herself was forged for the special fraudulent purposes of these claimants.

[The Court took a recess for half an hour.]

ON RE-ASSEMBLING,

Mr. Randolph said:—Before Mr. Benjamin resumes his argument, I wish merely to notice at this particular time an inference which might be drawn by the Court, or intended to be drawn by the Court, that I was at the time of making my argument acquainted with or cognizant of the rest of these articles in the newspapers, but did not candidly make my argument on that point. In answer to that, I have to say, that I was unapprised of any such matter existing at all, much less as evidence in this case. As to what the fact may be, I shall of course make observations in my own proper place. I merely rise now to disabuse the mind of the Court of any such impression, accidentally or intentionally conveyed.

Mr. Benjamin said:—I will answer for my brother Randolph. Nobody who knows him could by any possibility suppose that he would endeavor to impose upon the Court. No one supposes that he would have made use of arguments which he did, if he had supposed we could overthrow them in the manner we have done. And that makes my argument all the stronger, and shows him and the Court how dangerous is that style of argument in which my brother Randolph indulges all through this case; when he dwells upon certain prominent facts in general history which he supposes to be inconsistent with this particular case, without giving us any opportunity of explanation. We cannot explain the entire history of the country in this record. We cannot put in our record everything that occurred in Mexico and the United States for ten years, but must select only that which bears on this mine. And I trust that my brother Randolph will see himself, that arguments which he hourly draws from just such facts as these, have no more foundation than the argument which he sought to draw from the supposed omission of these papers to speak of Fremont's affairs; that his whole fabric of similar arguments has just as little basis as the one which has just been destroyed and is now abandoned.

Mr. Randolph—No, sir; I abandon nothing.

Mr. Benjamin—Now, may it please the Court, I take it to be a matter of some consequence, since the question has been raised, that there shall be no shadow of doubt left upon a human mind, that the brig "Hannah" left Monterey at the time we name, that she carried this correspondence, and had Lazaro Piña on board. I have not done with the proof.

The official documents which Mr. Randolph himself introduces, show the facts.

The facts are: that the "Hannah" is to leave on the 4th of March; that the Consul detains her until the 7th, and then sends correspondence by her. Now, we say that this vessel stopped at Santa Barbara on her way down; and we say, that although Consul Larkin had written by her before her departure from Monterey, other facts had occurred in the interval which rendered it necessary to dispatch a courier down to Santa Barbara to the "Hannah," and send communications to the American squadron at Mazatlan. My brother Randolph has read a letter of the 9th of March, written by Consul Larkin to President Buchanan, then Secretary of State. It shows that after the 7th of March Mr. Larkin got frightened in relation to Fremont's position. On the 9th, he had intelligence that a company were being raised to proceed against Fremont; and there was danger of this American officer's being overwhelmed by the enemy. It was this fresh danger that incited Larkin to send couriers to Santa Barbara, and to send double couriers to Fremont. And I call upon brother Randolph to tell me this: *How came Mr. Larkin to know in Monterey that there would be a vessel at Santa Barbara to take his dispatches down to Mazatlan?*

MR. RANDOLPH—He took his chances.

MR. BENJAMIN—Pretty chances. He took his chances to INTERCEPT the vessel by his messenger, because he KNEW that the brig "Hannah" had been dispatched three days before from Monterey for Mazatlan, intending to call at Santa Barbara.

Do your Honors want further proof? Look at page 3020, at the letter of Mott, Talbot & Co., which I read:

U. S. PORTSMOUTH,
Port of Mazatlan, 1st April, 1846.

T. O. LARKIN, Esq.,
Monterey,

Dear Sir: It has been hinted to us that this ship is bound to Monterey, and although the fact is doubtful, we avail of the chance to acknowledge receipt of your much valued favors of the 2d, 4th, and 5th ult., the former enclosing a remittance of a Russian bill for $2000 for account of Mr. A. B. Thompson.

Your letters have this moment reached us per "Hannah" and we are much obliged for their contents. You were correct in supposing that the destination of this vessel when she sailed from this was not known to us. We have only time to return you our sincere thanks for your kindness in collecting the debt of Mr. Thompson, which we frankly confess we had written off as lost. We are also much obliged for the fruit, which, however, as the Hannah has just anchored, we have not yet received.

Now here we have got this fact: the arrival of the brig "Hannah" at Mazatlan on the 1st of April, her arrival from Monterey, bringing down letters of the 2d, 4th and 5th of March.

My brother says that there is no proof in this record that the "Hannah" ever left Monterey from the 1st to the 10th of March.

There is Consul Larkin's original book with the entry of the 7th of March—"The 'Hannah' for Mazatlan." Here is Mott, Talbot & Co's letter, dated on board the ship Portsmouth on the 1st of April, announcing the arrival of the "Hannah" at Mazatlan, with letters up to the 5th; and what occurs on the arrival of the "Hannah." The "Hannah" has just cast her anchor. Suddenly the intelligence is spread among the persons on the Portsmouth, that that vessel is going up to Monterey. How is this? For what purpose is this sudden dispatch of the Portsmouth? It is not known publicly. It has been hinted to Mott, Talbot & Co., by some friendly officer. But this fact is known on shipboard, and preparations are at once made agreeably to the requirements of such a determination; and in twenty-one days from the date of the arrival of the "Hannah" at Mazatlan, the Portsmouth is in the Port of Monterey.

And the next thing that we find is a letter from Consul Larkin to Mr. Gillespie, who is sent out with confidential instructions. Gillespie is there in pursuit of Fremont, who is on his way to Oregon. Gillespie goes in pursuit of Fremont, and brings him back. The Consul, in the meantime, writes and sends a letter to Gillespie, telling him of the arrival of the Portsmouth, by reason of letters dispatched to the American squadron at Mazatlan, by the "Hannah." Now, brother Randolph says that the "Hannah" never left for Mazatlan at the date we fix, and that consequently it was not in reply to any letters sent by her that the Portsmouth made her trip to Monterey. You have the entry of her arrival here in the newspaper:

1st day of April, American hermaphrodite brig Hannah, of 89 tons, Captain Benjamin F. Thusum, crew 10 men, proceeding from San Blas, 5 days' sail, in ballast. Passengers, Anselmo, Latayada, and Gregorio Aguirre.

This gives ample time for the "Hannah" to leave Monterey, stop at Santa Barbara and receive further Government dispatches for the Portsmouth, which secured the prompt dispatch of that vessel to Monterey for the succor of Captain Fremont.

Now, let us look at what the Consul writes on April 23d, 1846, to Mr. Gillespie. It will be found on page 2674–5 of Transcript:

CONSULATE OF THE U. S. OF AMERICA,
Monterey, April 23, 1846.

SIR:—Captain Montgomery, of the "Portsmouth," being under sailing orders (the 1st or 2d instant) at Mazatlan, was waiting for the Mexican Mail, when Commodore Sloat heard, per brig "Hannah," of the situation of Captain Fremont near San Johns, and immediately dispatched the ship. She was twenty-one days from Mazatlan to Monterey. I send to you four or five New York, and one Mexican, newspapers—the former to 5th February, the latter of March. New York papers of Feb. 25 was in the hands of the Commodore. Captain Montgomery has not any certain information of Mr. Slidel's situation in Jalapa, in March; he says that the Commandaute General of Mazatlan had later news by six days than Commodore Sloat; that all Custom-House and other Government officers had left Mazatlan, taking away the archives and Government property, publishing in the streets that Commodore Sloat would in all probability declare a state of blockade the next day, thereby giving reasons to suppose they were aware of the cause.

The "Portsmouth" will remain here and in San Francisco some weeks, perhaps a month, according to circumstances. I have (as my opinion) said to Señores Castro, Carillo, and Vallejo, that our flag may fly here in thirty days. The former says, for his own plans, war is preferable to peace, as, by war, affairs will at once be brought to a close, and each one know his doom. I answered, without war he could make certain officers, and secure to himself and his friends fame, honor, permanent employ, and pay. He and others know not what to do or say, but wait advices from Mexico, per their Commissioner, by "Don Quixote." She ought to be here by the first of July.

In the meantime, Andres Castillero, on the 4th of April had gone down to Mexico, as Commissioner, on the Don Quixote. He had left on the 4th of April. The California local officers were awaiting his return by the Don Quixote, which was expected by the 1st of July, in order to determine what their own conduct should be.

Now, I want your Honors to look at another little circumstance, insignificant perhaps, but yet, when an entire body of papers and transactions—as in the present case—are stated to be forged, when the most trifling fact is denied, every little particle of proof serves as some guide, some indication by which you can be enabled to arrive at a conclusion satisfactory to yourselves.

How about Captain Paty's testimony? Let us look at it. (First volume of the Transcript, page 424.)

Brother Randolph has read from the memorandum—"Information for whom it may concern"—annexed to Captain Paty's deposition, a statement, which he says he took from his wife's journal, of what occurred upon that trip; and the third entry to be found on page 429, reads:

Third.—Don Andres Castillero and his servant, or companion ("Lazaro Piña" I think was his name), Mrs. John Paty, child and servant, were passengers on board.

Now, may it please the Court, this testimony of Captain Paty's was given on the 15th December, 1857. The name of the servant in the memorandum is put in with marks of quotation, and in parentheses, to indicate that it did not come from the original memorandum, but was an impression upon his memory; or, in other words, he did not here quote from his wife's journal, but he thinks that "Lazaro Piña" was the name of Castillero's servant. It is placed separately from this paper; evidently it is something interpolated—no doubt in perfect good faith—by Captain Paty. Having heard it remarked by some gentleman interested in this case in the City of San Francisco, that it was important for the Government to prove that Castillero had a companion by the name of "Lazaro Piña," he was induced to imagine that such was the name of a person who sailed with him as passenger on the trip referred to. They said to him, "Didn't Castillero have a companion on that trip?" "Yes," he replied. "Wasn't his name 'Lazaro Piña'?" Naturally enough he answers, "Yes, I think that that was his name." This is the whole story. The evidence of it is on the face of the paper. It is in the past tense, "*was* his name." Evidently somebody had been talking to Captain Paty, and had directed his memory as to the names of passengers, ten or twelve years ago, in this way. If this memorandum had been a contemporaneous entry, it seems to me that it would not have expressed any uncertainty as to the name of the person in company with Castillero.

And this is the only evidence, introduced by the Government, tending to shake the truth of the circumstances we allege as connected with the whole of these papers and Consular archives.

Now, my brother Randolph has got something like testimony. I congratulate him on that, and hope that he will make a proper use of it.

But the sixth item of this same memorandum:

Sixth.—I agreed to wait twenty days at Acapulco for Don Andres Castillero, and after waiting until the 16th of May, I heard that he wished to meet me at Mazatlan; consequently I left Acapulco on the 18th of May for that place and St. Blas.

I ask your Honors to remember that Castillero's bargain with the Junta in the City of Mexico was, that they should give him a draft on *Mazatlan.* He had to go there to get his money. He expected to receive his money at Mazatlan.

And here we find the evidence furnished by the Government itself, in an out-of-the-way place, in a memorandum of Captain Paty. Instead of waiting at Acapulco for Castillero, as the terms of his first agreement bound him to do, he got information on the 16th of May from the City of Mexico from Castillero, that the latter wanted him to go to Mazatlan to meet him; not to wait longer at Acapulco. This is in precise accordance with the arrangement made by Castillero with the Junta when he did get to Mexico.

Andres Castillero followed Lazaro Piña very closely. The date of the latter's arrival in Mexico does not appear. We know this, that on the 13th of April Herrera, the President of the Republic, communicated to the Junta extracts from two letters of Castillero's, dated late in February, 1846, which he said he had just received in relation to the quicksilver mine: and we know that Don Tomas Moral received also a letter dated 19th February, and we know that specimens did reach the City of Mexico early in April, and were placed in the Cabinet of Minerals, and were exhibited to and inspected by the professors. That we do know. How they got to Mexico it was not incumbent upon us to say. In order to prove the genuineness of our title, it is not incumbent upon us to prove every fact in regard to the passage to and from Mexico of letters, documents and minerals which the United States may think proper to question.

Really, it appears to me as if all legal principles were inverted in this case. Everything we have done is denounced as fraudulent; and when we produce documents and papers, and witnesses to prove our title, the Government of the United States seems, and actually is disposed, to call upon us to prove that forgery is an impossibility. Not to raise a doubt themselves, not to prove themselves that anything is wrong, but we are called upon to prove that it is *absolutely impossible* that any of our papers *could* be forged. Now that is a hard burden to place on any man.

I do not know how far the art of the forger has gone. I cannot tell what can or what can not be done by accomplished forgers. I only know that we prove these documents to be true. Whether they could have been forged by accomplished forgers I do not know, I cannot tell. Certainly it is no part of my duty to prove that it would have been an impossibility to have forged them.

Now Andres Castillero had reached Mexico, and commenced a series of communications with the officers of the Government. But, says brother Randolph, he did not get there until the 6th of May. Well, the argument upon that point was very much like the argument upon all the other points,—in relation, for instance, to the Fremont news not reaching Mexico until the 5th or 6th of April. Because an evening paper in the City of Mexico on the 6th of May, inserts in its issue, at the last moment before going to press, a statement that the Supreme Government had received news from Upper California; because that evening paper did not receive, or did not publish that news till the 6th May, my brother Randolph argues, as if he was perfectly certain of it, that Andres Castillero did not arrive till that date—that he did not arrive in the City of Mexico until that paper got that news. Now I would have drawn just the opposite conclusion. I would have supposed that a messenger sent by the Governor of California to the President of Mexico with special dispatches, would not first run to the office of an evening newspaper and tell its editor what tidings he had brought, before even going to see the President. I should suppose that the first thing which he would do would be to call upon the President of the Republic, or the Minister to whom his dispatches were addressed, and give his dispatches; and that after they had examined the dispatches, and consulted in relation to their contents, they would allow the dispatches to be published, or not, as they might judge best for the interests of the country. But it did not occur to me, that because the evening paper did not have this news before the 6th May, therefore it was impossible that Castillero should have arrived in Mexico before that day. I am willing to leave that point of inference to your Honors. I do not think that it is worth spending breath upon.

We find communications in relation to Andres Castillero on the 5th of May. That we know. We are therefore authorized in saying that he arrived on the 5th of May.

And here let me comment on a series of papers obtained from Mexico, which are most remarkable. We have again to thank brother Randolph for a most important piece of evidence.

We had sent to Mexico to obtain from the different public offices, copies of the proceedings in those offices in relation to this mine.

We applied to the Junta de Fomento for a copy of the espediente there. We applied to the Faculty of the College for a copy of their proceedings relative to the assay of this mineral; and what the result was, and the richness of the ore, was spread before your Honors. We applied to the Ministry of Justice, and all that appeared in this connection in its archives is brought here. We applied to the Minister of Relations, and obtained a record of the espediente existing in that office in relation to this matter; and here we rested our case.

But brother Randolph's acuteness put us upon a new track, inoculated as with a new and valuable idea—just as he cross-examined Chard, and proved that he was a native of New York—thus making his testimony exactly tally with the Consular archives. It was brother Randolph's acuteness that put this question to the witness [Couto]: "You say that all these things were done by the Junta?" "I do." "You know that they were done by them?" "I do."

Q. 173. Was there any record kept of the discussions of the Junta on the subject of such applications?

A. 173. The result of each day's deliberations was written down in a book, and signed by the gentlemen of the Junta. This record was called the book of the *Actas*.

Q. 174. Where is that book now?

A. 174. In my possession, in the office in Mexico. Those books are not allowed to leave the office.

Q. 175. Were not the final resolutions of the Junta on the subject of any application extended more formally than in the book of the *Acta*, or journal of their proceedings, to which you have referred?

A. 175. This book of the *Acta* is a formal record. The first and last sheets in the book are stamped. The book is stamped in the stamp-paper office, and the pages are all numbered in that office before it is stamped, so that it is the same as if each leaf was stamped. The same amount of fees were paid to the stamp office as if each leaf were stamped. The *Administrador de papel Sellado* certifies on the first page of the book the number of pages in it. This book undergoes all this preparation before it is used by the Junta.

Q. 176. Is this the only book in which the final resolutions of the Junta are recorded?

A. 176. Yes.

* * * * * * * * * *

Q. 188. Have you brought with you a copy of the *Actas* of the Junta in reference to the application of Andres Castillero?

A. 188. I have not.

Q. 189. Why did you not bring a copy of the *Acta* of the Junta upon his petition, that being as you have said, a mode by which the Junta executed contracts and assumed obligations?

A. 189. Because my business here was only to prove my own signatures, and to prove that the dates upon the documents in the expediente of which I have spoken, were the true dates of those papers, and to prove that they had been in my custody. I had nothing to do with any other matters.

Now, brother Randolph has got us. We have been keeping back all these actas, which would demonstrate that the whole of our papers were forged, fabricated, simulated, etc.

But we send to Mexico, and we find not only the minutes of the Junta, but the original rough draft of the minutes, as taken down in brief, whilst the Junta was in session; and we bring up traced copies of these original borradores, and traced copies of the clean copies drawn from them. These copies of the actas and minutes contain the entire proceedings relating to the Mining Body of Mexico; contain thousands of facts and names, embodying all the doings of that office, and afford a complete picture of the whole mining business of Mexico at that time. We have here every possible indication which any human being could desire of the accuracy of all our statements in regard to the acts of the Mining Junta. Take up any one item in these thousands of statements. Here is a whole year's business of the mining body, their proceedings day after day and hour after hour. Every two or three days, everything that has been done during the two or three days previous, is entered carefully into a well-written volume, in order that the record may remain. We have done in this instance just what we did in reference to these newspapers. We have brought in files of the public newspapers of the day. You say that they are not true; that we have forged entries; that we have forged proofs in them which the true papers do not contain. We have brought in these entire files of papers which prove and vindicate themselves. We anticipated these charges. We bring in these minutes as we brought in the papers, and we challenge you to prove *one single fact* in them to be untrue. You have not brought a witness to disprove a single fact in them. If the newspapers are forged, can't you get a true copy? If this official newspaper is forged, we say, bring in a true copy. *Then we will give up the case.* The whole mining proceedings of the Republic for fifteen months are thrown open to you. Prove one statement forged; prove one single entry forged by us and we give up the case. You have had years to do it in.

Mr. Randolph—These papers don't belong to me.

Mr. Benjamin—Don't belong to you? No. They were offered in evidence. They are produced; the dates are given; the full record is there.

Mr. Randolph—We don't keep them.

Mr. Benjamin—No; but mark what you do have. The Department of State has duplicates of them. The Department of State at Washington keeps the official newspapers of all nations with which we hold any relations. Now, prove that we have forged or interpolated one solitary newspaper item, and we give up the case. You have got the evidence. You can prove that we are forgers, and drive us out of Court in shame. Make us hang our heads for our attempt to impose upon this Court. You can do it with the greatest possible ease if your charge is true.

They offer no evidence on these points. They offer no proofs in this connection. And it is because all the evidence proves the charges unfounded, that we have this regular, persistent determination that your Honors shall see nothing that they can keep from your view.

Hence the refusal at Washington, when we urged the authorities there to send down their own officers to Mexico at our own expense to examine the archives and witnesses there. Let them look at the records. If you do not feel authorized to draw on the Treasury for such an object, here is the money to pay for all expenses. Name anybody you please. Let them go down to Mexico, investigate, and come back and make their report. Take any men you choose. Send down such men as you sent before—such men as Henry May. Let them search the entire archives. Give them ample time and opportunity.

Now, we say to you, Mr. Black, if your object is to do justice—if you uphold in your office those traditions of the honorable men who have occupied your seat before you—if you have any regard for the memory of the Wirts, the Legares, the Berriens and the Johnsons, who have illustrated that office—pursue their example. Send out intelligent, skillful, scrutinizing, honest men to examine into the truth of this whole business, and if they satisfy themselves that our titles are forged, then we will admit that we are in the wrong, and we will do what we can to repair the injuries we have inflicted upon the just owners of this property. But Black exclaims, "No; *I* know that you are forgers, and *I* will not agree to your bringing any evidence into Court." "Then you won't send down a commission to look into this matter?" "No." Well, we will do one thing; there is one thing remaining which we can do. We think that under the law we have the right to take testimony in Mexico before the United States Consul. We are going to take our witnesses before the Consul, and we give notice to you what examinations we propose to make; so if you choose to take any steps by way of cross-examining the witnesses, or introducing evidence of your own, here you have an equal opportunity with us. The names of our witnesses are given. What is the answer to that?

The Attorney General runs post-haste to Secretary Cass, and gets him to issue an order to the United States Consul at Mexico to take no testimony for us. That is in the record. That is the way our Government does justice; with the plighted faith of the Nation, published to the world, that it would respect California titles.

Now, may it please the Court, these documents, these archives, are sworn to by witnesses by the dozen. Men have been brought up here from the City of Mexico, of high position. It has cost a great deal of money. My brother can make any thing from that admission which he pleases. We have been compelled to charter a steamer for the work. We have been compelled to send down agents upon her to Mexico. We have been compelled to get brother Billings to go down to Mexico and induce some of these officers to come up here and give us their testimony. He succeeded in his mission.

These men were not coming to a distant country without some indemnity, and no sensible persons would ask them to do so for utter strangers. They could not afford to leave their business and come up here on a chance of payment for their trouble and losses, or for a trifling consideration.

But there is one of these men who would not receive any indemnity, and to him I will again refer. All the rest were examined by us on this point, and they stated that they had received money for coming up and giving in their testimony, and made no secret of the amount.

Who were they that came up from the City of Mexico?

Manuel Couto. He was Secretary of the Administration of the Mining Fund.

The next man is José M. Bassoco, a member of this Mining Junta that made these very bargains with Castillero which we allege. He was not a Government official. He was elected by the creditors of the Mining Fund to sit in that Junta and watch their interests. He represented $5,000,000 of the creditors of the Junta.

The next man is Blas Balcarcel, Director of the Mining College.

The next man is Antonio del Castillo, Professor of Mineralogy, who divided out the specimens into different departments, and marked them. And this witness testified to the intense interest with which these specimens were received and examined.

Then comes Yrisarri, Fifth Clerk in the Ministry of Justice. Brother Randolph sneers at this witness, as "only a clerk." Clerks can't tell the truth. He is only a clerk, the same as any clerk in one of the Federal Departments, he says. Are our Department Clerks here incompetent to tell the truth? Does the fact that they are clerks effect their credibility?

Who else will you prefer to bring up to testify to a certain writing, rather than the man himself who did the writing?

Here is Mariano Miranda, a former Clerk in the Ministry of Justice. Here is Velasco, Clerk in the State Department. Then here is Castillo Lanzas, Minister of Foreign Relations. Then here is José Maria Lafragua, Minister of Foreign Relations, who made a report in 1846, speaking distinctly of the matters now in controversy. Then here is Francisco Villalon, Notary Public, who put his signature and seal to a large number of the instruments introduced in this record from Mexico.

Every official in the City of Mexico, whom we thought could be of use in explaining and verifying any of our papers, was brought up here and examined before the Court—and, need I say it, *cross*-examined. These cross-examinations are a part of the judicial history of the country—a remarkable portion. Upon that cross-examination I may have something hereafter to say. It is something totally unprecedented in the whole history of jurisprudence. Six or seven hundred questions to one witness are what is termed, in familiar language, a mere flea-bite. That constitutes a mere commencement. Every possible thing which human ingenuity could devise, in the shape of a cross-question in such a case, is brought up and propounded.

Men are not only asked their age and birth-place ; their entire domestic relations are inquired into. They are asked where they went to school, who were their school-mates and *school-masters*.

They are asked whether they are the men they profess to be, *and if they have brought proof of their own identity!*

Castillo Lanzas is called upon the stand. My brother Randolph asks him : Who are you? What is your name? Can you *prove* that you are the person you say you are?

Well, I am Castillo Lanzas, is the reply. Mr. Buchanan knows me. I knew the whole diplomatic corps in London, when he and I were there together. It did not suggest itself to me to come here with certificates of identity. "Who knows you here?" inquired brother Randolph. "I think that there is a gentleman here by the name of Arce, who keeps a tobacco store, who knows me, and to whom I gave a passport in London."

"Well," says brother Randolph, "I give you permission to summon this witness, in order to prove that you *are* Castillo Lanzas." Now, my friend Mr. Peachy plays rather a scurvy trick. Brother Peachy pretended that he was very apprehensive that Castillo Lanzas would be proven to be not Castillo Lanzas. At least I take this to be so, for he objected very strenuously. Brother Peachy would not allow of the introduction of this tobacconist on the stand. Brother Randolph then insisted upon his own right to bring him forward, and summons him in behalf of the United States. Well, in comes Mr. Arce. Castillo Lanzas is pointed at, and the question is put: "Do you know that man?" "Yes." "Who is he?" "That is Castillo Lanzas." "Where did you know him?" "I knew him in London. He was Mexican Minister there." "Well, now," exclaims brother Randolph, "who are you, sir? Who knows you?" That is the question next put to the witness.

[The Marshal had to call "order" in Court, the laughter and demonstrations of applause being very loud.]

The moment the man said that he knew Castillo Lanzas,

brother Randolph wants to know who knows the witness that he himself had summoned!

Now, may it please the Court, I will do my brother Randolph the justice to say, that toward the close of the examination, the mild conduct, the gentlemanly bearing, the evident respectability of the witness upon the stand, was such that my brother became somewhat repentant of the mode in which he had conducted the examination; and he testified his admiration of the witness and his respect for his character, by asking him for his photograph, *and has his thanks for the portrait* carefully recorded in the Transcript.

Now, there is Castillo Lanzas. I suppose we have afforded sufficient proof of his identity. Mr. Head, who likewise was called by Mr. Randolph, knew him at the Mansion House, Philadelphia. He remembered Castillo Lanzas.

"But," says Randolph, "Who knows you, sir? Who are you?" I presume that the result of all this must have been that my brother Randolph became satisfied of Castillo Lanzas' identity.

MR. RANDOLPH—No; I am in doubt yet.

MR. BENJAMIN—You don't believe it yet! May it please the Court, my brother still remains in doubt on that point!!

Well, may it please the Court, so it happens with all the rest of our witnesses.

This large body of unimpeached witnesses, men peculiarly competent to testify in this case, against whom not a solitary word of reproach has been produced, have sworn that these papers are genuine, written by themselves, and bearing the date at which they were actually made.

How are all these witnesses disposed of? Brother Randolph disposes of them with the facility peculiar to himself: "I say that these papers are all forgeries, and every one of your witnesses is a perjurer." That disposes of them effectually. Not a syllable in the evidence, not a syllable in the Transcript, going to show that any one of these gentlemen is capable of telling a falsehood; but the whole of them are disposed of by a single statement to your Honors—that they are all perjurers, and that all they have sworn to are forgeries.

Now, amongst these witnesses there was one old man whom I have never seen, but who has been spoken of to me with affection and veneration by gentlemen in no way connected with these claimants' interests, who made the old gentleman's acquaintance while he was in San Francisco.

Francisco Martinez Negrete has been represented to me, by gentlemen outside of this case, who know him as a model of a Castilian merchant of old, as a man above suspicion or taint of reproach ; as a man whose every action was always open as the day; who, to save the life of himself, or the lives of the members of his family, would not utter a syllable of falsehood. This man, in 1845, was a merchant in Guadalajara ; he is yet living in the same place, in his sixty-second year. Mr. Peachy received a letter from him a few days ago, which he was kind enough to show to me.

In 1846, owing to some circumstances disclosed in the record, he was in the City of Mexico. He was an old correspondent of Alexander Forbes of Tepic. Alexander Forbes of Tepic, had just made a bargain with McNamara, the agent of the owner of the mine, that he would supply this mine for sixteen years on certain terms. Forbes writes to Negrete in the City of Mexico, informing him that he wishes to buy some mining shares, and begs him to act as his agent.

Negrete and Eustace Barron were old friends, and when in the year 1858 or 1859 Eustace Barron, in the City of Mexico, was on his dying bed, Francisco Negrete, being then in the City of Mexico, was requested by him to go up to San Francisco, and vindicate the character of his old friend, by proving that he himself bought these barras from Castillero in Mexico, in December, 1846; that he paid for them then; that he received the two-league grant, and forwarded it to Tepic; and to produce such evidence as was in his power to vindicate Eustace Barron from the aspersions thrown upon his character in the Courts of California. And it was because Eustace Barron was dying that Negrete promised to do this. He died, and that Castilian merchant complied with his pledged word, and did come up to this city to give his testimony in this case. And it was because he was performing a friendly service, acting from his love and affection to a dying friend, that he agreed to make this journey, and give his vindicating evidence at great personal inconvenience and without pay or reward.

He is asked, "What did you get for coming up here?" "I got nothing," was the reply. "If anything had been offered me, I would not have come." He did not come up here as a matter of business. He came here as a matter of duty to a dying friend, whose character had been aspersed, and whose last hours were embittered by charges of fraud brought against him. And this dying man told his son to spend the last dollar of their fortune, and never to compromise this case until his memory was cleared from these foul assaults; and it is for that reason that no compromise has ever been made. All the wealth that was left by Eustace Barron, and it is large, will be expended in this case until his character is vindicated. Many offers have been made to buy us out, and proposals made to secure the withdrawal of the injunction by compromises. All these efforts have been made. All propositions for compromise are useless and idle. We expect and ask for nothing but justice from your Honors.

And now we say that we repel this charge of fraud made by a body of speculators who have succeeded in deceiving the Attorney General of the United States, and cause him to issue some of the most extraordinary productions that ever emanated from any public officer; bringing the whole power of the Government to bear against the claimants; suppressing, so far as lay in his power, all testimony which was suspected to be of a tendency favorable to our interests. We repel the charge and defy the proof. Finally, at a cost of nearly $200,000, we have obtained the evidence in this case now before the Court; and this we never could have done if we had not had large means wherewith to bear up against the whole power of the Government. Against persons of only moderate means, this vile conspiracy against rights would have succeeded.

Well, all these witnesses come up here; and amongst them is old Francisco Martinez Negrete. Here we had in our possession the letters which he had written to Alexander Forbes, in 1846. Alexander Forbes left their correspondence behind him in Tepic, when he departed for England. He is now in England, eighty odd years of age. These letters were produced here. Negrete at once recognized them as the letters written at the time at which they are dated. He gives his testimony calmly and consistently from the beginning to the end. The whole of his transactions with Andres Castillero are spread before the Court. Castillero's first letter is received. He offers Negrete two shares for himself. His reply is, that he don't wish to go into the mining speculations. This prudent old merchant had never done anything but attend to his mercantile business, and the idea which he entertained of engaging in carrying on Mexican mines was, naturally enough, that it was like throwing a fortune into a bottomless pit. He could not believe that anything was to be realized from this mine. To use his own very expressive language, "his very flesh quaked" at the thought of a mine.

Now I want to refer your Honors to the letters of Negrete, which give a picture of what occurred in Mexico at the time of this purchase.

I do not intend to detain you by reading largely from the deposition of Negrete. I merely wish to read his letters, produced here in the original, and inquire if, from their bare inspection, they do not in themselves rebut the possibility of forgery. Indeed, the pretense that they are forged is not in the record. Only my brother Randolph says they are forged.

I will read from pages 2484–5–6–7. The first things that we have are three checks. Negrete procures them from his banker in Mexico, and brings them up with him. Here is the first check:

Señor Don DONATO MANTEROLA:

Mexico, December 18, 1846.

My Esteemed Friend: I beg of you to deliver to Don Andres Castillero the sum of four thousand dollars, for account of D. Alexander Forbes, charging the same to the account of your sincere friend.

FRANCISCO MARTINEZ NEGRETE.

Received—Andres Castillero.

At the bottom of this check we read "Received—Andres Castillero." There is the payment. This comes out of the banker's vaults. Check No. 2 reads:

Señor Don DONATO MANTEROLA:

Mexico, December 22, 1846.

My Esteemed Friend: I beg of you to deliver to Señor Don Nasario Fuentes the sum of one hundred and thirty-seven dollars and twenty-five cents, for account of the Sres. Barron, Forbes & Co., charging the same to the account of your sincere friend.

FRANCISCO MARTINEZ NEGRETE.

Received on the above date—Nasario Fuentes.

We read at the bottom of this: Received on the above date, "Nasario Fuentes."

There is the order on the banker to pay the Notary his fees for drawing up the papers; and this paper also he procures direct from the banker's vaults.

Again, on the 2d of February, 1847, as appears on the face of these papers, Forbes wrote to the City of Mexico, and asked Negrete to send him a second copy of the contract. Under the Mexican law, Negrete was obliged to go before a Judge and request him to order the Notary to make out this second copy; otherwise, the Notary could not furnish one. The Judge's order is here. The petition to the Judge is here. The copy delivered by the Notary is here. And here again is the check on the banker for payment of charges on this second copy:

Señor Don DONATO MANTEROLA:

S. C. Mexico, February 6, 1847.

My Esteemed Friend: I beg of you to have the kindness to order the payment of the sum of seventy-five dollars to the Señor D. Nasario Fuentes, charging the same to the account of your sincere friend.

FRANCISCO MARTINEZ NEGRETE.

Received—Nasario Fuentes.

Now, not only are these banker's checks here, showing the payment by Negrete at that time, but there is a letter here showing the whole transaction precisely as it occurred.

On the 5th of December, 1846, Negrete first acknowledges the receipt of orders from Forbes to buy for him:

Señor Don ALEJANDRO FORBES:

Mexico, December 5, 1846.

My Esteemed Friend:—Pursuant to your favors dated 24th ultimo, I have spoken to Señor Castillero in the terms which you express in one of them, and he has promised to think over the business and bring me his propositions to-day, and also the document of the ownership of the mine which he has in his possession. It is now eight o'clock at night, and he has not brought his proposals, from which I understand that he has not been able to finish them, and should it be so I shall send them to you by next mail.

Your affectionate friend and servant,

FRANCISCO MARTINEZ NEGRETE.

POSTSCRIPT.—After having signed the above, Señor Castillero has come in and has brought me the contract of partnership in the mine, with a copy of the original, which I transmit, as I have not time to make a second copy. You will observe the simplicity of that document, and that it is wanting in explanation. By article second, none of the associates can alienate his share. Señor Castillero says that he drew up this article to prevent any of his partners from disposing of any share (barra) to the North Americans. He has also handed to me the letter, the original of which I send you, for the purpose that having informed yourself of their contents, you may communicate to me your orders in the terms you may think proper. According to Señor Castillero, the half of the proceeds belongs to the supplier (aviador) of a mine. You, in view of the contents of the contract, and of the proposals which he makes in the letter addressed to me, can decide upon what is most expedient, and I assure you that on my part I will obey your orders to the best of my ability. Said Señor Castillero places much value on the Island of Santa Cruz, which is his property. He says it is thirty-two leagues in circumference, with a good harbor, timber and fresh water. He is disposed to sell it rather than have it taken from him by the Yankees. It is distant four leagues from the port of Santa Barbara, and he considers that it would be of much importance to England, should the United States take possession of the Californias. If you are inclined to purchase, you can inform yourself of all this leisurely.

The mail is about to close, and without time to say more, I remain your affectionate friend and servant,

FRANCISCO MARTINEZ NEGRETE.

Now, here is a letter (page 2486) dated Mexico, December 9th, 1846:

Señor Don ALEJANDRO FORBES:

Mexico, December 9, 1846.

My Esteemed Friend:—Don Andres Castillero delivered to me, just as the last mail was leaving, a copy of the papers of the mine in Upper California, and a letter addressed to myself on the subject of the terms on which he would agree to enter into a contract with you. That you may carry this business into effect, it is requisite to make it sure that it be done in a manner which shall leave no room for disputes afterwards; and as the said documents sent to you in my letter of the 5th, furnish some information, it would be conducive to this end to submit the business to the judgment and opinion of a good lawyer, who will draw up the conditions in conformity to the ordinances of mining and the existing laws. With this and the instructions you may give me, I can do everything else here which I may be ordered. Señor Castillero appears to be a sedate person, and he manifests a great interest that this negotiation may be placed under the protection of the English flag, as being the only means of securing that property, in which he owns twelve shares (*barras*), having ceded the other twelve out of consideration to the persons who now own them.

I had written thus far when I received late through the English Minister your esteemed letter of date 28th ult., and in consequence of its contents I have spoken with Señor Castillero again on the subject of the negotiation of the quicksilver mines, and find him entirely disposed that the half of the mine which belongs to him be supplied (se refaccione ó habilite,) or to sell four shares (barras) as may appear best, on account of the interest which he has in defending this negotiation under the protection of the British flag. Seeing him so well inclined, and desiring to acquire information regarding the value of the shares, I have resolved not to enter into any complete arrangement before speaking first with Father MacNamara; neither has it seemed to me proper to divulge what I know about the contract you have made there, until the arrival of the Father and I have a conference with him, and it be seen whether it will not be better that he himself inform Castillero of said contract, which I have read carefully, and find that it has not all the formalities prescribed by the laws, because, in the power of attorney granted by Señor Castro should have been inserted the power which the other two parties interested gave to him, that is, the Father Real and Robles. I would have been glad that you had fixed some starting point for me, as I find myself in the dark and unable to form any opinion as to the value of each share; but let it be what it may, and whatever may be the opinion of Father MacNamara, I will not go beyond the price of fifteen hundred dollars, or at the utmost, seven thousand dollars for the four shares which Señor Castillero is disposed to sell, as appears from the letter which he has addressed to me to-night, a copy of which I inclose. I shall inquire of said Father what price Don Diego Forbes gave for the two shares which you mention, and be guided by the same; and you will please observe, that should said price be more than two thousand dollars each, I will suspend making the purchase till I receive further orders from you; as in truth, not being a miner, nor inclined to that branch of business, my flesh quakes at only thinking upon the precarious nature of such negotiations and their results, etc.

I am your affectionate friend and servant,

FRANCISCO MARTINEZ NEGRETE.

Now, what is it that Castillero wrote to Negrete, in the City of Mexico, and which the latter forwarded by copy to Alex. Forbes:

Señor Don FRANCISCO MARTINEZ NEGRETE.

Home, December 9, 1846.

Dear Sir:—To the proposals which this morning you were pleased to communicate to me, regarding the quicksilver mine, I will say briefly that you may assure Señor Don Alejandro Forbes of the sale of four shares, and that to conclude the contract I only await the arrival of Father MacNamara, that, in view of his authority, more validity may be given to the instrument, and we may make arrangement of all the matters of the company. The Board of Encouragement (Junta de Fomento) has offered me one distilling apparatus for quicksilver, *of the two which it has ordered from England*, and two thousand flasks which it has in Tasco, which is important news for our business.

I am, etc., ANDRES CASTILLERO.

Just here I stop a moment. Brother Randolph, in his cross-examination, refers to the translation of these words:—*La Junta de Fomento de los dos apparatos que ha mandado traer de Inglaterra para la distilacion de azogue me ofrece uno.* That is to say, that the Junta, of the two apparatus for distilling quicksilver which it has ordered from England, offered him one.

Now, by reference to the actas of the Mining Junta, we find an entry corresponding with this very matter. It appears that when Castillero made his bargain, which was broken off because the Government consumed in their war operations all the funds in the public treasuries, he proposed to the Junta that they should give him some iron retorts which it had, and some flasks from Tasco. The Junta agreed to this proposition. The negotiations were broken off. Nothing was done. There was no habilitation of the mine. Castillero remained in Mexico. Alexander Forbes leases the mine; and in December Castillero tells Negrete that the Junta *offers him one* of the distilling apparatuses for quicksilver recently ordered from England. Now, where is the entry in relation to the distilling apparatus?

At page 1495 your Honors will find, in Spanish, the acta for the Session of the 10th of September, 1846; and the last entry but one in that acta reads:

> There was read a petition from Don José A. Nieto, requesting that there should be sold to him, at cost and charges, one of Doctor Ure's apparatuses which the Junta has bought, to put up at Guadalcazar. The Junta acceded to the request, asking Sōr. Nieto whether it would suit him to take the apparatus at Havana.

Now, we find, by following on these entries, that the Junta sells, on petition, at cost and freight charges, one of these distilling apparatuses.

The Junta enter into an agreement for the disposal of one of these apparatuses, provided that it might suit him to receive it at Havana. It had not yet got to Mexico.

At page 1529 you find that on the 26th of November, 1846, the second entry is:

> From Messrs. Baring Brothers, in London, 31st October, advising having placed in Havana, at the disposal of Messrs. Picard and Alven, the two distilling apparatuses ordered from them, with the corresponding plans to put them up, and with the invoice of same.
>
> Let the receipt of said communication be acknowledged, and let the amount remaining in their hands, of £270 0*s.* 5*d.*, be drawn for. For this purpose Mr. Bassoco is commissioned.

Here, then, in every little circumstance of their proceeding, nothing ever issues from the pen of Castillero, or any of these parties, that does not receive some emphatic confirmation in these actas.

Now we have all these actas brought in. What are the entries?

Here is Castillero's letter of the 9th of December, noticing the offer made to him by the Junta of one of these distilling apparatuses. Here are the records of the Junta selling one of these apparatuses in September. These papers descend into all these little details of circumstances. Let the United States prove one of them false or forged.

Negrete, on the 12th of December (page 2488), again writes:

Señor Don Alejandro Forbes.

Mexico, December 12th, 1846.

My Very Esteemed Friend:—I have given orders at the stage office to inform me immediately when the Father MacNamara may arrive. I do not believe that he has arrived by to-day's stage, because if he had they would have sent me word. As the business which you have on hand is of great magnitude, as I conceive it to be, and as its success depends entirely on Señor Castillero, I intend to submit the arrangements to be made in conformity with the requirements with the laws, to a lawyer in whom I have all confidence; as in such cases it is better to spend at the beginning five hundred dollars, than to have to incur afterwards a lawsuit when the negotiation is fully established and productive. In whatever way it may be done, this business will be managed in the best possible manner.

I am neither a miner, nor do I like the business. From the beginning Señor Castillero offered me two shares (*barras*), which I declined. Reckoning, with your consent, I might perhaps take one share for some friends of mine, but this would come in, like all the rest, in the contract of supply (*avio refaccionario*). In this manner, if it should be that the holders did not gain any great advantages, they would know that at most they would lose only what the share had cost them.

Wishing you good health, I remain your obedient servant,

Francisco Martinez Negrete.

There are two young men who had got it into their heads that a share in this mine would be worth a good deal. One of them was a relative of Negrete. Castillero told Negrete that he might have two barras. Negrete cannot consent to enter into a mining company; but he has two young friends who would like one share between them. Then Negrete writes to Forbes.

That share was taken; and of these two young men, one of them holds his half-barra. He lives in Mexico, and is owner of a half-share to this day; and Negrete gives his name, and tells his place of residence. This we offer in evidence for the benefit of the Government, who may prove Negrete a perjurer in this particular instance.

On the 16th December Negrete again writes:

Señor Don ALEX. FORBES.

Mexico, December 16, 1846.

My Esteemed Friend:—The Presbyter MacNamara arrived here on the 12th, by the stage, and yesterday, the 15th, he was in my house for about a couple of hours, when we talked particularly about the b siness of the mine in California, and what should be done as regards Señor Castillero. To-day at eleven o'clock we met again, together with Señor Castillero and a lawyer in whom I have entire confidence, that he might be informed of what was to be discussed, and draw up the articles of our agreement in conformity to law, or in the best manner possible. I have instructed him regarding all the essential points of the transaction, by means of extracts of your letters, that is, of such portions of them as favor our position, so as to treat on the conditions which may be most favorable to your interests, considering the present condition of the business, and I will inform you of the result of the whole, either in this letter, or separately; but, I advance these lines in case I should not have time.

The measures you have adopted in regard to the vessel which carries the cotton, are, in my opinion, very proper, and were even indispensable.

By to-morrow's mail I expect to receive accounts from Madrazo and Palacio of the result of the fair, and although all complain bitterly, I am sure that they must have sold all the stock from the factory notwithstanding the Yankee wagons which came from Santa Fé, which were allowed to enter San Juan without any difficulty. The Supreme Government is inclined to prosecute the war, and it is believed that the national representation is of the same mind; and such being the case, there will naturally follow a formal declaration prohibiting all communication with the North Americans, and even expelling those residing in the country, as was done with the French. The English courier has arrived; till now I have received no letter from my *compadre*, although I have received other letters. I hope to get it to-morrow.

I am, your affectionate friend and obedient servant,

FRANCISCO MARTINEZ NEGRETE.

And here is a postscript:

P. S.—Señor Castillero desires that you should maintain possession of the island, and for this purpose he has addressed me the inclosed letter, which I recommend to your attention.

And here is the inclosed letter:

[LETTER REFERRED TO IN THIS EXHIBIT.]

Señor Don FRANCISCO MARTINEZ NEGRETE.

Home, December 16, 1846.

Esteemed and Respected Sir:—I would thank you much if you would recommend to Señor Don Alexandro Forbes, that on taking possession of the quicksilver mine, my island of Santa Cruz, situated in front of Santa Barbara, may also appear as an English possession. Being distant only four leagues from the coast, it is a very important possession on account of the abundance of water and timber, and its good harbor. To English vessels, and even to the company, it may be of service for the absolute independence in which it is of all the country. Excuse me for repeating my request, and I remain always your most obedient servant,

ANDRES CASTILLERO.

Now, may it please your Honors, in relation to the question made by the U. S. counsel about this mode of passing property from under the flag of Mexico to that of some other power, in order to save it from spoliation by the United States, then at war with Mexico, I desire to say a word. Why? Was it not the duty as well as the privilege of a Mexican, when his country was at war with the United States, to injure the United States as much as he could? Was not this according to the laws of war? And if he is the weaker party, is it not his duty, is it not his right, under the law of nations, to hide his property from the Government of the United States, and save it from spoliation? When the war's desolation breaks upon a land, does not the shepherd hasten to the field, and gathering his flocks and herds, fly with them to some secret place of safety? Does not the prudent housewife take her treasure and bury it in the earth, safe from the hand of the despoiler? Does not the merchant send his vessels and stores of goods to distant lands, where they will be secure from seizure by his country's enemies?

But the land cannot be moved. Then every possible stratagem is employed, and legitimately employed, to hide it from the power and use of the common enemy of the country. Does the gentleman suppose we are going to offer excuses for the Mexicans who avail themselves of this undoubted privilege? Far from it. Is not the instinct of self-preservation implanted in every created being, animal as well as man?

The little fish in the sea, when pursued by a superior foe, will eject an inky fluid, and conceal its flight 'neath the murky cloud. Every living creature is born with this instinct. Go out into the field, and observe the little bird hopping away from the traveler, simulating a wounded wing, until it has turned him from the place where its treasures lie concealed, and then, when its pretty stratagem has succeeded, it will rise buoyant in the air and carol forth its thanks to God, who gave it the instinct by which it preserved its offspring.

All this is natural and proper. Why excuse these Mexicans for doing this very thing? We do not offer any excuse. It was proper. It was justifiable. It was the duty of any Mexican when at war with the United States. So much for that point.

I proceed with Negrete's statement.

On the 16th December, 1846, another letter was written, and this letter first speaks of the success of the transactions, and the sending of the papers:

Señor Don ALEJANDRO FORBES:

Mexico, December 16, 1846.

My Esteemed Friend:—At last, to-day, a little after eleven o'clock, the Presbyter Don Eugenio MacNamara, Don Andres

Castillero, myself, and Señor Don José Antonio Romero, met together to arrange the final result of the business of the mine, and, as we had to deal with matters of law, I considered it proper that Señor Romero should attend, and certainly in this I was not mistaken, as he was of great service to me in the conference and in the illustration of some points—and he also sustained and elucidated my ideas with the judgment, wisdom and experience which he has acquired from his practice in the management of affairs—so that, had he not been so well posted on the subject relating to which we had already held various meetings, and fixed the starting points, and noted the flanks on which we might be approached, probably we would not have obtained the desired results.

Señor Castillero, then, conforms to the contract executed in Tepic by his associates, and takes part in it also for his twelve shares in the mine—and, after various debates, he decided to sell you four shares. Having agreed upon this, I requested him to cede to me another share for two friends (who are your friends also), and he acceded, stating that my mediation and the regard he professed for me, were sufficient motives for him to accede to my request. Thus it is, that I have contracted for the five shares, at $800 each, making in all $4000, which I shall pay over as soon as the writings are made out—it being understood that, although the $800 of the share for the two friends shall be paid by them, this shall be included in the avio agreeably to the contract which you made in Tepic.

Besides this instrument, there will be executed another for the supply (avio) for the part belonging to Señor Castillero. Señor Romero has taken a note of all the points, so as to order the instruments to be drawn up in such a manner as to prevent disputes afterwards. Señor Castillero engages that his associates will consent to the sale of these shares; and, if they do not, he will pay you their cost out of the proceeds of the twelve shares belonging to him.

He has in his possession the title paper of the Government, by which is granted to him in the mining district (*mineral*) two leagues of land in circumference, which he has ceded for the benefit of the negotiation for the sixteen years of your contract. He has assured us that the Board of Encouragement of Mining has entreated him to sell it some share, but that he declined, and I have told him that for no reason nor consideration ought he to alienate any more shares, as in such case you should be preferred. I believe that I have acted and worked for your best interest, and if after this you obtain the advantages you propose in this enterprise, my gratification will be doubled.

By next mail I will send you the writings, and give you any other details which at present I may have forgotten.

I am, your affectionate friend and obd't serv't,

FRANCISCO MARTINEZ NEGRETE.

And here follows the letter of the 19th of December:

Señor Don ALEJANDRO FORBES:

Mexico, December 19, 1846.

My Very Esteemed Friend:—Conformably to what I expressed to you in my last letter, there have been drawn up the writings of the treaty between Don Andres Castillero, as the owner of the half of the quicksilver mine, and myself, as your representative, the originals of which were signed yesterday by both, after having acquainted ourselves thoroughly with their contents, and in the presence of Father MacNamara, who was of opinion that they were expressed in the terms best for your interests, and for this purpose our lawyer, having taken note of all the points, superintended the drawing up and other matters connected with it, because being myself no lawyer, I desired to have the assistance of one, as this would prevent differences and disputes afterwards. I have been promised the legalized copies (testimonios) to-day, and shall transmit them with this letter, together with the instrument which you executed there, and the document showing the grant which the Supreme Government made in favor of Don Andres Castillero for two leagues of land (two ranges for neat cattle) at the place where the mine is situated, and which he cedes for your benefit for the term of sixteen years.

After having made the contract with said Castillero, and not having been able to obtain from him more than four shares for you, I requested him to cede to me one share for two friends of mine, and he acceded to my request immediately, as is shown by the inclosed letter addressed to you by said Castillero. I wish to have this share for my nephew, Don Francisco Maria Ortiz, and Don Martin Lapiedra, in consideration of the first having indicated to me that a small interest in this business would make them happy, and I have desired on my part to gratify them in the belief that you wish to do the same, as I know the regard and other considerations which you have for them. As a matter of course this share is to be included in the supply (*avio*) of the company, conformably with the contract made. If we are to believe the statements of Castillero, that mine is the richest in the world, and he surely says so in sincerity, and is to me a man of strict morality and probity. He expresses himself to the effect that the contract is very favorable to you, and says that my mediation overcame everything in the way of it, so that he would not have done so much with any other person. Be this as it may, after having done my duty, my satisfaction will be much greater when I know that by this means your fortune, and that of your respectable house, has been increased.

Señor Castillero has promised to give me a certified copy from the Mining Tribunal of all the concessions which it made to him, when on the proofs given by the Professional Board (*Junta Facultativa*) it engaged to favor the mine, but which it did not do, because the Government seized upon the funds.

Now, may it please your Honors, that paper was never seen by any of the owners until 1850, and they did not know that it had the slightest interest for them. They had not the slightest idea that it could be of any interest for them. What cared they for any agreement which Andres Castillero might have made with the Junta, when it was known it had not been carried out, because the Government seized the funds? That bargain had been replaced by a new one. They now themselves furnished the funds; and it did not enter into their brains that the old bargain that Castillero made with the Junta was of any value to their interests.

It was not then until 1850, when Castillero came to Tepic, and saw these letters of Forbes, which were shown to him, saying that he ought to get a "ratification" from the Supreme Government, that he said: "I have got one; I have got my grant; the Government gave me a 'ratification.' What do I want with another?" And then he and Eustace Barron upon this statement start for Mexico, where they both resided, promising to get a certified copy of the document showing the ratification of the mining possession. Where is it found? Not in the dispatch from the Minister of Relations to the Governor of California. Nobody dreamed where else it could be. But the Government communications were two. One says to the Junta de Fomento that the President had approved of the bargain made with Castillero for the working of his mine on the terms proposed; and the other says to the Minister of Relations, give him a grant for two leagues. Now the very language of the Government to the Minister of Relations induced the belief that there was nothing contained in the terms proposed but some bargain about working the mine; and men might have gone on under that conclusion, and looked over these papers forever, and never have dreamed that under these very words was concealed a ratification. Amongst the *terms proposed*, was one providing for a ratification of the mining title. Castillero tells Eustace Barron in Tepic that there is a "ratification." They go on together to Mexico and find it, and sent it on instantly. This is the paper which Castillero was to have given Negrete in 1846. It was the paper which proved the perfect folly and uselessness of the suggestion of James Alex. Forbes. If James Alex. Forbes had known that amongst Castillero's propositions there was one to ratify the mining possession, he never would have suggested the forgeries which he did. He proposed to forge titles because he did not know that amongst these propositions so made, was one involving the ratification of the mining title.

And the best possible proof of this fact is contained in James Alex. Forbes' letter of the 20th December, 1849. It will be found at page 846. This man is proposing to forge titles; he is proposing to forge them because he does not know that they already exist. Here is his letter of the 20th December, 1849. I read from foot of page 847:

I know not what conditions Castillero may have proposed to the Supreme Government of Mexico, but, whatever they may have been, they, of course, cannot be fulfilled in California by reason of the change of government.

It never had yet suggested itself to any man, that amongst Castillero's proposal for working the mine, was one in which he asked for a ratification. The people in Tepic knew nothing of it. James Alexander Forbes knew nothing of it. James Alexander Forbes had written down letters urging forgeries. Castillero arrives in Tepic from Lower California. James Alexander Forbes' propositions were spread before him. He exclaims, "Why, what in the world is he writing about? My mining possession was ratified when I went to Mexico in 1846. The ratification is there in the archives. What does he want now with another ratification?"

And then what is the answer written to James Alex. Forbes by the house of Barron, Forbes & Co.? "Your letter has been received. In reply, we would state that Castillero is now here, and he says that his ratification was given him in Mexico in 1846."

On the 7th of January, Castillero was still in Lower California. On the 3d of February, 1850, Alexander Forbes writes to James Alex. Forbes: "Castillero has returned, and is also here." * * * "He says such approval was given, and that on his arrival in Mexico he will procure a judicial copy of it."

Now, this is the first time that it is ever suggested by anybody that the propositions for working the mine contained a condition that the mining title should be ratified. This is the very first thing that Castillero says when he is shown Forbes' suggestions.

From 1846 to 1850 there is a separation. War intervenes. Peace is declared. Castillero has been in Lower California. When he arrives we will show him James Alex. Forbes' letters and see what he will say. He does come over. The first words he utters are: "There is already a ratification. I got it in Mexico in 1846."

This brings us back to Negrete's letter of the 16th December, 1846:

Señor Castillero has promised to give me a certified copy from the Mining Tribunal of all the concessions which it made to him, when on the proofs given by the Professional Board (*Junta Facultativa*) it engaged to favor the mine, but which it did not do, because the Government seized upon the funds.

Now, he goes on to say:

At the end of this letter is a statement of the capital and expenses of the whole, with the only difference that the Señores Ortiz and Lapiedra will make good to you the eight hundred dollars which they have to pay for their share.

I have not received the instrument which I have mentioned, but I remit the other documents.

He had been in hopes that the Notary would give him copies

of the two acts—the power of attorney is sent to Tepic, as well as the grant of two leagues which Castillero gave him to send on.

Here, then, this grant of two leagues is traced directly from the Mexican Government to the hands of Castillero; from the hands of Castillero to Negrete, and from the hands of Negrete it is dispatched—as we have seen in this letter—to Alexander Forbes. Did Alexander Forbes receive them?

I read from page 2496:

Señor Don ALEJANDRO FORBES:

Tepic, Mexico, January 9, 1847.

My Very Esteemed Friend:—By your favor dated 29th ult., I noticed that you had in your possession the before mentioned documents, and I believe that on a close examination you will find them to be in order.

I shall mention your indications to Castillero, and according to his opinion the principal supplies for the mines should consist of goods of the country, mostly domestic cottons and brown sugar, which articles he noted in the report which I transmitted to you in one of my former letters.

About the end of September there sailed from New Orleans, bound for San Blas, a ship loaded with cotton, and I do not know whether the cargo belongs to Rubio or Drusina.

Without any other particular matter to communicate,

I remain, your affectionate friend and obedient servant,

FRANCISCO MARTINEZ NEGRETE.

This is from Negrete, acknowledging the receipt of Forbes' letter, advising the receipt of the report forwarded by Negrete.

From the top of page 2493, I read:

I have paid to Señor Castillero $4000 yesterday, which I have charged to your house, $120 exchange, at 2 per cent., making $4120, of which my house in Guadalajara will give your office advice.

I will send you, by next mail, a note of the other expenses.

Your affectionate friend and obd't serv't,

FRANCISCO MARTINEZ NEGRETE.

On December 19th, Negrete forwarded another little note:

To Don Alejandro Forbes:

Mexico, December 19, 1846.

My Esteemed Friend:—Immediately after having put into the post-office the package addressed to you separately, there were delivered to me the two instruments mentioned in it, and which I have the pleasure to transmit to you herewith.

Your affectionate friend and obedient servant,

Francisco Martinez Negrete.

Now, we have got *four* documents forwarded,—the two copies of the contract of avio, and the sale of the barras. Castillero's title for two leagues of land, and Castro's power of attorney, all leave Mexico for Tepic, and all reach Tepic and are acknowledged as received by Alex. Forbes on the 29th of December, 1846.

Now, these same papers are brought forward into Court, and they are exhibited to Negrete, who recognizes them all. He recognizes the Castillo Lanzas document as the very one which he at that date sent to Alex. Forbes in the City of Tepic.

But observe, before these papers went to the City of Tepic, the two-league grant was used in making a notarial act.

When Castillero told Negrete, "I have got two leagues of land adjoining this mine; I have got a grant of it for cutting wood for burning in the mine; and as you are going to take the mine for sixteen years you will want it, I throw it in; I don't give it to you; but you may go and cut wood upon it, just as you need, to supply the mine. Here is the title; take it, and send it to Alex. Forbes."

The Notary is then required by Negrete's lawyer, Mr. Romero, to put this two-league grant in the deed of avio. The Notary does this. For what purpose? That it may be inserted in the copies sent out of this deed. They wanted them all to go out together.

And in this connection we have a little incident indicative of the good heart of this old merchant. He will not engage in the mining business, but he reserves a share for two young friends, who would be made "happy" on obtaining such a possession. He will divide the cost between them. The mine *may* turn out well. At least this share will make my young friends happy. I don't want anything for myself. I take this share for these young men, not for myself. That is the tenor of these letters on this subject.

In February, 1847, Negrete presents himself before a Judge in Mexico, and petitions for an order requiring the Notary to make him out a couple of additional copies of these papers. The Judge grants this order; the copies are made out; and then Negrete writes as follows:

Señor Don ALEJANDRO FORBES:

Tepic, Mexico, February 6, 1847.

My Dear Friend:—I have written you a separate letter, and with this I have the pleasure to transmit the inclosed writings, and also the receipt for their cost, which amounted to the sum of seventy-five dollars, which you will credit to my house in Guadalajara.

I am, your affectionate friend and obedient servant,

FRANCISCO MARTINEZ NEGRETE.

This closes the Negrete correspondence.

[END OF FIRST DAY.]

SECOND DAY.

THURSDAY, Oct. 25th, 1860.

MR. BENJAMIN—May it please your Honors, in the course of my argument yesterday I called the attention of the Court to the fact, that upon the last examination made in the Mexican archives, some papers had been discovered—the existence of which was previously unknown to the claimants—those papers being a correspondence between the Governor of California and the Mexican President at the date of the discovery of this mine, and of this denouncement. And I called attention to the fact that upon the return of these papers from Mexico, the keeper of the archives in California was requested to make his search, and there found the papers which corresponded with the archives that had recently been discovered in the City of Mexico.

Among those archives is found a statement that the dispatches from California to Mexico, were taken by a special commissioner, Mr. Covarrubias. In addition to the testimony disclosed by the archives themselves, and by the testimony of witnesses who state the fact of the transfer of these communications from California to Mexico at that date, we have what I had omitted yesterday to call to the attention of the Court. In the index [holding up a small pamphlet], which I have prepared for the convenience of your Honors, are stated two entries from the "Diario Oficial." One states the arrival of the brig "Juanita" in Mazatlan, on the 2d of March, 1846, from San Diego, in twelve days. That is at page 2354 of the record. And then, on the 12th March, by the marine news at Mazatlan, published in the papers of the day, the brig "Juanita" leaves Mazatlan for San Blas. The paper contains a list of her passengers; and amongst them is Mr. Covarrubias, the messenger from the Governor of California. This fact I omitted to call to the attention of the Court yesterday.

I have stated to your Honors what the archives in the City of Mexico are. I have detailed to you what they contain, and refer you to the testimony of the witnesses who prove the genuine character of the documents relied on by us.

All the testimony to which I have hitherto adverted, is the testimony of Mexican officials. We desire now to invoke, in behalf of these claimants, the testimony of a witness sent from California for the purpose of examining and comparing these archives; and that of the United States Minister, Mr. Forsyth, who was then in the City of Mexico.

Mr. Brodie, a gentleman well known in this city, of unimpeached, and I imagine, unimpeachable character, was employed by the claimants for the purpose of taking the copies to the City of Mexico. He had long been a resident of the Mexican Republic; knew the people, knew their language, and knew some of their public officers, as he states in his deposition. He took the traced copies which had been sent from Mexico back to the City of Mexico, and went to the archives there for the purpose of comparing them, and ascertaining their accuracy. His testimony is conclusive as to the entire accuracy of the copies, and their concordance with the originals there found. Upon the cross-examination of Mr. Brodie—whose testimony is at page 1059—the counsel for the Government propounded to him two questions, to which I will call the attention of the Court.

The first question is at page 1067. The witness had stated that he went with Mr. Forsyth, the American Minister, for the purpose of making this examination of the archives and comparison of the copies with the originals.

Ques. 27.—Did you explain to Mr. Forsyth that the papers that you wanted were to be used in a certain case in California, in which Eustace Barron, Escandon, and other *foreigners*, were claiming the New Almaden quicksilver mine against the United States?

Ans. 27.—I did not explain to him anything at all. I did not know that Escandon had any interest in the matter; do not know it yet. In conversation with Mr. Forsyth, after the papers had been examined, we talked about the purpose to which these copies were to be put, and he understood perfectly well that they were to be used in the United States Courts in California.

I call the attention of your Honors to this remarkable question, as exhibiting the entire theory of the Government in this case; that it is the duty of the officers of the United States, under no circumstances to lend any aid whatever in the taking of testimony, or in affording any facilities for bringing before the Court the evidence on which these claimants base their rights:

The next question is (*Ques.* 28): "Were not those traced copies, Exhibits G. H. I. and K., all traced in California?

Ans. 28.—I say no; impossible, unless they had the originals to trace them from here.

Now those originals were in Mexico. Let us see what Mr. Forsyth says upon this subject. On page 1112, he states the facts on his examination.

How came Mr. Forsyth to go to the public offices of Mexico for this purpose? It was because the Government of the United States had refused to send any agent, or to authorize or facilitate any investigation whatever; for that reason, and in order to show the entire good faith of these claimants, Mr. Eustace Barron, of the City of Mexico, unwilling that it should be supposed that he was doing anything clandestinely or secretly, prevailed upon the representative of the United States Government to go to the public offices of Mexico, and there satisfy himself, by actual inspection, of the authenticity of the originals, of which copies were to be produced.

Mr. Forsyth says (page 1112):

In the months of July and August, 1858, I was Minister of the United States of America to the Republic of Mexico, and residing as such in the City of Mexico. In the months of July and August, 1858, at the instance of Mr. Eustace Barron, of Mexico, who was one of the parties claiming an interest in certain quicksilver mines in California, known as the New Almaden quicksilver mines, I went in company with the British Consul in Mexico, Frederick Glennie, Esq., Mr. John P. Brodie, of California, who had come as agent for the claimant of the mines, and the Licentiate Emilio Pardo, first to the office of the Junta de Mineria.

And then he describes the production of the originals by the proper archive keepers, one by one, and the careful examination made by him of those archives—by reason of which he is enabled to testify to the truth of the copies. He says that these copies were precisely alike in every respect; and to these copies he gave his certificate. At the foot of the page he says:

These original documents were found in the several offices where they appropriately belonged, and were produced by the officers having the custody of them; and I saw nothing whatever to cause me to doubt their being genuine originals. I, as Minister, certified each of the copies hereinbefore mentioned, and the facts set forth in those certificates are true, and the certificates are in accordance with the laws of Mexico.

The Government of Mexico will not allow the great seal of Mexico to be attached to copies of such documents, nor will they allow the originals to be withdrawn. And the manner in which the copies herein mentioned have been authenticated, is the only way in which such copies can be authenticated. The copies, so as above mentioned certified by me, are not now before me, but the printed volume hereto attached as a part of my deposition, and marked Exhibit A, and purporting to contain printed copies of those copies, I believe to be correct.

Now, having produced these copies, with certificates sworn by the United States Minister in Mexico to be the proper certificates for authenticating them, and the only ones known to the Mexican law by which copies can be authenticated, I shall not detain your Honors by referring to the law book, because you have familiar in your memory the provisions of the Act of Congress constituting this very Court under which you are ordered to determine these cases according to the usages of Mexico, and yet your Honors are invited by the counsel for the Government to take these certificates thus made according to the laws and usages of Mexico, carry them to the window, tear them up and throw them out, as impertinent matters in this case.

They are good, valid, legal certificates. They are proof conclusive of that which they purport to state. They are good certificates in any Court of Justice in Christendom. They are good by the very words of the law under which your Honors are bound to decide the cause.

Now, it is said: True it is that all these documents are there in the archives. True it is that they purport to be original.

True it is, that the Minister of the United States, charged with the defense of their interests in a foreign country, has examined them, and can find no reason whatever to doubt their authenticity; but I, the Attorney General of the United States, instruct the special counsel of the United States to say in Court that they are not true originals; that they are antedated, fabricated, forged and fraudulent.

Your Honors are called upon to say this, without a shadow of testimony in support of the assertion. What semblance of probability is there in this assertion? If our object had been to forge papers by which we could show to the satisfaction of this Court a grant of this mine and of this land—if the gentleman [Randolph] gives us credit for sufficient subtlety and ingenuity to carry on, without leaving a trace behind us, this long series of forgeries, why does he not then give us at least credit for sagacity enough to make papers that would be undoubted grants? For, he says, the papers we have fabricated, and that we have taken all this pains to forge, are no grants, and are not legal conveyances of the property. If we wanted to fabricate papers to prove our title, the papers themselves, on their face, would have admitted of no two constructions. But he says our papers do admit of two constructions. He says they are doubtful, vague, uncertain, no title papers at all. And yet we, forming this scheme to place in the public archives of Mexico, a body of documents which are to put our title beyond dispute and contradiction, fabricate titles open to all the objections as to their true construction which his legal ingenuity has suggested to the Court! He says the paper we rely on as title to two leagues of land, on its face does not purport to convey it. He says the papers we say are a ratification of our mining possession, are not such in law, nor do they purport to be such on their face.

Now, if we wanted to fabricate something, we would not fabricate something that admitted of these doubtful constructions; but we would have put in the papers a direct grant of what we claim—a direct and undeniable grant of the land to us, without any necessity of referring to local officers in California to put us in possession, when it is known that no possession was ever had. All these papers which we produce are such as he says do not carry out the object. We had skill, we had sagacity; we could put papers in the Mexican archives by the bundle; we could do anything we pleased; and yet it did not please us to forge plain and unequivocal papers—but only papers that are liable to all the legal objections which he has suggested to the Court; liable to double constructions, not purporting to be grants, and if grants, requiring further proceedings to give them effect according to their true interpretation!

What a strange system of forgery is this! All this risk, all this skill, all this labor and ingenuity, exhausted to forge papers that are not, he says, good titles; which, he says, are not good, if true! What probability, may it please your Honors, in all these wild assertions?

But the gentleman goes farther, and tells us that it is absolutely impossible that all these transactions should have occurred in the City of Mexico at the time that we say they did occur there; and why not? With his vivid imagination, with his artistic power, he has drawn a picture to the Court of the condition of Mexico at the time. He represents the American armies as thundering at the gates of the city; the entire Republic as in alarm; the whole mind of its public officers as intent on but one thing—the defense of the country. He says, that if a thousand witnesses came forward to swear to the fact that all these transactions about the mine occurred, the inherent improbability of the assertion is such that your Honors ought to reject it as an idle fable. And then, in the fervor of his imagination, he furnishes us with a comparison: he says, we may as well believe that such an act could have been performed by the officers of the beleaguered city, when Cyrus, with his Persian army, was thundering at the gates of Babylon. Does not my brother Randolph remember what was occurring in Babylon that night? With Cyrus thundering at the gates, did not Belshazzar and his Court that night engage in riot and in feasting? So far from the inhabitants of that doomed city keeping watch and ward, or fearing danger, mad orgies filled that night in Babylon. Cyrus was entering the gates, yet the people did not know it, so confident were they in their security. Vainglorious and presumptuous were the Babylonians; nor did they suppose it possible that their well fortified city could be taken.

Will my brother also call in doubt that relation, and say *that the dread handwriting on the wall was also forged and antedated?*

Mr. Randolph—Certainly not.

Mr. Benjamin—Yet he will see at once how very unfortunate has been his allusion to ancient history, for the purpose of satisfying the Court that all Mexico was ringing with cries to arms, and that it was totally impossible that any other business could have been under consideration. Let him look at the newspapers, which it is said we have forged, and which are here offered before the Court. All the daily operations of life were going on; the theatres open, people going to their usual evening amusements; the administrative acts of the country spreading over so vast an extent of territory as was then embodied in Mexico, going on in daily routine. And all this theory that it was impossible that the Government of Mexico should have been occupied with anything else than the war is utterly annihilated by the facts which we know to have occurred. Besides which, the gentleman's picture is overdrawn. At the time when these things were going on, it is true there had been some frontier battles on the Rio Grande; true that the American army had dispersed the Mexican army at Palo Alto and Resaca de la Palma; but the news of these events had not reached Mexico.

Mr. Randolph—The news reached the city on the 19th of May.

Mr. Benjamin—These things all occurred on the 5th, 6th, 7th and 12th of May.

Mr. Randolph—The grant is dated the 20th.

Mr. Benjamin—It is proven that all the arrangements were completed previously, and needed nothing but the President's "*acuerdo*" to make the whole complete. All the preceding forms had been complied with. All the operations, the bargaining, and the communications between the different Ministers, had taken place before even the news of the battles had reached the City of Mexico—so that my brother's argument is false in theory as well as unfounded in fact.

My brother Peachy recalls to my memory the ball in Brussels on the night before the battle of Waterloo, when the British officers were called from their dancing with the ladies to fight that memorable field.

But, if your Honors please, another ground is suggested. It is suggested that the demand for the land, alleged to have been addressed to the President in Mexico, could not in all probability have been acceded to, because upon the very face of the papers produced by Castillero himself, it was stated that his mine was upon private property; that it was upon the land of the "retired sergeant, José Reyes Berreyesa." Now, if the mine had actually been upon private property, this argument would have no force whatever, for the reason that there is no pretense that the land of Berreyesa extended for two leagues all around this mine; and, therefore, the grant of two leagues contiguous to the mine, and upon the mining possession, would of necessity have been considered by the local authorities, when giving possession, as subject to the rights of Berreyesa, and would have been so located as not to interfere with him. The very purpose of the Spanish law, in requiring the delivery of a juridical possession, was to take care that the grant—which is made very frequently in the neighborhood, and sometimes on the very boundaries of, private land—shall not intrude upon that private land, but that the land (given by the grant) shall be so laid out by the public authorities as to respect prior possession and prior title. So, under any circumstances, the authorities of California, upon the production of this title, would have laid out the land without encroaching upon the possession of Berreyesa. But all this is explained away at once by some documents that have since appeared in the case.

At page 2729, your Honors will find Berreyesa's title in the Spanish; and on the next page, in the English. The title purports to be issued by "the citizen Manuel Micheltorena," then the Governor of California. In the 4th article, the grant states: "The land of which donation is made is two square leagues, a little more or less, as is explained by the respective sketch."

That paper, purporting to be a grant of two square leagues, was handed by Berreyesa himself to Castillero. It is in Berreyesa's handwriting, as shown by the evidence. The fact that it was his handwriting is proven by the testimony of two witnesses.

Well, now, Berreyesa, as is admitted on page 2728, died about the month of June, 1846; so that this paper, which is a copy of Berreyesa's title, made by himself, was made prior to June, 1846.

Berreyesa, then, prior to June, 1846, furnishes in his own handwriting what he says is a copy of his title, and which purports to be a conveyance of two square leagues.

Now, if your Honors will look at page 542, at a letter written on the 7th of February, 1848, by James Alexander Forbes to Alexander Forbes, at the time that they were making the survey of the two-league grant through the agency of the United States Surveyor, you will find Forbes says: "My opinion of the title of the widow of Berreyesa, or rather my first view of that title, and the information given by her sons, was that the grant was for two leagues; but I have been at her house on my return hither, and I find that the title was given for *one* league, or sitio, and that the word one had been converted into the word two."

Now, when Castillero first applied to the Governor of California for a grant of two leagues, in the papers which have since been found here, and in which his petition is described, he asked for two leagues *adjacent* to his mine, believing his mine to be on Berreyesa's land; believing it by reason of this forged paper of Berreyesa's, in which he had erased the word "one" and inserted the word "two," and added an "s" to the word "sitio." Therefore, when he asks this land, before leaving California, he wants it *adjacent* to his mine. He goes to Mexico and determines that he will ask for the grant there directly from the President.

What occurred in Mexico? Will your Honors look now at page 3049? On that page is a letter written from the City of Mexico by Castillero to Alex. Forbes on the 14th of January, 1847, just after he had in the contract of *avio* leased his two leagues of land for sixteen years, and had sent the title to Alexander Forbes:

Esteemed Sir:—The Sōr. Don Francisco Martinez Negrete, has informed me of your willingness to place my Island of Santa Cruz under the protection of H. B. Majesty's Government. I thank you as well as Mr. Negrete for this favor, and to that end I enclose a letter for Sōr. Don Antonio Aguirre, a citizen of California, who holds my power of attorney and a document which proves my property.

It has seemed to me proper to enclose you a copy of the title of possession of Don José Reyes Berreyesa, in which is situate the quicksilver mine; in my opinion it is not within the possession of said Berreyesa, but is situate on vacant land. I state this to you because I was so informed by several adjacent proprietors (colindantes). The copy which I send you herewith is given by himself, and bears a falsehood on its face; for Governor Micheltorena gave him only one league (sitio) of land, as that gentleman himself has informed me. I remark this to you, so that you, by an agent of yours, may examine Berreyesa's title in the archives office at Los Angeles, so that you may take possession of the other two which belong to the company and which the Government of Mexico has granted.

Now, compare Castillero's application to the Mexican Government, after he has gone to Mexico and learned this fraud, with that which he made before he left California. When he leaves California, under the impression that Berreyesa has a title for two leagues, he asks the Government here for a grant of two leagues *contiguous* to the mine. When he reaches Mexico, and finds that that title (Berreyesa's) is a fraud, and that the grant was indeed but for *one* league instead of *two*, he applies to the President of Mexico for a grant of two leagues *upon* the mine.

Is that plain? Is that natural? Is that consistent? There is nothing on the face of these documents that does not carry out our theory to the fullest extent. He (Castillero) wanted his land as near his mine as he could get it, evidently. It was for the purposes of his mine. In California he thinks his mine is on another man's land; he asks for land which is not *on*, but *near*, the mine. In Mexico he finds his mine is *not* on the land of another; he asks for two square leagues on the land of his mining possession. So much for land on another person's property.

Your Honors will find in the record in this case, the espediente of the Berreyesa title; he has left out the Micheltorena grant, and he has put his claim upon a grant from another Governor (Alvarado) to one league. Mr. Lewis proves that there is enough land to cover the one league for Berreyesa, without at all infringing upon this mine, or this mining possession. Castillero's impression in Mexico was that it was vacant land, and as such he asks for it. But whether vacant, or not, was a matter of very little consequence to the President of Mexico, for the reason that Castillero had to be put in juridical possession by the local authorities; and that juridical possession would necessarily respect the boundaries of any preëxisting proprietor.

There is nothing in the record contrary to the testimony of these witnesses, that this copy of Berreyesa's grant was given by him to Castillero in his own (Berreyesa's) handwriting; the fact is sworn to by two witnesses, not controverted by any. You will bear in mind that Berreyesa's handwriting is here in different original papers; that it has been the subject of comparison by experts, and no attempt has been made on the part of the Government counsel to show that that writing was not Berreyesa's own handwriting.

Well, now, what is the next objection? The next objection is this, weighed upon and commented upon, detaining the Court for hours: That in all the early conveyances of shares in this mine, between the parties, the instrument of copartnership is referred to, instead of the registry and denouncement, as the title of the different individuals who sell their respective barras. Well, I am somewhat surprised at this, because I cannot imagine what possible title they had other than that instrument of partnership.

Where had Padre Real got a title to a barra in this mine, except under Castillero's grant to him? Where had Teodoro and Secundino Robles got a title in this mine, except through the grant by Castillero to them in the instrument of copartnership? The original denouncement and juridical possession gave no title to these other parties, because those stood in Castillero's name. The parties selling were not those who registered as denouncers. The registry and denouncement are in

Castillero's individual name. What, then, more natural; what else could be done when Robles sold his shares, and Castro sold his shares, and Padre Real sold his shares? What else could be done to show a title, than to show the grant by the discoverer to them of one-half the mine? What other title had they? None other.

And then it is said that these parties, all throughout, speak of three pertenencias, while the grant is of a mine of three thousand varas as a mining possession. From the very origin, the gentleman has harped upon this point; although it was answered in his own first examination of James Alex. Forbes, who gave him over and over again the whole theory of all the owners of this mine from the origin, in language too plain to be mistaken.

A *discoverer's* mine, under the ordenanzas, is three pertenencias; therefore, when these parties were selling out shares in the mine, they sold out their shares in the three pertenencias. That was the mine.

Now, the Alcalde says, after giving possession of the mine, "I have thought proper, ('*he venido en concederle*,') I have come to the conclusion, to grant three thousand varas, as a *gracia*." The Alcalde had no earthly power to do it. It gave nothing. It was of no value as a grant or concession, but still it was there on the face of the papers. And what is that *gracia*, or mining possession? A tract of land, upon which the *hacienda* —the reduction-works—can be established, and upon which the mining workmen may live; land for stables for their cattle, and for every operation that is necessary in the working of a mine.

Now, the gentleman, after James Alex. Forbes had been introduced as a witness in favor of the Government, by every possible means endeavored to make him say that there was something inconsistent in these purchases and sales of shares "in three pertenencias." But James Alex. Forbes persistently told him it was all right. I refer your Honors to Forbes' deposition, at pages 446, 455 and 870. Page 446:

Ques.—Of what did he give you the possession?

Ans.—Of the mine itself, *the hacienda*, the mining utensils, and the ores that had been extracted from the mine.

Ques.—What do you mean by the mine? Do you mean the spot only where the ores were dug, or that spot with a definite extent of land about it; and if the latter, of what definite extent of land around the mine did he give you possession?

Ans.—I do not recollect that any definite extent of land was specified, other than that given by the Alcalde. It was understood that the *mine* contained three pertenencias at the time.

[The latter part of this answer objected to.]

Ques.—By "pertenencias" do you mean the quantity of land known as such in the mining ordinances of Mexico?

Ans.—I do.

There that stops. Now we will turn to page 448:

Ques.—Will you answer yes or no, whether he gave you possession of a definite tract of land about the *hacienda?*

Ans.—There was no definite tract of land stated. The possession of the *hacienda* was comprised in that of the mine.

Again:

Ques.—When you delivered possession of this mine and *hacienda* to Robert Walkinshaw, agent of Alex. Forbes, of what did that possession consist? Did you deliver possession of any tract around the mine and *hacienda*, or either of them?

Ans.—I delivered to him that possession which I had received, together with a considerable quantity of ore which I had extracted. At the *hacienda* I delivered what I had received, together with some utensils purchased by me. I did not deliver to him possession of any definite tract about the mine or *hacienda.*

Ques.—When you recovered possession of the mine and *hacienda* on your return from Tepic, of what did you regain possession?

Ans.—I did not state that I recovered possession on my return from Tepic; I said I received it by virtue of a power of attorney.

Ques.—Of what did you receive possession on your return from Tepic?

Ans.—I received possession of the mine, a large quantity of cinnabar ore, and *of the hacienda, comprising all the works* erected by Mr. Walkinshaw."

The gentleman is not yet satisfied; and goes on, at page 455, when Forbes again attempts to make him understand what he means:

Ans.— * * *. There was only one act of possession which I understood to have been given. *This embraced three pertenencias, so far as regarded the mine. Three pertenencias*, and *also lands about the hacienda*, I understood to have been given to Castillero in 1845.

Ques.—How much land do you mean to say was granted about the hacienda?

Ans.—I understood there were three thousand varas.

Ques.—When do you mean to say you understood this?

Ans.—I think I learnt this about the time I received possession from Real. I am not certain, it may have been subsequently.

Ques.—When you bargained and bought of the Robles two of their four "barras," did you not obtain from them the half of all their interest in the mine and its appurtenances?

Ans.—I did.

Ques.—How is it, then, that your deed, drafted by yourself, expressed only two of their barras in "each of the *three pertenencias* of that mine?"

Ans.—Quite correct. I purchased two barras, and as a *matter of course the deed expressed according to mining custom all the pertenencias supposed to be comprised in a mine.*

Ques.—How is it, then, that there was no allusion to their interest in the tract of three thousand varas around the hacienda, which is one mile from the mine?

Ans.—The terms of the deed of sale comprise everything; all their right, title and interest to *lands* and mine.

Ques.—Show what words in that instrument convey an interest in the three thousand vara tract around the hacienda.

Ans.—It is that clause which commences, "all their rights and shares in each one of the three pertenencias," etc.

By Mexican custom *a sale of barras in a mine includes an interest in the hacienda.*

Now, what plainer explanation can I give the Court than this witness has given—for this witness does not lack intelligence? He has only too much of it. If he had a little less, or if one-half could be carried to the credit of honesty, he would be a very considerable man. How am I to explain to your Honors, better than the gentleman has explained, in his own testimony, that, according to the mining customs, all these deeds speak of the *mine* as a *mine* of three pertenencias, although there were other lands comprising the hacienda; and which were considered as *appurtenant* to the mine, and consequently needed no mention in the conveyances. James Alexander Forbes so understood it, when he took a title in the year 1847, and when he bargained for his title in 1846; and it never occurred to him that the three thousand varas were not included in the sale of the three pertenencias. Yet the gentleman stands up here day after day, and calls on your Honors to infer forgery, from the fact that in their deeds of sale no special reference is made to the three thousand varas, which, according to Mexican custom, went with the pertenencias, and required no mention.

Mr. Randolph—You argue, then, that the Junta, misunderstanding this document of Castillero's, supposed it to be for additional pertenencias, and as such recommended its confirmation.

Mr. Benjamin—Certainly.

I now come, may it please your Honors, to a paper somewhat celebrated in this record, called the Vejar certificate.

This Vejar certificate appeared to me, when I read the opinion of his Honor the District Judge upon this evidence, to have precisely the signification which his Honor attached to it. I read it over; and whilst I could not see what bearing it had upon the genuineness of these papers, I *could* see that it would certainly have a very prejudicial effect upon the mind of the Judge to have a certificate brought before him purporting to be a certificate of facts, known not to have existed.

Now, nothing surprised me more than that there should be a false certificate from a Mexican Notary. It surprised me for many reasons. When a student-at-law, I passed a two years' apprenticeship in a Notary's office in Louisiana. I was familiar with the old Spanish certificates in the notarial offices there; I knew the notarial mode of doing business of the old French and Spanish notaries; I knew the excessive and almost ridiculous particularity with which they carried out the forms prescribed by law—but in the whole course of my experience I have never seen such a certificate called into doubt or question, nor one of them forged or fraudulent. I might have looked anywhere else for false and fraudulent certificates; but the

Notary is so hemmed in with pains and penalties, the punishment so severe and exemplary both under the French and Mexican law (the Mexican law being derived directly from Spain), that it would require some most extraordinary temptation to induce such an officer to make a false certificate.

Yet this certificate (Vejar's) did upon its face apparently bear the construction which your Honor of the District Court had put upon it. But when I read the record through, and found a dozen other certificates of the same nature, and noticed the particularity with which Vejar made up each certificate in turn, then I began to be shaken in my belief of the correctness of the translation, or of the correctness of the meaning attached to the words of this instrument.

In the first place, the certificate, with the meaning which at first blush would be attached to it, appeared to be nonsense. I could not understand how a Mexican Notary could say that he had seen insertions of this instrument in acts done by executive officers in California. How could a man see an insertion of an instrument in an executive act? I did not understand that exactly.

Then, I did not see how the Notary in Tepic could certify that he had seen executive acts in California; nor could I see the slightest sense in the world in bringing up a certificate from Tepic to prove that something had occured in California, with the remotest idea that any Court would admit it in evidence! because if any official act had occurred in California, the evidence would be in California. Why should we go to Tepic for the evidence of official acts in California? I did not understand it. I then took up Vejar's certificates in the Transcript, and examined them, and the whole thing became plain and clear as day, for it turns out that on every occasion throughout this whole record, in which he certifies to a signature (except when the parties had passed the signatures before himself in the usual form), he had stated the grounds upon which he certified it. He says: "I know such and such facts, and for that reason I say, that in my judgment that signature is true." I found that his (Vejar's) certificates did him honor for their scrupulous regard to facts; and I found that this very certificate, so harshly attacked, contains within itself the proof of the scrupulous regard of this Notary to truth and honor.

I propose now to call the attention of the Court to those certificates according to their date, to show how the language of each certificate changes according to the facts which come by degrees to the Notary's knowledge; how scrupulously he abstains from certifying a signature when he does not know it, and how, when a signature is brought before him to certify, that he does know, he leaves it to those who read it to ascertain whether the grounds upon which he attaches his certificate are sufficient, or not, in law.

The first certificate in this record, of this Notary, is on page 511. It is a certificate to the power of attorney, of the 23d of April, 1849, given by Alex. Forbes to James Alex. Forbes. This certificate is signed by Jesus Vejar and Panfilo Solis. It is to the signatures of Alex. Forbes and the two witnesses, Joaquin Andrade and William E. Barron. What is the language of that certificate:

The undersigned, Notaries Public, certify that the foregoing signatures are those of Don Alexander Forbes, Don Joaquin Andrade, and William E. Barron, the same that they use in all their business transactions. In testimony of which we have given the present in Tepic on the 19th May, one thousand eight hundred and forty-nine.

(Signed) JESUS VEJAR. PANFILO SOLIS.

There is Vejar's reason for certifying those signatures. He says, they are used in all their business. It was not signed before him, and he does not say it was. But his reason for believing the signatures is, he says, because they are the same signatures they use in all their business.

On the 13th of November, 1849, he was called on to certify a copy of this Castillo Lanzas decree. Your Honors will find his certificate at page 546.

Now, there was the time, if anything fraudulent was intended, to certify that the original had the true signature of Castillo Lanzas. But the Notary did not know, had never seen, Castillo Lanzas' signature. He did not know whether the signature was his or not, and he would not certify it. Let your Honors look at the particularity with which, when this instrument is first presented, he certifies the copy.

"The above is a copy taken,"—not from "*the* original;" to use which words would imply he knew the original signature; but—"from '*its* original.'" * * * The certificate reads as follows:

The above is a true copy from its original, which was presented in this office by Messrs. Barron, Forbes & Co., of this place, and the same being returned to them marked with my rubric, the present is given at their solicitation for the appropriate use.

In testimony of which I sign and attach my signet to these presents, in Tepic, 13th November, 1849.

JESUS VEJAR.

A paper is presented to Vejar. It purports to be this Castillo Lanzas decree. He is asked to certify a copy. He will not certify that the original is signed by Castillo Lanzas; he will not certify that it is an original document; but he *will* certify that the copy he makes is a copy from "*its* original," which was handed him by Barron, Forbes & Co., and returned to them by him. Whether that original be true or not, he will not certify. He returns them a certificate that it is the original *of the copy*. That is all. There is no statement there that it was signed by Castillo Lanzas. He knew nothing about *that*, and he would not certify it.

At page 556 is the next certificate by this Notary; given on the 3d of December, 1849. He was called upon with a copy of the instrument of *avio*, which had been passed in Mexico in December, 1846, and which contained a copy of the Castillo Lanzas decree, authenticated by the signature of three or four Mexican notaries, and the seal of the National College of Notaries. That instrument is now brought to him, and he is requested to certify; not to the Castillo Lanzas decree—that has not come up yet—but to a copy of this instrument of *avio*. He says:

At the verbal request of Alexander Forbes, and for the appropriate uses, I give him the present, signing the same in Tepic, the third of December, one thousand eight hundred and forty-nine, having used common paper, there being no sealed paper of the fourth class in the office where it is issued, which I certify. JESUS VEJAR.

Now, why was that paper taken to Vejar to be certified? Why was not the copy from the City of Mexico sent up for a certificate? The answer is plain. On the 3d of December, 1849, California belonged to the United States. The original copy from the City of Mexico, certified by the notaries of Mexico, without an authentication of their signatures, would not be admitted in California. Therefore, Jesus Vejar certifies the previous copy given by the Mexican notaries, and then that copy is certified by the British Consul in Tepic. It was supposed that, with the certificate of the British Consul in Tepic, it would be admitted here in our Courts, or used for any purpose required as an authentication of the notarial copy from Mexico. A notarial copy, sent from Mexico, being intended to be used in *Mexico*, required no authentication; but here he was asked to make a copy to be sent to *California*. Making that copy, he saw the notarial certificate of four notaries of the National College of Mexico attached to a paper containing the Castillo Lanzas decree, "treating it with deference as genuine, respecting it, obeying it," as he says in his certificate of that day.

Then, on the 15th of March, 1850 (page 801), he was called upon to give a certificate in relation to another signature. On that day he was, in fact, asked to certify three different signatures. Now, I call your Honors' attention to the manner in which each of the three signatures was described and certified to by this Notary. I want your Honors to see the scrupulous care which he takes with each of the signatures. He certifies each, giving what he knows about the facts, and leaving the Court to determine from those facts whether his certificate is to be treated as sufficient proof, or not.

He is asked to certify to the signature of Jas. Alex. Forbes who was in California. What certificate does he give?—

I, Jesus Vejar, Notary Public, hereby certify and accredit, that the last preceding signature of Don James Alexander Forbes, is the signature of that gentleman which he is accustomed to use, it being so known to me, when I knew him in his transit through this city on his way to Upper California, by various acts which he executed in the house of Barron, Forbes and Company. And at the request of those gentlemen, I sign and execute this certificate at Tepic, on the 15th day of March, 1850. JESUS VEJAR.

That is one thing he is asked to certify—the signature of James Alex. Forbes: "I certify it to be his signature; and I certify it because I *know* it, having seen him use that signature, when he passed through this city, in various acts. That is the way I know that signature." He is there asked to certify the copy of the *escritura*, or instrument of partnership.

On the same day he is asked to certify another instrument. Your Honors will find his certificate on page 3132. These papers were presented to him solely for him to certify *signatures*. This (page 3132) is again to the signature of Jas. A. Forbes.

Now, the Castillo Lanzas document is presented to him. We want him to certify this signature too. What does he say? He is asked to certify the Castillo Lanzas signature to that document which had been sent to him before for a certificate, and which he *then* declined to certify as being Castillo Lanzas' signature. What does he say *now?*—(Page 68 of the record):

I, Jesus Vejar, a Notary Public, hereby certify and attest that the foregoing authentic instrument, signed by his Excellency the Minister of Foreign Relations, Government and Police, Castillo Lanzas, has been respected under that signature, and obeyed by the Mexican authorities that governed in Upper California in the year eighteen hundred and forty-six, according to insertions which the said authorities made of the said instrument in acts which they passed upon the subject of which they treat, and which I certify to have seen, and for this reason that signature in the said instrument should be esteemed as authentic, and signed in the handwriting of his Excellency the Minister; and as also by proceedings which have passed under my observation Sōr. Don Andres Castillero recognized it.

The certificate is not offered for the proof of any executive acts in Upper California; it is offered to prove a signature. The Notary says: "I say that is the signature, and here is my reason: my reason is, that I have seen it respected in insertions of that document in acts passed in which that document has been respected and treated with deference by Mexican authorities."

Now, may it please your Honors, if two meanings could be given to this certificate, would you select the one which would brand this man with falsehood, perjury, and official misconduct; or the one which the act naturally bears on its face? Observe, too, he certifies to this certificate for two reasons; the latter of which you may judge to be sufficient, or not, as you please. "These are my reasons," he says: "First, I have seen acts containing insertions of this document in which this signature has been respected and obeyed, as being a genuine signature, by the Mexican authorities; Secondly, by proceedings that have passed under my observation—Castillero recognized it." That may be a bad reason, or a good one. It is his reason. Take it for what it is worth. The grant purports to be made to Castillero. Vejar says he has two reasons for believing the signature to be a genuine signature: one is, that Castillero recognizes it to be a genuine signature; the other, that the Mexican authorities have inserted it in acts passed before them which "I have seen." We produced the acts in which he had seen it, and in which it was inserted.

Now, what is there against all this plain and logical sequence of facts? These words: "that governed in Upper California in the year 1846." Now, let us put ourselves in the Notary's place for a moment. He has given his certificate to the genuineness of an instrument which is to go out of the country; he certifies, therefore, that the signature has been respected by Mexican authorities; but it strikes him that, the fact that it has been respected by Mexican authorities gives it no authenticity in California, which is an American possession; it is therefore necessary to state that it was at the time that Mexican authorities were governing in Upper California, to wit, in the year 1846. For the purpose required by these parties—to send it into the United States—it won't do to simply certify that it was respected by Mexican authorities; that does not help it in the United States; but it must be at a time when Mexican authorities governed in Upper California. I must state the date when it was so respected, the date when the Mexican authorities were governing in Upper California, to wit, the year 1846. The ideas passing through the Notary's mind become apparent to any individual who will read the certificate candidly.

It was suggested by his Honor the District Judge, that those words (*actos que autorizaron*) might refer to other officials besides notaries; and it is asked what reason had we for restricting "actos" to notarial instruments in this particular case?

To that I answer, first: That it is true that those words "*actos autorizados*" are Spanish law terms which do apply to other instruments, as his Honor observed, but that those words are particularly and almost invariably applied in the Spanish law to notarial instruments. In the Mexican Febrero, vol. 2, page 532, the author, speaking of the various kinds of public and private instruments, says: "There are various classes of public instruments that the laws of the Partidas recognize, such as documents authenticated by the seal of the Pope, King, Prince, Archbishop, Bishop and others; but among them special mention should be made of those authenticated by the signet of a Notary, who is the officer created by law for this purpose."

In the whole of these notarial certificates scattered through this volume, you find these words "*actos autorizados*" used by every Notary who attempts to certify a signature. And, although it is true, as his Honor the District Judge observed, that those words could be applied to other authorities, yet when we find about one hundred certificates in this volume, in all of which the words are applied to notarial acts, the presumption is that they are in this instance applied also to notarial acts, and nothing else. It is a Notary that is certifying; he is using notarial language; all through the record you find notarial instruments *certified in these words and in no other.*

Everywhere else the meaning is undisputed. Why is that meaning changed for this particular certificate? Why should it be so changed? And why should it be so changed for the purpose of establishing that the Notary certified a falsehood in regard to a matter of which he was not asked to certify at all? The certificate on its face states its purpose. He is not asked to certify facts in California; he is asked to certify a *signature;* that is all his certificate purports to be—an authentication of a signature.

Now, whether he gave good or bad reasons for his certificate, is to us a matter of indifference, for he states his reasons—such as they are—and your Honors can judge of their weight. All we asked of him to do, and all he could do, was to certify the signature to three or four papers—which certificates he made out as he thought proper. *This* signature he certifies to be James Alex. Forbes', because "I have seen him use the same in various acts;" "*that* I know for other reasons." I shall bring to your Honors' attention some other certificates where he uses these words "*por esta razon.*" "*This one,*" he says, "is true because I have seen it inserted in various acts." We produce the acts in which it was inserted, and those very acts prove the statement to be true that he had seen the document inserted in them, and treated as genuine by Mexican authori-ities. Who made these insertions? Mexican authorities. The California authorities were designated as *local* authorities. All through these books of record, when California authorities are spoken of, they are called *local authorities*. That is not what this Notary has seen. He has seen the signature respected by Mexican authorities; that is, by authorities of the City of Mexico.

I think, may it please your Honors, that this amounts to absolute demonstration. We were all deceived in our opinion of the meaning of the paper; which I admit, upon its face, appeared to be that which the Court attributed to it, namely, that this Notary was certifying to something which had occurred in California, when he had not the remotest idea of certifying to anything in California, at all, and could not know in Tepic what occurred in that Department.

Shall I give your Honors more of his certificates?

On the 18th of March, 1850, he was called to certify to some other papers. It is another certificate in relation to the signature of James Alexander Forbes; and given for the same reasons as before. James Alexander Forbes' ratification of the *avio* was presented to him. What is his certificate to this? (Page 549):

I certify and give this certificate, that the preceding signature of James Alexander Forbes, which is subscribed to the foregoing power of attorney granted by Don José Castro, is the same that he is accustomed to use as Vice Consul of Her Britannic Majesty in Upper California, under his seal of office as it appears; and in such character of Vice Consul such signature is recognized by foreign mercantile houses in this city in giving faith and credit to authentic instruments which he has issued—and for this reason I certify the signature.

That is the reason why he gives that certificate.

On page 2475, this Notary is called upon to certify the certificates of notaries of the City of Mexico, on this same day (18th of March, 1850). In so doing, he gives a different reason. He certifies their signatures, and gives his reason for it.

It appears to me, then, may it please your Honors, in the highest degree unjust to this Notary to give to his instrument the signification which was formerly given; a signification with which we have nothing to do. He was asked to certify Lanzas' signature, amongst others, and if he gives bad reasons for believing it to be his genuine signature, it is nothing to us.

MR. JUSTICE MCALLISTER—Taking your construction to be correct, do you say that the certificates of a foreign Notary, made in a foreign country, should be taken as evidence by this Court?

MR. BENJAMIN—Not at all. I simply wish to prove *good faith*, so far as the certificates are concerned.

Now, how is Jesus Vejar's certificate further assailed? By a very simple method; by saying he was a piece of *office furniture* of Barron, Forbes & Co. Well, so he is, just as my brother Randolph is a piece of office furniture of any of his clients, and as I am of mine. If it suits him to call me at present a piece of office furniture of Barron & Co.—

MR. RANDOLPH, (interrupting)—By no means. I shall never do so, sir.

MR. BENJAMIN—Why not? They are my clients. Barron, Forbes & Co. were clients of Vejar. Notaries have clients as well as lawyers. Each Notary has his clients. Those clients do all their business in his office. You might ridicule any set of certificates brought from any Notary, any where, as signed by a piece of office furniture, because all the business of the client is done in the same office. I will say, that so far as my professional business is concerned, I do all my client's business, or none. There would not be a particle more of propriety in the epithet being applied to any gentleman of the bar, because he was the legal counsel of his client, than in applying it to a Notary, because he was the Notary of a commercial house.

I now come, may it please your Honors, having reviewed some of those facts which are spoken of as inducing doubt in relation to the genuineness of these papers—a genuineness that is now established beyond all controversy—to inquire into the origin of this charge of fraud.

How came it to be made? Whence did it derive its origin? and who were its respectable progenitors?

A man by the name of James Alexander Forbes—who has been often mentioned in this case—became bankrupt. Not yet thoroughly bankrupt in character, though that was rather bad, he became utterly bankrupt in fortune. He came to these gentlemen, who are the claimants in this case, and asked them for money, and they would not give it. He asked them for a loan of $10,000. They had already lost money by his bankruptcy, and declined giving him any more just at the moment when he had become bankrupt. Animated by motives of revenge, and by a desire to swindle somebody out of some money, having failed in his efforts on Barron & Co., he bethought him that there was a body of men here who were busy in claiming this valuable property, pretending that it belonged to them under agricultural grants, who had the ear of certain Government officers, and who, if they could get rid of the New Almaden Mine owners, might perhaps defraud the Government out of it for their own benefit. How were these New Almaden Mine owners to be gotten out of the way? Up to that day not a human soul had breathed a suspicion of the genuineness or validity of these papers which had been spread upon the records of the Courts of this country for years.

"I will go," says James Alexander Forbes, "and get hold of a shrewd Frenchman here by the name of Laurencel, and satisfy him by forging certain papers, and putting them in connection with certain other true papers, that there is something wrong in this title. I won't let him have anything until he pays me $10,000 cash down ; and I will run no risks. The way I will avoid risk is this : I will stipulate that these forgeries of mine are not to be used except to effect a compromise ; and if Laurencel can get a compromise by their use, then he shall give me $10,000 more. If he cannot effect a compromise, why, the papers are not to be used for any other purpose, and I run no risk. I will, at any rate, get my $10,000 in my pocket, and may be these gentlemen, the owners of the mine, will be frightened by this conspiracy into some compromise, and will give Laurencel some of their property—in which case I will get $10,000 more. If they do not, why Laurencel will be left to suffer, and I will have pocketed $10,000."

Forbes proceeds immediately to carry out this notable scheme, and he finds a willing auditor in Laurencel. Laurencel, in his testimony, admits this bargain ; Laurencel, under stress of the orders of the Court, is forced to answer the question put to him relative to it. He declines answering it at first.

Mr. Randolph—Excuse me ; you fall into an error, Laurencel testifies first, that he gave a large sum of money for the papers ; and the amount of the money was the only thing which afterwards it required the order of the Court to induce him to divulge.

Mr. Benjamin—Certainly. (Deposition of Laurencel, page 377.)

Ques. 18.—What amount did you pay ?
Ans. 18.—I paid a large amount ; I decline to say the amount.
(Page 378). *Ques.* 21.—What did you pay Mr. Forbes for those papers ?
Ans. 22.—I have already declined to answer that question.

[The Court requires the witness to answer.]

I paid Mr. Forbes twenty thousand dollars for the interest that I was acquiring in depositing these papers, and the engagement on his part to go to Mexico, when I should request him to do so, and procure for me the proof of the fraud committed in this case, which he assured me he could procure. I was also to pay, if necessary, his traveling expenses.

"Was that your bargain?" "Yes, it was." "Was it in writing?"

After some equivocation on that subject, which I do not care about dwelling on, Laurencel finally answers: "Yes, it was in writing." Where is it? On page 380, he answers that—"*I have destroyed my counterpart;* it was in duplicate; *I destroyed it by the advice of counsel;* I did not preserve a copy of either; *I thought it would be inconsistent with the object with which I destroyed the papers.*"

What moral leprosy was there in that paper which his counsel would not suffer even to appear in a Court of Justice? Was it that the infamy of the transaction had so penetrated into the very substance of the paper on which it was written as to taint the paper itself? That paper must never see the light!

He says now, that Forbes' promise was to procure him proof, proof which he said existed of the fraud. Procure proof! Was that all? They *say* that such were the contents of the paper burnt up!

What a field is left for the imagination, as to the infamy of this transaction, when the man who enters into it is told by his counsel, Burn up your paper, sir! It is *infamous!* It must never be seen!

What was this bargain? Was it that James Alexander Forbes was to procure other witnesses who would join him in establishing his forgeries?

Was it that he was to go to the City of Mexico and endeavor to get witnesses from there to swear that these papers, containing our title, do not exist in the archives, were not the original papers, or were not placed there at the time that the dates purport?

What was this infamous bargain which polluted the very paper on which it was written? No man knows but the parties to this conspiracy; to others, it will remain a secret forever.

They have disclosed to the Court whatever they chose to disclose. They have destroyed the evidence which would disclose the rest. That, I suppose, is locked up in their own bosoms, to be carried into effect between them, on the principle that, as regards themselves, there is honor between thieves.

The next that we see of Mr. Laurencel, he is inducing Benito Diaz to come in and back the testimony of James Alexander Forbes. He shows him papers; induces him to examine them; has conversations with him—which poor Benito Diaz, who has a bad memory, says he cannot exactly recollect; and then comes into Court with him to back the testimony of James Alexander Forbes in carrying out the conspiracy between Laurencel and himself (Forbes).

What else do we know of Mr. Laurencel?

Having in his possession, and under his control, or rather having access to the records of the United States Consulate of Monterey, containing in their pages the proof of the departure of certain vessels—

Mr. Randolph, (interrupting)—You refer now to a book which was at the time in my keeping? I told you the other day that that book was not within the access of Laurencel. For all that concerns that book, I alone am responsible.

Mr. Benjamin—If Mr. Randolph states that Mr. Laurencel did not see that book, of course I have nothing further to say.

Mr. Randolph—Mr. Laurencel, or any other gentleman who chose to see it, could see it. Further than that, I had charge of the book which was loaned me by Mr. Larkin; and I alone am responsible for its keeping.

Mr. Benjamin—I say nothing to the contrary. My brother Randolph does not know what I have reference to. I say that Laurencel had access to certain books containing certain facts.

Mr. Randolph—If you refer to the letters to the Department of State, they were there. If to the other letters, they were not there to the best of my knowledge and recollection.

Mr. Benjamin—Of course my brother cannot suppose I refer in the remotest degree to him. I am referring to Laurencel's acts.

MR. RANDOLPH—When referring to a book in my keeping, you are referring to me in a great measure.

MR. BENJAMIN—I will drop that part of the subject, as Mr. Randolph seems sensitive about it.

MR. RANDOLPH—No! I request you to go on.

MR. BENJAMIN—I was simply going to state this, that Laurencel had access to these records. It is a fair presumption that he there ascertained that certain vessels had left California at a particular date; yet he brings up evidence to sustain my brother Randolph in the fact that no vessel left. I do not for a moment suppose that my brother covered up the evidence. Of course he only got such evidence as Laurencel brought.

MR. RANDOLPH—The records to which you refer were brought in evidence. It is all in evidence. But not one single line or word is put in evidence, and you are speaking altogether in regard to a private document, without warrant from the record.

MR. BENJAMIN—I proposed to abandon the subject. You said, go on! When I go on, you object. I will, therefore, leave all that concerns Laurencel and the departure of those vessels aside. I will leave aside the fact that he (Laurencel) produces witnesses in Court (whose memory is eulogized by Mr. Randolph) to prove, as they did most emphatically, that they remembered perfectly well all the vessels that left Upper California in the first half of the year 1846. Yet here is the proof in the newspapers and marine news of the day, of other vessels leaving at that time—facts which they, with a memory stated to be extraordinary and unimpeachable, had no recollection of when they were brought up by Laurencel.

What else does Laurencel do?

He sends James Alex. Forbes down into Lower California to corrupt General José Castro by an offer of a bribe of $10,000. I say Laurencel did this, because no man can suppose the United States Government did it. Forbes said to Castro, to be sure, that he came in behalf of the Government. But, in truth, Laurencel had bargained with him that he should go. Laurencel states that was his bargain.

(Page 2616.) The witness (Castro) called upon the stand, is asked in relation to a visit paid by James Alex. Forbes to him in Lower California. He first states the date of the visit, as April or May; and then before his testimony is complete he states that the data given by him previously is erroneous. It was in January or February, and not April or May. Now, what was this visit?

Q. 18. Just before you left Lower California on that occasion, had not James Alexander Forbes been there on a visit to you?

A. Yes, sir; he was there about the latter part of April, or the early part of May.

Q. 19. Please to state what was the object of Mr. Forbes' visit to you, as disclosed by himself, and what passed between you in your interviews; and under what circumstances Mr. James Alexander Forbes left Lower California.

A. At my residence (the Sausal de Camacho) I received a letter from James Alexander Forbes, dated at Don Juan Bandini's rancho, in which he stated that he had a matter of great importance to consult me about, and requested me to come at once to see him. I did so, and took my Secretary with me. When we arrived, after the usual salutations had passed, Mr. Forbes and myself walked out together from the house, and he then stated to me that he was authorized by a powerful company of speculators (empresarios) in San Francisco to furnish me with the necessary means, if I would consent to detach Lower California from Mexico, to make it independent, and that subsequently it should be annexed to the United States. I listened to this calmly, and told him to go on. He then said, there is another matter of importance to you. If you will give your testimony against the owners of the Almaden mine, the Government will pay you a considerable sum—more than ten thousand dollars. I answered: "With regard to what you first proposed, I am an officer of the Mexican Government, and will not be guilty of treason; and as to the other matter, I cannot be bought with money to do an infamous action; say nothing further to me on the subject." We then returned to Mr. Bandini's house.

I then went home, leaving Mr. Bandini an order, directed to Forbes, stating that my duty, and his own personal safety, required that he should immediately leave the country. A few days afterward, I heard from Mr. Bandini that Forbes had left the day after he received my order.

After asking some other questions, he answers to the twenty-second interrogatory thus:

A. I did not give him an opportunity to enter into further details. He was speaking about being dissatisfied with the house, etc., as I stated before, when I stopped him, saying that I did not wish to hear any further explanation about his affairs.

Q. 23. Did he mention the names of any members of this powerful company of speculators in San Francisco, who were desirous of detaching Lower California from the Mexican Republic, and of annexing it to the United States?

A. He gave no names.

Q. 24. Did he tell you by what authority he spoke for the Government, when he promised you ten thousand dollars and more, if you would give your testimony on its behalf, and against the owners of the mine?

A. He did not state how he was authorized to make the offer—he only said the Government would pay me.

Q. 25. Did he exhibit to you any written authority from the Government to treat with you about this matter?

A. No, sir.

Q. 26. Did he tell you who defrayed the expenses of his mission to Lower California?

A. He said that there was a fund here in San Francisco, out of which his expenses were paid.

In the cross-examination, he is first asked how he came to recollect and change the date of the visit from April or May to January or February. He says, in looking over his testimony it occurred to him. (Page 2621.)

Q. 40. About this date of your interview with James Alexander Forbes in Lower California, what had happened in the two or three days which had elapsed, to enable you on yesterday to remember more accurately what the date was?

A. Because I thought the matter over and came to that conclusion in my own mind; but I am not even now absolutely certain about the date.

Q. 41. Has not Forbes been to see you, and told you that he could prove your statement was false, and that he was not in Lower California at all at the time that you have mentioned?

How did the special counsel for the Government know that Forbes had come here to have a conversation with General Castro, except through the omnipresent Laurencel?

Mr. Randolph—I will tell you. Forbes came into my office and told me so. I should have guessed *that.*

Mr. Benjamin—*I* should have guessed the contrary. If he came to your office, I have no doubt but Laurencel sent him there. Laurencel had previously been sent for by Forbes, as stated by the latter, to consult with him *as to the proper manner of sending a paper to the United States Attorney.* He was sent for, and had to go all the way to Santa Clara to consult about the proper manner of sending a paper to the District Attorney! *That* certainly must have required a *very* grave consultation! However that may be, as soon as Castro on the stand is forced to divulge this attempt to corrupt him by money, and to induce him to give testimony against the owners of the Almaden mine, Forbes is on the spot. (Page 2624.)

Q. 41. Has not Forbes been to see you, and told you that he could prove your statement was false, and that he was not in Lower California at all at the time that you had mentioned?

A. It is not true that Forbes ever told me that any statement of mine was false. I allow no one to tell me personally that I am a liar, and I have never received such an insult from any one in my life. Forbes came to me in a friendly manner, saying that he regretted I had revealed what had passed between us, that it was a secret, and so on, and endeavored to make excuses about it. I told him that being under oath I had to tell the truth.

Now, if Castro's version was not the truth, why was not Forbes brought up again? Why, the answer springs instinctively to the lips: the Government counsel were ashamed to bring him again into Court. This vile creature, with his lips steeped in lies and his soul blackened with perjury, had been paraded day after day in the Court-room, until the very precincts of justice were filled with the stench of the corruption which dripped from his every pore. The Government did not dare to bring him in again. But if Forbes had not in fact been down to Lower California—and no other motive for his trip is suggested anywhere—it was open to proof by other persons than he that he had not been down. No such attempt was made.

Who sent Forbes down? Laurencel tells us that he had made a bargain with him that he should go; and Forbes admitted that he had made the bargain. The papers which stated the terms of the bargain are burned up—burned up by those who wrote them, as too infamous to be produced in Court.

There then, your Honors have the names and the characters of the conspirators who are brought forward to rob the owners of their property in this mine because it is a rich mine. In the quaint language of old Gamboa, "suits do not arise about poor mines; but whenever a mine is rich there are plenty of suits."

Now we come to the famous letters in which this conspiracy "cropped out,"—I believe I use the appropriate Californian mining language. Before that the matrix had been in the bosom of the earth. Parties had been simply "prospecting," but finally they found the "cropping out" of this mine, and it is produced in Court; and what do we find?

A series of letters, produced by one whom the Government's counsel admits to be proven a forger. He ridicules us for impeaching the testimony and showing the perjuries in the deposition of this man—James Alex Forbes,—"for," he says, "it never entered my brain that you would take that trouble. Of course he is a forger; I bring him in as a forger; I do not pretend to defend him."

Very well; there is one step. Then, a man known to be a forger,—whose testimony is not, the counsel says, relied on at all by the Government in anything that cannot be proven by somebody else,—that is the man who is in constant confidential communication with the man who bought his testimony. The man who is to make money by that testimony is riding day by day with this known forger, who is not to be believed on oath, and whom the Government counsel does not pretend to defend in any way or shape. They go from rancho to rancho; they seek the old Spanish and Mexican settlers. They concoct, they combine, they conspire day by day, and hour by hour,—how can this mine be stolen? How can we rob this mine from the owners? I will tell you how you can steal the property, says James A. Forbes,—"There are some letters containing expressions which can be made to give resemblance to a charge of conspiracy to forge titles; take them, use them; blackmail these parties into a compromise."

The letters are brought; they are exhibited to the claimants; "How much will you give?"

How much *did* John Parrott give? How much did William E. Barron give?

They told these people: "Take your forged documents, and go away with them; don't insult us with your base proposals."

MR. RANDOLPH—Did they ask Hall McAllister?

MR. BENJAMIN—I speak to the record in speaking of those to whom the proposals were made, and by whom.

Why did not John Parrott buy up all these *proofs* about these forgeries? He is a shrewd man; he has the reputation of knowing where his interests lie. I believe people don't generally get much the better of him in bargains. Why did not John Parrott buy up these papers which were going to destroy his title? He handed them back to the parties: "Take them! Take them! Go away! No compromise about such matters here!"

Well, that won't do. John Parrott won't buy. But James Alex. Forbes has got the money ($10,000), the documents are deposited in the vaults of a bank here (Davidson's), under a contract that they shall not be brought out without his (Forbes') order; and he goes off whistling about the country, contented in the knowledge that he has defrauded Laurencel out of $10,000 without any risk run.

But Laurencel is not to be defrauded that way. He comes into this Court, and procures a *subpœna duces tecum* to bring the papers into Court. He uses the Government as an instrument to break his bargain with Forbes. He makes a cat's-paw of the Government to break his private contract. So the Government does lend its aid to Laurencel, who undoubtedly must be worthy of its protection. Still, it does strike me as curious, that it is precisely Laurencel's counsel who is so vehement against my clients for being foreigners. I have sometimes thought it would be an interesting puzzle to find out how the Government of the United States got interested in foreigner

Laurencel against foreigner Barron. That, however, must be one of the secrets of State, not proper for us to know.

MR. RANDOLPH—That is not the case at all. My learned friend (Peachy) next me, for a year or two—I know not how long—was engaged in opposing the case known as the Fossatt claim. He acted as the counsel of the United States in that particular instance, because of his employment by the New Almaden Company in another instance. His endeavors to procure testimony were untiring, and a vast amount of damage was done. In that situation the counsel for the Fossatt claim followed the example, and did the same thing as the New Almaden Company.

MR. PEACHY—One word. My connection with the case has been alluded to, and I desire to say this: that I appeared for the Government to prove that this mine did not belong to Laurencel; that was all. I never did contend that this property belonged to the Government.

MR. BENJAMIN—I was going to say, further,—if Mr. Peachy had not desired to make his personal explanation—that Mr. Randolph's explanation lacks one little particular.

MR. RANDOLPH—It was not an "explanation" but a statement of facts.

MR. BENJAMIN—Well, facts. One fact is left out. Did the Government employ Mr. Peachy, or pay him, or was he employed by private clients to appear for them in that Government suit? What I am trying to ascertain is, what the secrets of State are that induce the Government of the United States to employ and pay counsel for the purpose of having Laurencel proved the owner of this property, instead of Barron; Laurencel being as much a foreigner as Barron is. If foreigners have no rights to justice in the Courts of the United States, how is Laurencel any better off than Barron? If, on the contrary, as has hitherto been believed by people in this country—and will be hereafter, unless your Honors decide otherwise (which I doubt)—foreigners, by law, have a right to own property, then I cannot see what difference it makes whether Barron, Forbes & Co. are foreigners or not. However, in all respects I say Laurencel and Barron stand on precisely the same footing—both being born out of the United States.

I have said that this contract being completed remained a mystery between the two rogues who made it; the papers which would show what it was being destroyed.

A failure to obtain witnesses from Mexico, as undoubtedly had been promised by Forbes to support his testimony, having occurred; the papers being in a bank vault; Forbes having his money in his pocket,—Laurencel uses the Government as a cat's-paw to enable him to break his bargain. Instead of a joint order of Forbes and Laurencel being taken to Davidson, the banker, to produce the papers that had been deposited in his vaults, the United States Marshal is procured, through the agency of Laurencel, to be sent to take the papers by violence and by stress of law, which Laurencel had bound himself to Forbes should not leave that depository without his (Forbes') consent.

They are brought into Court. I propose to examine them. I propose to show your Honors, in addition to what Mr. Peachy has said, that a large number of the documents produced from that case are forged; that nearly every letter purporting to be written by James Alex. Forbes, as produced by himself, is a forgery; and I do this with a full knowledge of the fact that in the answer in chancery William E. Barron swore that he believed them to be true. In spite of that belief at that time, I will show the forgery on the face. I will show it so that no man can doubt. It does so happen, that William E. Barron being called upon suddenly, and within a very brief delay, to file an answer in chancery, a large body of correspondence was brought before him, and he was asked whether it was genuine. He saw that one document was undoubtedly a forgery, and that document purported to be a letter of Alex. Forbes. The remainder of the documents, which James Alex. Forbes had forged and brought in as his own letters to the parties, were forged out of true letters, by picking out paragraphs here and there. Without an opportunity of comparing the originals in our own possession; without an opportunity of looking at the letters; hurried to file an answer in chancery, William E. Barron said, "I believe they are true." But when we come to examine the vast mass of documents produced in the cause, we find the very original letters, of the very dates, and in the handwriting of James Alex. Forbes, from which he has forged and brought into Court what purport to be copies retained by him of the originals. I will follow him all through these letters. I will follow his track as the huntsman follows the game; and with the same unerring certainty.

Mr. Justice McAllister—On what page do the letters begin?

Mr. Benjamin—If your Honor please, these letters are scattered through four volumes, without order of date, sequence, or anything by which the Court can follow up my argument in relation to them, unless I furnish your Honors with a chronological list, which I will take pains to do, in order that your labor may be facilitated.

Before speaking of the letters produced in this case that were forged by James Alex. Forbes, I wish first to call the attention of the Court to a letter which is a true letter; having no reference, however, to this case, but brought in for the purpose of deceiving the Court, and since made use of by the counsel for the Government in a former argument. I think I can show that James Alex. Forbes put this paper into this bundle of letters for the purpose of deceiving Laurencel; that he has deceived the Government counsel and deceived the Court in so doing; that it was in relation to an entirely different subject, and had no reference to New Almaden at all.

Mr. Randolph—What letter is it you refer to?

Mr. Benjamin—The letter of the 19th January, 1848.

The letter is at page 386, and requires reference to but one other for any immediate purpose—which is at page 542. This is a part of this conspiracy of James Alex. Forbes to deceive into the belief that there was some fraud.

Now, I will show to your Honors in a moment or two, that after the discovery of this New Almaden mine, the whole neighborhood was aroused to the fact that there were veins of quicksilver in that range of the Sierra Azul, or Blue Ridge, as they term in Virginia a similar mountain range. The whole neighborhood was aroused to the fact that there was quicksilver in that range. It was supposed that other veins would be found; and the record teems with proof of denouncements by different parties, who thought they had discovered other similar mines. Amongst them were the mines of San Antonio and Guadalupe. In those two mines James Alex. Forbes had an interest. In those mines he was desirous that Alex. Forbes should also take an interest; and they had a correspondence upon the subject: and this letter (19th June, 1848) is a letter relating to those other mines, which, as I shall proceed to show to the Court, was brought into this case, having no earthly bearing upon it.

First, let me read the letter, premising to your Honors that only two or three months previous—as late as the 24th of November—Alex. Forbes had made a thorough examination of the New Almaden mine (*Vide* letter of the 24th November. 1847). After having been first alarmed at the prospect that the mine was, as he supposed, a mere *manto*, or horizontal layer—a mere superficial deposit—he found the vein, and he proceeded to inform his associates of it in the letter to be found at page 386:

MINE, 24th Nov., 1847.

For Mr. J. A. FORBES and The PADRE REAL, } Santa Clara:

We have at last found the vein or "cinta" of ores which we were looking for, so that I have now the pleasure to inform you and the good Padre of our luck as I promised I should do; but I fear the mine will be reduced to this "cinta," and the great body of it will be "Tepetate Muerto."

But perhaps this cinta may be wider below than it is above.

To see whether this is so or not has been the object of our labors, *since discovering the proper direction of the vein of the whole mine*, which discovery makes everything more plain.

This direction was before *entirely mistaken*, of which and other things we will have a great deal to talk about when we meet. When Mr. Walkinshaw arrives and takes a look at the mine, I think we shall take a turn to the Mission. I expect him to be at Bernal's Rancho this afternoon.

I may say now, that it is impossible we can go off the main vein of the mine, as it is entirely different from the walls (respaldos), they being of hard rock of quite different character, whereas the vein is quite soft and easily distinguished.

All we have to do is to look for the "cintas" which have got ores, which, in my opinion, will be reduced to one, *not very wide.*

A. FORBES.

(Endorsed) ALEX. FORBES,
Mine of Almaden, Nov. 24, 1847. No. 4.

Here, Alexander Forbes has discovered the vein of the mine. It is thoroughly distinguished. The walls are clear and plain. The ore itself is equally plain and manifest. This he is now communicating. No doubt now amongst these mine owners in relation to the extent of the vein of the New Almaden mine. As to the quality of the ores, that has been known for years. There is no necessity for testing the ores of New Almaden. They were selling the quicksilver; they were reducing the ores in their furnaces, as far as their furnaces would permit it; they were furnishing quicksilver for the market; therefore, they had no necessity, nor was it a time for making tests of the quality of the cinnabar. That time had passed.

Now, look at this letter of January 19th. When I first read it I did not understand it all. The idea did not suggest itself to me that it did not refer to this mine at all; and I could not make head or tail of it until I saw the endorsements. Then, following up the track, which my friend Mr. Randolph left, I found out what it meant.

(Page 386).—My Dear Sir:—I am very much obliged to you for your very prompt attention to the business in hand, and return the expediente immediately.

I am very much surprised at the result of your assay, and shall try what I have?

Is that the ore of New Almaden? Is it probable that in January, 1848, these two parties have each taken specimens of that ore, and that James Alex. Forbes is testing one of the specimens in Santa Clara, and Alexander Forbes testing another specimen at New Almaden? Does that paragraph seem at the start to refer to New Almaden ore? No! It looks to something new, something strange; something that Alexander Forbes had no idea would occur. But he knew the richness of the New Almaden ores. The New Almaden ores, from the original assay, afforded an average yield of 35 to 40 per cent. Some of it was pure cinnabar, pure sulphurets of quicksilver. He could, therefore, be surprised at no assay of New Almaden ore. But he has been sent the results of an assay of some ore by James Alex. Forbes. He says, he is greatly surprised, and adds, "I will try what I have." Does that mean New Almaden ores?

(Page 386).—"It will of course be better to say nothing about it, particularly as I have already written to Monterey that there is no mine." Is that New Almaden? Is there no mine there? "Nor does there appear to be any quantity of this kind of stuff."

Now, what is it that this is written about? It is some mine the ores of which surprise him intensely, and which he is immediately going to assay, to see if his specimen corresponds as to results with the other. New Almaden ores have gone up, in some instances, to pure cinnabar. No assay of *those* ores would surprise him. But this is something that does surprise him. Besides that, he has written to Monterey that there is "no mine." Now, the mine of New Almaden was known to the public officers, and to the whole world; and the quicksilver therefrom was being sold! Is that New Almaden, where there was "no mine?"

I looked on the back of this paper, and saw that it had been endorsed in pencil: "Suñol, 4½; Sainse, 4½; Pena, 4½; Narbaez, 4½; Padre, 4; Forbes, 2." I added that up and found it made twenty-four; just the number of barras there are in a mine. Well, that is not New Almaden, because these are not the names of the owners of New Almaden. This is an entirely different set of men. Yet that is on this letter. Then this letter is about a mine; the assay of some of the ores of which are then—1848—thought surprising; and about which Alexander Forbes has written to Monterey that there was "no mine;" on the back of which letter there is written in pencil an endorsement, showing that the owners are not those of the New Almaden mine.

Now, James Alexander Forbes was asked about this. I

want to omit this poor devil's testimony, and will simply state its substance. He was asked about this endorsement, which shows the letter is not a New Almaden letter at all. He says it *is* a New Almaden letter, and that this endorsement is put on the back of it, as he supposes, by mistake. (*Vide* page 489.)

Now, what is the face of this paper? On its face it is unintelligible as regards New Almaden. But that is not all. Look at the letter of the 7th February, 1848, page 542:

SANTA CLARA, 7 Febr'y de 1848.

ALEXANDER FORBES, Esquire:

My Dear Sir: By the Mason I send you some specimens of the new vein, which was duly registered on Saturday, and poso opened; I have not had time to make an assay. Your opinion respecting the abundance of ores of cinnabar in the vicinity of New Almaden is very correct. I have this morning caused another denuncio to be made of another veta! and on my return from it I have yet another to make.

Is that New Almaden? We all know that there never were such denouncements made about New Almaden; but we have in this record these denouncements made of other mines all round it. So you see, by positive archive testimony, that these passages of the letter of the 7th of February, cannot refer to New Almaden. They refer to denouncements and registrations of other and different mines made on a particular day which we have brought in and copied in the record.

Your Honors can see that this letter of the 7th of February is but a continuation of the correspondence begun a few days before on the 19th of January. Here you see, that at this very time he was writing about other mines, and he has written to Monterey that there was "no mine." That has been perverted by the counsel for the Government into a statement that he had written "*to the authorities*" at Monterey. There is not a word in relation to "the authorities" in the letter; and if he had written to the authorities at Monterey, he would have told them what the authorities at Monterey already knew—that there was "no mine;" that is, in the place he was speaking of.

How did we find that out? Why, brother Randolph turned the attention of the witness (Halleck) to the correspondence of

Governor Mason in this executive document [holding up a pamphlet], for another purpose; and he referred to a letter of Governor Mason relating to Walkinshaw's opinions about the right of a partnership to four or five pertenencias, under the Spanish law. Naturally I looked at that—as at everything he has brought forward in this case—for the purpose of informing myself, and it occurred to me that there was at this time a false denunciation made to Governor Mason of something that was at that time "no mine." I refer your Honors to Governor Mason's letter to Alcalde White, under date of 17th April, 1848, at page 551, House Executive Documents, First Session Thirty-first Congress, 1849-'50—generally known as the California Message and Correspondence. In that letter he refers to his visit to this very neighborhood; speaks of his examining this very locality in relation to the business of the mines, and finding that there was no mine, and that there had been a fraudulent denunciation. And who are the parties? Who are the owners? The very persons whose names are on the back of this letter—or several of them. Just the names that James Alex. Forbes penciled on the back of this letter of January 19, are found in Governor Mason's communication of the 17th of April, 1848. I must read some of the passages of that communication to the Court. (Page 551, House Ex. Doc. 1849-'50.)

SIR:—I received your letter of the 2d inst., late in the afternoon of the 6th, together with the papers which accompanied it concerning the quicksilver mines denounced in your office on the 17th of December and on the 5th of February last—the former by Cook, Belden, Abrego, Ricord, and others; the latter by Pedro Sainsevain and partners—two days previous to the reception of your letter, I received one dated on the 4th inst., signed by Messrs. Abrego and Larkin, though written by Mr. Ricord, as he informed me, in which it is stated that the mine denounced on the 17th December was denounced by Mr. Cook, and shortly thereafter transferred to their company, and that the one denounced on the 5th of February was denounced by Jose Suñol, Abrego and Larkin—"as for an executive order for Suñol to desist from digging within their proper limits, and that the right of Suñol to be put in possession be deferred until the creation of certain Tribunals;" and state that the denouncement of Cook on the 17th of December, and that the possession obtained under it on the 2d of February, were prior to Suñol's denouncement, which was on the 5th of February; the other party, those interested in the denouncement of the 5th of February, claim to be the lawful and rightful owners of that mine, and demand possession of it under the 4th article, title 6, of the Mining Laws, and assert that the party claiming under the denouncement of the 17th December have no right to possession, because neither their excavations nor measurements of possession were made in accordance with the requirements of the aforesaid fourth article of the Mining Laws.

I availed myself of the opportunity which a visit to your part of the country last week afforded me, to personally examine and minutely inspect and measure both of those mines, denounced on the 17th December, and the 5th February last; there were three excavations at the mine pointed out to me as the one denounced on the 17th of December, all three within a few paces of each other. In one of the excavations, over which was erected a windlass, a man named Cash was at work; and near the two excavations lowest down the hill, stood a small cabin and Mr. Taylor's tent; this was on the 11th inst.; the deepest of these three excavations, measuring in the most favorable manner from the upper side of the hill, did not exceed fifteen feet, though one of the interested parties states that one of the excavations was of the depth required by the mining laws when possession was given, on the 2d of February, but that it since "caved in." I examined that spot which is said to have "caved in," with great care; it had the appearance of having been purposely filled up; it was dug in rock, not in earth; and in my mind it was as much impossible for that rock to have "caved in," as it would be for an auger-hole to "cave in" that had been bored in a large solid block of timber.

So, your Honors see that there were places around this mine that were denounced as mines that were no mines. People in those days were getting specimens of good ores, and going around and representing them as found in neighboring mines, and trying to deceive the unwary into purchasing, or taking an interest with them in the pretended mines. Amongst them, it seems, there was some mine of which Alex. Forbes had written to Monterey that there was "no mine;" that he did not believe any existed there. He was afterwards surprised at the assay made by James Alex. Forbes, and said he would try what he had, but did not believe in the thing at all. That is the letter of the 19th January, 1847.

There, then, we find on the 17th of December, that a body of men had been engaged in digging a hole in the rock, and filling it up, pretending it was their *pozo*, and that it had caved in, when, as Governor Mason said, "it could not cave in any more than could an auger-hole in a block of timber."

But what, in the name of sense, has that got to do with New Almaden? To whom was it that Alex. Forbes was writing, stating that there was no mine? To James Alex. Forbes and Padre Real, to whom he had written six or eight weeks before, that he had found the "*cinta*," that the vein was rich. James A. Forbes is asked "Why, what in the name of sense is the meaning of this letter (19th of January) written to you, if about the New Almaden mine?" "Why, the reason Alex. Forbes wrote that is this: He did not want people to know the richness of the mine; he wanted to buy out shares." "But who had any to sell?" "Padre Real and myself." Well then, he was writing (on the 24th of November) to the very men he wanted to deceive, the true state of the case. Only a few weeks before sending this letter of the 19th January, he had written to them of the richness of New Almaden, the certainty that he had found the vein, that the ores were all that could be desired; and sends a special messenger with the pleasing intelligence. Yet James Alex. Forbes says, the purpose of this letter written six weeks later, was to deceive him by telling him that there was no mine at New Almaden, in order that the writer (Alex. Forbes) might purchase out his (James A. Forbes') interest cheap! What could possibly make the Court believe such a ridiculous story as that? Here is the paper, endorsed on the back with all the names of the partners in a different and distinct mine, clearly having no connection with New Almaden. By reference to the correspondence in the record, I have clearly shown that these parties did have correspondence in relation to other mines.

Your Honors will find, at page 541, that there had been previous correspondence of these parties in reference to other mines:

SANTA CLARA, 14th July, 1847.

I was presented yesterday with a splendid specimen of quicksilver ore, from a spot within or near the limits of the two leagues conceded to Castillero and Socios, but situated upon the land claimed by the American, Cook, of whom you doubtless have been informed. The person who brought the specimen to me, was sent by one of the discoverers, and informed me that in May, 1846, this new vein was discovered and denounced before the authorities of San José.

So of some other veins he speaks of.

And, finally, we have a letter which has recently been brought into the record, which the Court will find at page 3050.

In January, 1849, James Alexander Forbes writes to Alexander Forbes:

I have received proposals from the house of Jecker, Torre & Co. for purchasing "barras" in the mine of Almaden. I have declined entering in any negotiation of this nature, but I have made proposals for them to take the habilitacion of the mine of Guadalupe, which is still in the state in which you saw it. Herewith I accompany a copy of my letter to D. Ysidro de la Torre upon that subject, for your government, and if he should not feel disposed to take the habilitacion, it would be more agreeable to me for you to include this mine in the negotiation of Almaden, upon terms which may be advantageous to both parties. I may add, that it is possible to obtain the major part of the shares in the mine of Guadalupe by judicious management.

Having shown, may it please your Honors, that this letter cannot, by its subject matter, in any manner apply to the New Almaden mine—that it is without sense or bearing on that mine, but has relation to other mines concerning which Alex. Forbes was corresponding at that time with James Alex. Forbes—I say that he (James A. Forbes) in bringing that letter into Court, commenced the series of frauds by which he attempted to impose upon the Court, and induced it to deprive, for a time, these parties of the enjoyment of their property.

At the very time—I think on the same day—that letter of January 19th was written, in which he (Alex. Forbes) speaks of his belief that there was no mine, referring to this other mine, he presents a petition to the Alcalde, setting forth that the mine of Almaden is a fine mine in beautiful order, and wants the Alcalde to go there with the "*peritos*" (mining experts) required by the law, enter the mine, inspect it, and make a record of its good condition. Now, those two things could not have occurred on the same day, in relation to the same mine. That thing asked of the Alcalde was done at once, the record of it made and put in the public office. The whole matter of this letter of January 19th shows clearly that its production here by James A. Forbes was the first of his series of forgeries. I call it "forgeries," for it is as much forgery to bring in a true document and endeavor to palm it off on the Court with a false meaning, as it would be actually to forge the paper.

I come next to the forged letter of the 28th March, 1848.

In relation to this letter I would say at once, that it ought not to be necessary to make any remark, for I understand Mr. Randolph to admit it to be forged.

Mr. Randolph—No, that is a mistake.

Mr. Benjamin—You said in the opening of your argument, that all the papers in the City of Mexico had been fabricated there after the 2d day of February, 1848. I noted that at once, because I immediately thought it an admission that this letter was a forgery. The existence of the two facts—that all the papers in the City of Mexico had been fabricated subsequent to the 2d February, 1848, and that this letter of March 28th is a genuine document—is totally impossible; because this letter, written up here in California, purports to be a letter of Alex. Forbes to James Alex. Forbes in relation to his knowledge of the forgery of these papers, in the City of Mexico, by Castillero.

Mr. Randolph—If you are familiar at all—and you must be now—with the course of my argument in this matter, you are aware that I maintain that a document brought up by Robert Walkinshaw in the spring of 1847, was a forgery. Hence, if forged in California, or in Mexico—where I infer it to have been forged—it was forged prior to the 2d of February, 1848.

Mr. Benjamin—I wrote down—Mr. Peachy wrote down—

Mr. Johnson wrote down—in fact we all noted at the time what I have stated. Inasmuch as the Government counsel was to give me an opportunity to reply, I had a right to know what his theory was.

Mr. Randolph—I think at the time Mr. Benjamin was taken ill, when I, perhaps, had grown too wearisome to be longer endured, I was discussing that paper of the 28th of March.

Mr. Benjamin—I do not deny that Mr. Randolph may have said the contrary of what I allege, in another part of his argument. I cannot pretend to follow his argument, for he makes one assertion at one time, and at another, another. But, when opening his argument—of which I took careful note for the purpose of reply—he put this proposition before the Court: that none of these Mexican papers were in existence prior to February, 1848. I wrote that down; all of us noted it.

Mr. Randolph—I desire the gentleman to precisely understand what he is to reply to. I say that all the papers now in evidence, including the Castillo Lanzas document as now produced, in my opinion were forged after the date (2d Feb., 1848) to which you refer; and further, that the paper referred to in the letter of May 5th, 1847, is not the paper now produced and called the Castillo Lanzas document; but that that paper was forged *before* the 4th of May, 1847, and that there have been here two Castillo Lanzas documents, entirely dissimilar. That is my case from the year 1858 to the present time. The Court understands it.

Mr. Benjamin—As I have to reply, I want to understand it myself. You say Robert Walkinshaw brought up a forged paper in 1847. I understand you to allege that another set of papers were forged after February, 1848. The act of possession of the mine—that was forged, when?

Mr. Randolph—That may be taken to have been forged in the month of January, 1848; possibly a little before; in 1847.

Mr. Benjamin—I will endeavor to remember these dates.

Mr. Randolph—When more familiar with the case you will not fall into error.

Mr. Benjamin—My brother forgets in one part of his argument what he said in another part, I think. I may be mistaken in this instance, perhaps. I said I supposed my brother, in the point made in his opening, had admitted this letter of the 28th March to be a forgery. He says he does not; that he said some documents sent up here were forged in 1847: and that the *other* documents which we produce were forged after February, 1848. But this letter of March 28, 1848, says, or makes Alex. Forbes say—which is the most absurd of all possible suppositions—that he wants to save himself from the loss that would ensue, should it in any manner leak out that in fact the documents procured by Castillero in Mexico, as his title to the mine and lands, were *all* obtained long after the occupation of California by the Americans. They were *all* obtained, Alex. Forbes says in this letter, long after the occupation of California by the Americans; and Alex. Forbes knows this in Monterey on the 28th of March, 1848.

Let us take that as a start in the matter of this forged letter. Now let us take up this forged letter, and see the forgery step by step as it goes along. It is really amusing; and may serve to beguile the Court for half an hour, while we go along. But, before we take up the correspondence of the 28th March, it may be well to begin with two or three letters that just preceded it. Your Honors will find the first of those letters at page 954, the answer to it at page 956, and the reply to the answer at page 955.

That is the *true* correspondence, and as soon as we have gotten through with it, we will see how it tallies with the forged.

The letter at page 954 is the rough draft, in the handwriting of Alex. Forbes, left in Tepic when he went to England, being then from 72 to 75 years of age.

The forged letter was produced here on the 15th of July, 1858. The true letters were produced, in order to show the forgery, on the 31st of July, 1858; so that but sixteen days elapsed after the forgery had been produced in Court before we produced the true letters. Those true letters are admitted by the witness (James A. Forbes), in his cross-examination, to be in the handwriting of Alex. Forbes, but he attempts to deny their receipt. I will show that they *were* received.

Now let us, with that light before us, observe how utterly impossible it was for these letters to be made up by us, within sixteen days from the 15th of July; for old Alex. Forbes was in London, and had left the country for many years, yet these were his own rough blotters of his letters, and were produced immediately after the production of the forgery.

Now let us see what these letters say. The first, which I am about to read, the Court will find is answered by James Alex. Forbes two days afterwards; and that the expressions in the one are copied and answered in the other.

James Alex. Forbes had gone down to Monterey; there had had a quarrel with Alex. Forbes, and had gone home to the mine. On the 20th of March, Alex. Forbes writes him as follows (page 954):

[NOTE.—Those portions of this letter printed in italics, and inclosed in brackets, show the words and sentences erased in the original.]

MONTEREY, March 20th, 1848.

My Dear Sir:—I have made an arrangement respecting the Admittance, by which she is to go down to the coast of Mexico at the same time [*with*] as the William, which arrangement amounts to a sale if certain contingencies do not [*occur*] take place. I therefore will not leave [*this*] Monterey till the 27th of [*the*] this present month. I mention this that you may know the latest day on which I can hear from you, and I request the bearer to be sent from the mine, to enable you to fulfill your promise of informing me if the Robles, or you as their attorney, [*would*] accept of my offer of $6,000 for their two shares.

This letter is a rough draft, and we have printed in italics what was stricken out of it—the erasures. At the commencement of the sentence immediately following what I have just read, the word "as" is stricken out, and the letter goes on as follows:

[*As*] I repeat what I said to you here, that it will give me much pain to leave this country with any bad feeling existing between us; and however much disappointed I may be in the arrangement which I hoped to make, I have no other wish than to do by you what I think is right and friendly. I, at all events, will expect to hear from you by the bearer [*or by any other conveyance which you think will be more speedy. I must feel most severely the disappointment of not hearing from you before I leave the country; and no good can result from a resort to any legal or harsh measures, on either side, between us who of all others ought to go in concert.*

I now beg that you will fulfill your promise in writing me so as to reach this before I leave, and I have still a hope that you will make such arrangements as will meet both our views and tend to all our] who will wait your orders, unless you choose to send your letters by some other hand.

Now this letter was received, when? On the 22d of March. In the letter acknowledging the receipt the statement is, "I have this moment received your letter of the 19th instant." Whether the letter copied from the rough draft was dated the 19th, by mistake, by the clerk who copied it, or whether James Alex. Forbes made a mistake in calling the letter the letter of the 19th, instead of the 20th, is of no consequence, because the letter cannot be mistaken. James A. Forbes' answer copies the very language of this rough draft:

March 22d, 1848.

My Dear Sir:—I have this moment received your letter of the 19th inst., and answer it immediately by your messenger.

The previous letter had stated that he (Alex. Forbes) had sent a messenger and wanted a reply by that messenger:

I had already closed a package for you, including a letter containing my answer to your last proposition, and in relation to the whole of the shares in the mine of New Almaden, held in the name of Robles. I beg to refer you to that letter.

I desire you to accept my thanks for the expression of your regret on leaving California with any bad feelings existing between us.

There, the expression in the rough draft of the 20th—"It will give me much pain to leave this country with any bad feeling existing between us,"—is copied in the answer.

Mr. Randolph—That, no doubt, is the answer.

Mr. Benjamin—You think there is no doubt of that? Let us, then, take that fact as conceded. What, then, does this letter contain? It contains the answer to the propositions of the letter of the 20th. This answer is replied to on the 25th. Let us see if the letter of the 25th is really an answer to the letter of the 22d. If it is, it will refer to expressions contained in the letter of the 22d.

In the letter of the 22d, James A. Forbes says:

(Page 956.) I desire you to accept my thanks for the expressions of your regret in leaving California with any bad feelings existing between us, and also for your friendly and equitable wishes towards me. None of the first exist in my breast, and I pray God that all success may attend you. There is no need of praise or merit attached by me to my actions or operations in this affair of the mine of New Almaden; nor shall I allege any services rendered to you; but I shall spurn at those base sycophants, who being no real friends of yours, but wishing to creep into your favor, have endeavored to poison your mind with regard to one who never entertained a thought sinister to your interest; and who, notwithstanding the cold, repulsive reception occasioned by your preoccupied mind, did not hesitate to manifest a sincere desire to serve you. But in what manner have those feelings on my part been appreciated? You declare that I have arrayed myself in defiance against you, and that you will take measures to counteract all that I can do! This has caused in me a poignant wound to my feelings, knowing that it may be augmented by your incorrect view of my intentions in relation to the affairs of the mine. I am quite sure, however, that the right of the Robles, which I represent, is most incontestable and tenable in any tribunal of the civilized world; consequently, it never has occurred to me that any litigation could take place, except by an attempt to impair that right.

It is difficult for me to believe that you, or any other person of sound, moral, unbiased judgment, will say that I have done wrong in the trust reposed in me by the Robles. But I will trespass no longer upon your time. I went to Monterey with a sincere desire to effect an arrangement with you; but you formed a misconception of my principles and sentiments. I have made a last proposition to you with the intention of retiring from any further intervention in the affairs of the mine of New Almaden. It remains with you to accept or reject that proposition, and also to remain assured of the respect with which,

I am, my dear sir,
Your most obt. servt,
JAMES ALEX. FORBES.

Here is the answer of the 25th:

MONTEREY, 25th of March, 1848.

JAMES A. FORBES, Esq., Santa Clara.

My Dear Sir.—I have received your letter of the 22d inst., and would have answered it immediately, but the man who brought it has been ill of a boil on his leg, which is the cause of the delay in writing you.

I shall not trouble you with any remarks respecting what has passed between us on the subject of the sale of your and the Robles' shares of the mine, further than to say that, I firmly believe your proceedings in the whole affair have been according to what you thought right.

That is the answer to this: "It is difficult for me to believe that you or any other person of sound, moral, unbiased judgment will say that I have done wrong in the trust reposed in me by the Robles."

The letter of the 25th goes on:

(Page 955.) But unfortunately I find the difference in opinion still as to the value of the shares exists; but in order to make another endeavor to effect an arrangement, I have named Mr. Walkinshaw and D. Manuel Diaz, who will wait upon you as soon as possible, and who take powers from me to conclude an agreement. I would not have named D. Manuel Diaz if he had been one of those whom you suppose had attempted to prejudice me against you.

In reference to that portion of the letter of the 22d, where James A. Forbes spoke of false friends, who had endeavored to prejudice Alex. Forbes' mind:—"On the contrary, he has always spoken of you as a friend, and as such I name him, thinking it will be agreeable to you."

"I am most sensible of your [*great attention*] exertions to serve me,"—copies the expression of the letter of the 22d, where James A. Forbes speaks of endeavoring to serve him—"and of the great trouble you took to forward in every way in your power the interests of [*the mine*] that concern, and [*I can assure*] you may rest assured that whatever way the business now depending between us is settled, and whatever proceedings may ensue, I do not leave California with any other sentiment towards you personally than that of friendship and respect."

There is the correspondence; and what is it?

First. A proposition to buy the shares of the Robles, and the price named.

Second. A reply, refusing the proposition.

Third. An answer to the reply, stating that as the difference of opinion still exists, the writer (Alex. Forbes) has sent Walkinshaw and Diaz to treat for the purchase.

When does that correspondence take place? On the 20th, 22d and 25th of March, 1848. These letters are produced here by us.

Now, for the forged one. The forged one is dated on the 28th of March, 1848; and the first line begins with an apology for never having written on the subject! (Page 864.)

> My Dear Sir—I have to apologize for not writing you before this, as I promised I would, respecting the purchase of your shares in the mine of New Almaden.

Is there another word necessary to be said? Is it necessary for me to go on? Three days after the 25th! Six days after the 22d! Eight days after the 20th! The correspondence of the 20th and 22d admitted to be that which took place! That correspondence was not in Court when this forgery was brought in, but it is now produced! The forgery copies portions of the true letters; is made up of portions of two or three letters, with an interpolated phrase about antedated titles which does not exist at all in the genuine correspondence, and the forgery starts out, being dated the 28th of March, with an apology for not having written before, as promised, in relation to the purchase of the shares!

What shall I say in addition to this striking fact? Can any human being believe this letter possible, after the previous ones? How is this letter, the forgery of the 28th of March, made up?

It states (page 866): I have to apologize for not writing you before this, as I promised I would, respecting the purchase of your shares in the mine of New Almaden; but really, as your opinion of their value is so widely different from mine, I considered it almost hopeless to make you any further proposals.

I do not, however, leave this, without making the necessary arrangements to effect that object, and have authorized Mr. Walkinshaw and Manl. Diaz to wait on you with my final offer for the purchase of those shares.

Two days before, he had written the same thing, viz. (955): "I have named Mr. Walkinshaw and D. Manuel Diaz, who will wait upon you as soon as possible, and who [*will have*] take powers from me to conclude an agreement." The clause about D. Manuel Diaz not being one of the men who had attempted to poison Alex. Forbes' mind is left out, and instead of that sentence, a forged sentence about antedated titles is put in.

Here is that forged sentence (Page 864): Were I not already so deeply interested in this negociation, I would never think of investing another dollar in it, but this interest renders it necessary for me to have the control of all the shares." * * * Why? in order that I may dispose of the whole, whenever an opportunity may offer, and save myself from the heavy loss that would ensue, should it unluckily leak out, that in fact the documents procured by Castillero in Mexico, as his title to the mine and lands, were all obtained long after the occupation of California by the Americans.

The forgery is so clumsy that it reveals itself at every step. The party who made the forgery did not know that we had the answer to the letter from which the forgery was principally made up. We have the answer to that *true* letter, answering everything that was contained in it, but saying not a word about this antedating.

Jas. Alex. Forbes would of course be startled by being told in that letter, if any such had really been written, that all the titles are forged and antedated! Some of the barras belonged to him. Yet, he holds on to them for years afterward—until he can sell them for $54,000. He is not disturbed about this antedating. I said we had the answer to this letter, or rather to the true one of the 25th, from which this was made up. What is it? (Page 957.)

SANTA CLARA, 28th June, 1848.

ALEXANDER FORBES, Esq.

My Dear Sir:—Soon after your departure from Monterey, Mr. Walkinshaw and Don Manuel Diaz made proposals to me in your name, for the purchase of the four "barras" or shares in the mine of New Almaden de Sta. Clara, of which I am the agent on the part of the Messrs. Robles. The offer made by your agents for those shares not being equivalent to their present or future value, I have declined entering into any negotiation for their disposal.

Why, how is this? Where is the "future value"? If it should unfortunately leak out that all these titles had been forged, where is the "future value" then? He is informed that all these titles *have* been forged! What has he to say to that? Not a syllable in this whole letter about it! Nothing but an answer to the letter of the 25th of March can be found in this letter of the 28th of June. There is no answer to the letter of the 28th of March there. No important part of the letter of the 28th of March is answered in the letter of the 28th of June.

MR. RANDOLPH—It was no news to James A. Forbes.

MR. BENJAMIN—My brother says it was no news to James Alex. Forbes. Is that so?

MR. RANDOLPH—Undoubtedly.

MR. BENJAMIN—Then why was it written to him as news. Why then was he informed of it so correctly in this letter? What is the object? Do men put down facts of this kind which may leak out and have the effect of destroying their title in letters to persons who know it already, and to whom, therefore, a slight allusion would be sufficient? On the contrary, this seems to be the communication of news. Either he (J. A. Forbes) knew it before—and then it was totally unnecessary to commit it to writing, and pass it through the mails—or he did not know it, and it was news to him; if so, why does not he answer it? Important as that news would have been to him, he did not even refer to it. No other answer to the letter of the 28th of March than the letter of the 28th of June is produced.

Go further into this letter of the 28th of June, and what does it contain? It contains bitter revilings. It contains violent personal reproaches about everything else—not a syllable about the news communicated by the letter of the 28th of March. It shows the existence of an embittered feeling between these parties which was going to bring Jas. Alex. Forbes into Court against Alex. Forbes. It shows also that he (J. A. Forbes) did not hesitate to write the grossest lies to those he corresponded with; for observe:

"I have declined entering into any negotiation for their disposal."

That is written by him on the 28th of June. The counsel for the Government produced here in evidence his sale to Walkinshaw of those very barras in the previous April. In other words, the first piece of treachery committed by Walkinshaw towards Alex. Forbes, was going to Jas. Alex. Forbes, as the agent of Alex. Forbes, to purchase certain shares for him, and buying them for himself instead, and after having bought them for himself and paid for them, Jas. Alex. Forbes, who has already sold those two shares to Walkinshaw, and who has himself bought the other two shares of the Robles, writes to Alex. Forbes: "I have declined entering into any negotiation for their disposal!"

The whole quarrel between Alex. Forbes and James Alex. Forbes, was in relation to the Robles' shares. Litigation is threatened. Alexander Forbes is told to send up some man to represent him in California; he is going to be sued. All in this letter of the 28th of June, is a plain, clear answer to the letter of the 25th of March. Not the slightest allusion is made to the antedating spoken of in the forged letter of the 28th, or to the facts contained in the forgery.

This forgery is brought into Court, pieced up with expressions from the true correspondence. It is made up by James Alex. Forbes, for his friend Laurencel, after they had broken down in their first attempt, and all their inventions had been scattered to the winds by the production of the original correspondence up to that date. After James Alexander Forbes had sworn in Court everything that could be required of him in relation to his first set of documents—had sworn that he knew nothing about these titles to the mine, and that he had never seen them, up to a particular date—brother Peachy pulled out of his pocket letters of a prior date signed by himself (James A. Forbes), that proved every word he uttered to be a lie.

What is to be done now? He had sworn that he had not seen certain documents up to 1848.

What is the document he speaks of in his letter of 1847? When asked, he equivocates, denies, tergiversates—won't answer. He is drawn back to it again and again, and finally he answers: "Well, it is the title. I saw them. I saw the copies he had brought up with him. I saw them in 1847."

He had previously sworn over and over again that he had never seen them at all.

He is driven out of Court in disgrace. The conspirators have lost their money. Laurencel had lost his first ten thousand, given for the papers to be used for blackmailing; he had lost his second ten thousand dollars, paid Forbes to give his testimony. Forbes had testified, and his testimony was routed. How much he was now paid for the forged letter I do not know. We may well surmise that these little services among these men were not rendered for nothing. Although there is said to be honor amongst thieves, James Alex. Forbes was too cautious to rely on promises for services he was rendering, and without doubt required cash in hand; but how much he received for this forged letter we do not know.

Forbes' testimony, when at an end, required bolstering up. My brother Randolph says forgery is natural to this man. The Government knows, Laurencel knows, who will forge papers, and at once they go to him. He is a man from whom any forged papers can be produced. He brings out this forged paper from the mine of true correspondence he has, from which any quantity, any description of forging required can be manufactured by him.

This is the only letter in this entire correspondence in which the remotest hint or suggestion of antedating is found. James Alex. Forbes again and again writes, urging these parties to forge titles. Again and again he says to them:

"You ought to have forged such and such papers. We want other titles than those we have here got. The titles we have got are an order to the Governor of California to put us in possession. That never was done. That is not a good title. We want a title so manufactured that it shall be a complete title to us without an order to place us in possession. We want a complete title at once. We have got a registry and a possession, given by an Alcalde. That is not a good title. We want a ratification by the Government in Mexico. We want that forged. Now you must have all those papers forged, and send them up to me."

That is the constant burden of James Alex. Forbes' letter to different parties, at different times.

He then says: "Walkinshaw has got our true papers. Walkinshaw has got our true titles. Walkinshaw is a villain, and is trying to cheat us out of the mine. I am afraid he will burn up the papers. I am afraid of every possible machination that villain can devise, by which to rob us of the property."

He (Forbes) never once hints, however, that the papers were forged or antedated. It is never again referred to in the entire record. The fact—spoken of in the letter of March 28, 1848, —that these things *had* been done, is never again referred to.

Again; why is not the original letter produced? James Alex. Forbes was asked: "Sir, you came here before. You sold papers for money. You sold papers for the purpose of blackmailing these parties, as you say. Your blackmailing project having failed, you afterwards got money to come and testify. You never came here—as you admit—until you got $10,000 for entering this Court and testifying. Now, sir, why did you not bring this letter with you? Why is it we have never heard of the original?"

He tells six different lies on that subject. I have marked them down. But the subject is too disgusting to follow out.

First, he says he lost it; next in sequence, he says it did not tally in date with the other letters he had sold; next, he says he kept it back on purpose, because he knew Mr. Peachy had gone to Washington to get some law passed, under which he supposed some new testimony would be brought in from Mexico, and that letter was to vindicate him; next, he says we were not hard on him in the cross-examination—"I would have brought it in if you had gone hard on me; when you were hard on me, I brought it in."

Time and time again he tells this story as the reason why the original was not brought in at first.

Again: can your Honors suggest—I cannot—what would make a man sit down at home and make a copy of a letter he has got in his drawer or desk? I never did such a thing in my life, and cannot see any reason for doing it. If you had a letter from a man, would you amuse yourself, sitting down and copying it, and then keep the original and the copy in your desk? What for? What did this man make a copy for unless he was getting ready to have the original stolen? He takes care to provide a copy, in order that he may have the original stolen, and bring in the copy.

And what is the story about the stealing of this original? Why, it would be a miserable thing to put into a comedy! People would hiss the play, and drive the actors off the stage, disgusted with the silliness of such an invention. It does no credit to Laurencel and Forbes.

Brother Billings, you and Barron went to the hotel and stole the paper out of James Alex. Forbes' carpet bag! You then said you did not know anything about the letter, and I will never associate with you again. This thing is proven on you, and James Alex. Forbes almost swears it.

Mr. Randolph—For the reputation of the profession, I must ask what is there in the evidence to show that Mr. Billings had anything to do with this theft?

Mr. Benjamin—If you will look at Laurencel's friend Birnie's testimony, you will see his name brought in for no other purpose that I can perceive, but to hint that Barron and Billings stole the original letter.

Mr. Randolph—I have nothing to say as regards Barron.

Mr. Benjamin—It is true Mr. Billings was at that time in Vermont; but that is nothing. It was just as easy for him to steal letters here, when in Vermont, as if here! It is true the doors were all locked, the windows closed tight, and the carpet bag fastened. But what difference does that make? You [tapping Billings on the shoulder] had false keys for all. It is true that Forbes had carried this letter all day upon his person, and just left it in his carpet bag for a convenient half hour, so that it could be stolen, and the copy he had prepared produced. But that does not make the thing improbable! You, brother Billings, must have known he intended to leave it in the carpet bag during that half hour, and you were all ready to go in and take it from thence. I am astonished at you!

Brother Billings, you went to Mexico, too. You found a big book there, You forged two pages in it. Not only that; you found on the paper "L. G. 15, fo. 140," and some other hieroglyphics, and got the clerk to swear that that meant the General Book, page 140, etc. You brought up the clerk to swear that that was the book of the Ministry of Justice, containing part of the original archives of Mexico. Now that is very astonishing again. Nobody did it but yourself. Nobody found it until you got there. You found it—and it is very easy for a man to find everything where he puts it, and I believe you put it there first. So of the carpet bag, the locked room, and the comedy; I believe you had a hand in it all!

James Alex. Forbes having taken a copy of the original, with an eye to its being stolen, and having put the original in his carpet bag for half an hour, so that it could be stolen, now brings the prepared copy into Court; and necessarily, as it is HIS testimony, it must be believed. Your Honors could not, for a moment, deny your credit to a copy, produced by James Alexander Forbes, himself, to an original he pledges his *honor* he once had! Once his *honor* pledged, the Court *must* be satisfied.

There is another witness for whom Brother Randolph has more respect than he has for Forbes. I think he does not admit that Birnie purjures himself for the mere pleasure of it, nor that he is a forger. Therefore, that not being admitted, we had to prove it. To prove it we produced only nineteen witnesses.

MR. RANDOLPH—On the other side we produce sixteen.

MR. BENJAMIN—We are three ahead of you. We have nineteen witnesses to prove that this man is a common and notirious liar and perjurer. We have all the notables of the place where he lives, who swear they would not believe him on oath.

MR. RANDOLPH—We have more notables to prove the contrary.

MR. BENJAMIN—Then, of course, your notables do not know him. And I say, if you had ten thousand witnesses to prove a man credible, and had I ten who would swear they would not believe him on oath, the presumption would be that your ten thousand did not know him, and that my ten did.

Who is the first witness, and what does he say about it? The first is Colonel Coffee. What does he know about Birnie? I will state to the Court that every one of these witnesses brought to impeach Birnie, were asked the legal question only, viz., "What is his general character for truth and veracity? Would you believe him on oath?" Every one says he would not believe him on oath, and that his character for truth and veracity is very bad.

But the principal points of his character came out on the cross-examination by the Government counsel.

MR. RANDOLPH—I was absent at the time. Had I been present I might not have asked the questions.

MR. BENJAMIN—I don't think you would. You know too much for that.

Now what does this Colonel Coffee say about Birnie? He says that he deceived him in a business transaction:

A man that would deceive me in a business transaction, I would not believe on his oath, and beside that, I have heard his reputation very frequently attacked in conversation.

The next witness is Warmcastle, County Judge, and member of the Legislature.

Ans.—From his reputation in the community, and my acquaintance with that reputation, I could not believe him, unless I was satisfied that he had no motive for misrepresentation.

Ques. 9.—Does not this rule apply to every person, and please state why particularly to Mr. Birnie?

Ans.—No, sir; there are many persons in my community, whom I would believe on oath, when it would be for their interest to testify falsely, but Mr. Birnie is not one of that class of men, from my knowledge of his reputation; my impression is derived from my acquaintance with Mr. Birnie, through public rumor, for a number of years.

Next, we have David Boss, who is a farmer. What does he say?

Ans.—When I find that a man, whose word is not good, whose note is not good, I would not believe him on oath. I regard Mr. Birnie as a dangerous man, who would take all sorts of advantages. I would not trust him out of my sight. He is a worthless man, who lives on the labor of other people. For my part I have made money by hard work.

His brother-in-law, William Welch, told me that he stole his name, and put it down on a piece of paper, without his (Welch's) leave. I understood him to say that Birnie had forged his name, and I believed that Welch told the truth about it at the time.

That is the cross-examination to sustain Birnie.

Nathaniel Jones is the Sheriff of the county. He is cross-examined.

State some acts of Mr. Birnie's which has caused the bad reputation which you speak of among the community in which you knew him.

A. It is not from any single act of his, but it is from his want of occupation, and the universal belief among people that he swears falsely, and procures false evidence in land cases. He is utterly worthless, and he is a loafer. If I were to start out to find a man who thought well of his character, I should not know where to go to find him?

That is the cross-examination which is to sustain Birnie, by testifying and sifting the evidence.

Again he is cross-questioned.

What land cases do you refer to?

A. Anywhere he could get a fee for swearing falsely, and procuring false evidence.

Now, we come to B. R. Holliday. What does *he* know about him, and what does *he* think of him?

After speaking of his reputation, and being asked if it did not arise from some lawsuits, the witness said:

A. Perhaps they did in part—not all. Shortly after his marriage, his pecuniary credit became bad, and in the community he got such a reputation that anything that was stated as coming from Birnie would not be believed.

It was enough that it came from Birnie.

The next gentleman, Nicholas Hunsaker, who was the Sheriff, was asked if there was anything outside of the lawsuits in which Birnie had been engaged, in which his character was bad?

A. Outside of lawsuits, it was said he employed a man to steal barley, which was in the custody of the Sheriff; this is the only act outside of litigation that I remember; with regard to those lawsuits, I have heard a great many people state that they would not believe him under oath.

The next witness, Smith, on his cross-examination, being asked, says:

A. I have no particular feeling against the man; I would believe him on oath if I could, but I could'nt do it.

So, it goes on. I will not detain your Honors by reading the rest of this testimony, with the exception of the evidence of one more witness, Mr. Elam Brown.

He says, when asked when Birnie's reputation commenced to be bad:

A. I had known him two or three years before anything occurred to cause me to lose confidence in him. His reputation became bad by degrees; he was charged with doing one bad thing here, and one bad thing there, until finally people began to say that they would not believe him under oath. This is what I have heard, not what I know of my own knowledge.

I am on no unfriendly relations; nothing has occurred between us. I am sorry to be here to-day testifying against him.

I consider him a dangerous man, and I believe him deficient in those principles of honesty which would make him a reliable witness.

I would observe to your Honors, in relation to the testimony taken upon Birnie's character, that that taken by us was from witnesses from Contra Costa, where he had been living five years.

Now, may it please the Court, what becomes of the forged letter and all the superstructure built upon it? These are the witnesses you have got to prove the existence of the original. You have paraded before you by the United States, James Alex. Forbes and Robert Birnie; to prove that he made a copy at home, before he brought it here, you have James Alex. Forbes; to prove that he made a copy also, you have Robert Birnie; they provided two copies in advance—one for Birnie and one for Forbes—before the letter was in a condition to be stolen. That the letter was put in a condition to be stolen, you have no evidence but that of James Alex. Forbes. Nobody saw him put it in the carpet-bag but himself. You have no proof but his own, which, I need not say, the counsel for the Government admits is utterly unreliable. In relation to the theft you have no evidence but his. When he left his room to get a cup of tea, he put the letter which he had always previously kept on his person, in his carpet-bag. Somebody got into the room and took it from the carpet-bag without breaking any of the locks of the apartment, or that of the carpet-bag. That is the story that is brought forward to bolster up a forgery patent on its face. On that your Honors are asked to believe that there is forgery, fraud and fabrication in this case. But I leave that letter.

Observe, if your Honors please, that I wish to deal candidly with the Court. I do not say there were no propositions to forge. I do not mean to say that all the parties who had an interest in this title, repulsed those propositions to forge with the indignation with which they should have been repulsed. That subject I will come to presently. I do not expect to gain any case by attempting to hood-wink the Judges as if they were children.

I am now engaged in endeavoring to show that the original conspiracy against our title was based on the forgery of nearly all the papers.

I will now come to the next forgery. Your Honors will now refer to the forged letter of the 28th of October, 1849, at page 392. The true letter of the same date is at page 844. Do your Honors want any further proof of forgery than this?

S. F. Oct. 28, '49.

WILLIAM FORBES, Esq., Tepic.

My Dear Sir:—I have been detained here until the present moment, occupied in carrying out the arrangements explained to B., F. & Co., in my letter to them of yesterday's date.

Here is the forged letter:

S. F. Oct. 28, '49.

WILLIAM FORBES, Esq., Tepic.

My Dear Sir:—I have been detained at this place until the present moment, occupied in completing the arrangements explained to B., F. & Co., under the date of yesterday.

Is anything further necessary? Two letters of the same date, to the same man, beginning with the same sentence, both marked private! Now, I say that letter was forged when he had the original. Do your Honors want the explanation? He had forged a memorandum, which he had said he left in Tepic the previous May, and he wanted to show that that memorandum had actually existed there. In order to bolster up the forged memorandum, he had to forge a letter that would refer to the memorandum; and as the true letter did not refer to the memorandum, he took the first sentence of it and forged the remainder for the purpose of referring to that document.

MR. RANDOLPH—Which letter do you like best?

MR. BENJAMIN—I like neither best. I want the Court to look at the true one. The true one is as bad as the false. It was not necessary to forge the false letter for any other purpose than that which I shall afterward show. These words "forgery" and "perjury" are hard words, and they come back to roost sometimes.

Mr. Randolph— * * * New Almaden.

Mr. Benjamin—I have no doubt of it. How about your friend Laurencel? These things come back to roost. We will see as the case goes on. Here, then, is the true letter of the 28th October, 1849; and here is the forgery of the same date to the same man, also marked "Private," and beginning with the same words. The one brought in by us after the other had been produced by James A. Forbes, as being a copy of his letter which he had presented, is the true, original letter. Now, in the true letter there is just as much adverse to the case of the claimants as in the false one; because the true letter calls attention to the forgeries he wanted committed, as well as the false. But in the true letter the statement in regard to the forgeries is this:

(Page 846.) "I now desire to call your attention to the following important matter."

Now, then. He is now initiating a proposition for forgery: (Page 846.) "In order to secure the possession of the land, you must forge and antedate titles."

But when he first came into Court it was necessary to carry the date further back. So he forged a memorandum, which he said he had left in Tepic; and then forged the letter of this date (28th October, '49). Instead of the statement in the true letter, "I now desire to call your attention to the following important matter," he inserts in the forgery, "I must *again* call your attention to the subject of the documents comprised in the memorandum left in Tepic."

And now, let me call your Honors' attention to a remarkable circumstance throughout the whole thing, to enable you to perceive the true state of that correspondence as originally brought before you: That in not a solitary letter throughout is that memorandum spoken of, except in letters brought in by James Alex. Forbes, as being copies preserved by him of letters written by him to the parties. There is no answer, no allusion in our correspondence to any such memorandum. It never appears in the whole correspondence, except in letters which James Alex. Forbes brings in and swears to be copies of letters which he wrote.

This is the first one of the copies of the letters he swears he wrote. It is a forgery. Two letters, of the same date, with the same beginning, and to the same man! In one of the two —that brought in by the claimants—Forbes is initiating the subject. In the true letter he says, "I now call your attention;" in the false, "I must *again* call your attention"—changing the word "now" to "again," and putting the initiative back to the date of the Tepic memorandum, which no man had ever heard of until it was brought in in this case. It is awful, awful, to have all this mass of forgery brought in here and put in evidence by the Government of the United States against a claimant.

MR. JUSTICE HOFFMAN—What do you suppose is the object in the production of this letter by James Alex. Forbes?

MR. BENJAMIN—His theory was, that there was a certain date for antedating. When first brought in, he had a theory to maintain. He could not show all his letters to Laurencel, because, if he did, he would show him the fact of the existence of our title. He had to pick out certain of our letters, take his answers and pick out passages, and present them to Laurencel as proof of antedating and forgery. He remained at home, gathered his correspondence together, excluded everything referring to the existence of the titles. Whenever his own letters referred to anything favorable to the titles, he cut the passage out. Pretending that the copies he made were true copies of the originals, he sold them to Laurencel—but not to bring into the Court. It was never intended by James Alex. Forbes that those copies should be brought into Court.

MR. JUSTICE HOFFMAN—The suggestions of the Tepic memorandum appear in this letter never to have been acted on. The whole letter is a reproach on that account.

MR. BENJAMIN—What motive influenced his conduct with Laurencel, it has been impossible for me to divine. But certain it is, that in this correspondence it is not one or two, but half a dozen letters, that he has forged in this way.

What his purpose was, God only knows. He never meant

the letters to be examined; never supposed they would be examined in Court. Again and again he urges in his cross-examination, that he never intended them to be produced here.

I am going to take up the true effect of the correspondence, and will show your Honors what I suppose the true theory to be, not denying any inference that may be drawn, but treating the Court with entire candor as regards the subject.

On page 393, you have the letter of the 30th of October, 1849. It is from James Alexander Forbes to Alex. Forbes, the sole purpose of it being to again speak of his having explained all this matter at Tepic. Nothing in relation to Tepic is in the original correspondence anywhere. But James Alex. Forbes left his tracks on the face of this letter. He forged it from his letter to William Forbes, of the 28th October. After forging it and making it up as addressed to Alex. Forbes, he put on the back the date of the true letter (which he copied from the true letter), while indorsing it with a wrong date, thus showing the origin of the forgery. Here on the back is the date (28th October) of the letter to William Forbes, containing, substantially, what James A. Forbes put in the forged letter addressed to Alex. Forbes; and the letter to William Forbes is dated October 28th; that to Alexander Forbes—which is made up from the former—October 30; and when James A. Forbes came to indorse his forgery with the proper date to bring into Court, he indorsed it October 28, copying from the letter of that date. There is the track of guilt forgers leave behind them. The true letter is addressed to William Forbes, under date of October 28. Out of that is concocted a false letter to Alex. Forbes, on which the date—October 30—is put; but when the forger goes to indorse the date on the back, that it may be brought into Court, he forgot he had made a change of the date, and writes on it that of the original, of which it is a copy. The true letter is at page 844.

You observe that every letter that I have shown you to be forged, is a letter which he brings in purporting to be a copy made by him of his correspondence.

Here comes another. Page 403; letter of February 26, 1850. Now, how was this letter produced in the original papers? Look at it! See how it deceives! The letter is brought in as

O. H. No. 9. It is followed by O. H. No. 10, O. H. No. 11, and O. H. No. 13,—all of which seem to have reference to its contents. Now it never was received, if ever written. But it never was written.

Mr. Randolph—How do you know that it was never received, if ever written?

Mr. Benjamin—This letter of the 7th of April announces the receipt of letters. As late as that date nothing had been received from James A. Forbes, posterior to the 19th of Feb'y. Page 405: "I wrote to you by the California, dated the 23d of Feb'y, and since then, have received by the 'Oregon,' yours of the 19th of that month." Where is the letter of the 26th?

Mr. Randolph—He had not received it yet.

Mr. Benjamin—Why had not a letter leaving California on the 26th of February reached Tepic by the 7th of April? Whether it ever was received or not, by the face of this correspondence, no man can tell. But it is introduced into this correspondence to make it appear that this is the letter to which the letters O. H. No. 10, O. H. No. 11, and O. H. No. 13, were answers. Upon examining those letters, we find that at the date of their writing this letter had never been received. Yet, it is arranged by Jas. A. Forbes in this manner, and presented to Laurencel for purchase. Time was not to be afforded for examining all of this purchase. Time was not to be given Laurencel to ascertain how this correspondence fits in. He is not allowed to do anything beyond making an inspection of the papers, until his money is paid. He is not to use them except to black-mail. And when at last brought into Court, this man Forbes, with $20,000 in his pocket, is put on the stand. Then, all these papers coming from this source are produced by the Government of the United States to break down our title.

Mr. Randolph—Forbes is one of the claimants in the record.

Mr. Benjamin—No! no! You have proven that he sold

out his interest, and then bought him to come in and give evidence against his partners—such evidence as we have seen in this case.

Now, may it please your Honors, from all this elimination of everything that is forged in these papers, I come to state to the Court what my true theory in relation to these letters is, and to place it before your Honors for your appreciation. These parties owned a valuable property in California. They owned it by titles derived during the existence of the Mexican Government. They had spent a fortune upon it. It is in proof in this cause, offered by the United States, at page 215 of this record, that these parties had spent upon this property, up to May, 1852, $978,114, and had gotten back $553,000, and no more.

Mr. Randolph—Where do you find that?

Mr. Benjamin—In the sworn answer in chancery introduced by you in this case. In these amounts are not included any allowance for interest on capital expended by the defendants. They were then $440,000 or $450,000 out of pocket. They had spent close upon a million of dollars upon a barren peak in the mountains.

Their liberal expenditures have turned out to have been judicious. If the mine had been poor, according to good old Gamboa, it would not have been (*muy codiciada*) very much coveted. Turning out to be rich, a host of harpies gathered around. These obscene birds of prey were befouling everything they touched. First came a body of people and said the mine was on the Berreyesa Rancho, because it had been so described in the original papers; Berreyesa having made a fraudulent alteration, and in that way defrauded Castillero into declaring it was on his land.

Next comes a body represented by my friend, the special counsel of the Government. Laurencel and his associates have been depicted with the feeble powers that I possess, and I have stated to the Court as plainly and as impartially as possible what their true character is.

The New Almaden Co. had sent up their titles through an

agent—an agent they had taken up and invested with their confidence—Robert Walkinshaw. He was their agent, and had all their documents of title. He, also, turned on them, with an attempt to defraud them out of shares of the mine, to get a larger portion for himself. He formed a combination with others to denounce the mine as having been abandoned, to represent the original title as bad, and to share the property with others, thereby getting a larger share for himself. In that he was playing a certain game. If he succeeded, he got a bigger share: if he failed, he still retained what he had. He had nothing to risk, something to gain. In those days, with people combining all around them to defraud them out of their property, Jas. Alex. Forbes—who had written to them of Walkinshaw's intentions—commenced afresh, in 1849, the suggestions made in 1847. When the suggestions were made in 1847, he was not their agent. They were not responsible for what he chose to write, but treated his propositions with entire disdain. This is best proven by his own letters, since brought into the record, to which I now refer your Honors, dated January 13th, 1849. (Page 3050.) In this letter to Alex. Forbes, he says to him:

I have much to say to you upon other points, but I cannot write too much upon the subject to which I have alluded, contained in my letter above mentioned, and which I fear has escaped your memory.

What is the letter above mentioned? Observe that, when that letter was written, the mine of these parties was in the possession of Walkinshaw; yet, Jas. A. Forbes, writing this letter to persons who were part owners of the mine, got no answer to his proposition. He had their interests in his charge, yet they did not answer it. No answer is produced or pretended to have been sent. On the contrary, in January, 1849, he (J. A. Forbes) writes that he fears that letter has escaped the memory of Alex. Forbes. Nothing has ever come of it.

Here you have in January, 1849, this: that his letter of May, 1847, had been forgotten. Nobody had ever paid any attention to his counsels. But, towards the fall of 1849, (and here I come to a point suggested by your Honors when you asked me what the difference was, whether it occurred in May or November), for the first time the propositions are made to Barron, Forbes & Co., who, in the interval, have taken possession of this mine, in behalf of the avio formed in Mexico for supplying it. At page 390, you find the first connection of Barron, Forbes & Co. with this mine.

TEPIC, 11th April, 1849.

JAS. ALEX. FORBES, Esq., San Francisco.

Dear Sir:—We beg leave to refer you to Mr. Alex. Forbes' letter of the 9th inst., respecting the arrangement of the affairs of the Mine of New Almaden, and beg to recommend that negotiation to your best care and management until we can forward the necessary instructions for your government. You may now rely on this mine being worked to the utmost of its capabilities of production and sale of quicksilver on the arrival of the apparatus, and we hope to make up for the delay which circumstances have hitherto prevented this important concern from being productive. We shall soon have the pleasure of sending you a list of the company of which the "Habilitadores" are composed. The House of Jecker, Torre & Co. of Mexico and Mazatlan, and our own, are chiefly interested, and as Don Ysidoro de la Torre has gone to Europe, he will concert with Mr. Barron everything which can tend to the successful development of this enterprise.

Now, fix your Honors' attention on this. The first proposition in May, 1847, never answered and nothing done, complained of in January, 1849, as forgotten. In April, 1849, for the first time, Barron, Forbes & Co. take an interest in this matter. They enter into it as *aviadores;* they furnish funds. They join with them Jecker, Torre & Co. of the City of Mexico, and the common funds furnished by these parties are to be used for putting this mine into active and successful working operation. Up to that time the funds had been insufficient. Probably there was doubt among the owners as to the amount of funds they were willing to expend. But now a new and vigorous impetus is to be given the enterprise. Alex. Forbes, long at the head of the business, then on the point of retiring and going to England, gives up his charge to the new company and they begin to operate for the supplies of the mine, necessarily under the influence of this old gentleman who had probably reached his three score years and ten nearly.

That is the condition of things in April, 1849. Now James Alex. Forbes goes down to Tepic, and succeeds in getting the house to make him the agent of the mine. What he told them about Walkinshaw, what means he used for that purpose (securing the agency), none knew better than himself. *We* know not. He comes back, turns Walkinshaw out of the mine and takes possession; then comes warfare; then Walkinshaw and James Alex. Forbes have a long and bitter litigation; then James A. Forbes begins to dread the destruction of the titles; then he writes again and again to his principals, "This man Walkinshaw is a villain; I do not know what I shall do; he is going to steal away the original registries from the Alcalde's office; he is going to destroy the documents of title which you gave him to bring up here, and which he refused to give me; he is buying up the rights of the neighboring proprietors that he may bring suits; he has denounced your mine as abandoned, and claims that he has a right to take it under the mining ordinances; there is no species of villainy which you may not expect of this man; I am struggling against him; I am desirous of defeating him; I have your interests at heart, but I do not know how I shall succeed; in my judgment and opinion, your titles are informal; I told you so years ago, yet you won't send me other documents that I wanted fabricated."

"On the other hand, I am acquainted with the machinations of the neighboring proprietors, who say that their titles extend over this mine; I want help; that help consists in your obtaining for me fraudulent, fabricated and antedated papers, which must bear the date of the Castillo Lanzas decree. That is what I want, and what you must send."

This is what he writes, for the first time in any original letter produced to Barron, Forbes & Co., or William Forbes, in the fall of 1849.

Now, under these circumstances, what are the answers? This property is in a foreign country, far from those men, in the hands of agents whom they distrust, one of whom has turned his back upon them, and is endeavoring to swindle them out of their property; the other of whom is writing them propositions for forgery. Perhaps you will say, and perhaps a man of rigid morality would declare as his judgment, that they ought instantly to have repelled the propositions with contempt and scorn, and driven him (James A. Forbes) out of the mine. But where would their property have been then? Then it would have been necessary for some one else to go and take charge of it in his place. They were not ready for that. The partners were abroad. The proposition amongst them is, "Let us temporize; let us wait a month or two, until one of our partners comes home, that we may send him to take charge and put an end to the scene of fraud and corruption going on up there." To that end Alex. Forbes conducted the correspondence in his own name, but of course for Barron, Forbes & Co., in his letter, and appears to accept the proposition.

Perhaps I am going too far in saying that it is apparent on the face of the papers, that if he could have done so safely he would have been willing to join James Alex. in the forgeries. But recollect another man existed in the firm, the man who controlled the house, and who was on the eve of his arrival in Tepic.

On the 3d of February, Eustace Barron returned from Europe. Your Honors will find the announcement in the letter from Tepic of the 3d of February, 1850, at page 400.

Eustace Barron has arrived in Tepic on the 3d of February. What is the result of his arrival? Within thirty days this man James Alex. Forbes is turned out of the mine; Ysidoro De la Torre is sent up here to take possession of it, and when James Alex. Forbes is turned out it is put in the possession of Captain Halleck in behalf of the owners.

I have already read to your Honors the letters of the previous spring, in which the statement is made that De la Torre will confer with Eustace Barron, who is in Europe; that is in April, 1849. Castillero was in Lower California; De la Torre had gone to Europe; the owners were scattered; their interests were in the hands of strangers who were endeavoring to swindle and defraud them.

Eustace Barron returns on the 3d of February, as shown in this letter. Castillero has just arrived, and the parties meet. What does Eustace Barron find? Walkinshaw endeavoring to swindle the partners; James Alex. Forbes proposing forgeries; the vast interests belonging to his house apparently in the hands of thieves and swindlers. That is the condition of things on the 3d of February.

Now, here is the letter of the 2d of March, page 404. Eustace Barron had not then been in Tepic a month:

TEPIC, 2d March, 1850.

JAMES A. FORBES, Esq., New Almaden:

Dear Sir:—We duly received your letters up to the 29th of January, per steamer Panama, which have had our best attention, and as our friend, Dn Ysidoro de la Torre of Mazatlan has been appointed, and has consented to proceed by this steamer to California, with full powers to act in behalf of all concerned, it is needless to enter into any particulars respecting the various matters contained in your letters, as you will be enabled personally to communicate your views to him; and to arrange everything in the best manner possible. Mr. De la Torre came to Tepic to meet Mr. Barron and the others concerned in this negotiation, and it was deemed necessary that some of the partners in the "Habilitacion" should proceed to New Almaden in order to consult personally with you, and to arrange respecting the future operations of this enterprise, and Mr. De la Torre has been prevailed upon, at much inconvenience to himself, to undertake the present charge.

We are sure that no one could be named more agreeable to you, than Mr. De la Torre, and have no doubt but that his presence will be most useful in sanctioning and arranging a plan of future operations, and of assisting in adjusting any difficulties which now exist, particularly as he has the full authority of the association to act, as to him shall appear necessary. Mr. De la Torre takes up with him Dr. Tobin in the hopes that he will resume his labors, and act in conformity with his duty.

Mr. Barron and Don Andres Castillero are about to proceed to the City of Mexico, and will attend to what you have recommended.

That recommendation was, as stated in the previous letter, to get a "copy of the ratification which Castillero says exists in the archives," unknown previously to the associates.

Now, at page 328, you have the deposition of Captain Halleck. In March or April, 1850, Captain Halleck takes the charge of this mine. De la Torre does not temporize for a moment after he gets here. James Alex. Forbes is at once turned out, and Captain Halleck put in in his place.

Mr. Justice Hoffman—Just read the letter of the 29th of January, 1850, from James Alex. Forbes to Alexander Forbes.

Mr. Benjamin—Page 399:

I have received the copy of the contract of habilitacion, and as you request me to address myself to B., F. & Co. on the affairs of the mine, I have now written to them upon this particular subject to which I request your earnest attention, not as regards the habilitacion, but another document which you know of.

Now, what is the answer to that letter? What are "the documents?" In the very next letter, dated February 6th, 1850, Barron, Forbes & Co. write to James Alex. Forbes (page 543):

Dear Sir:—In reply to your private letter of the 20th December, respecting the two sitios of land, we have to say that we had hoped the document lately sent for this grant to Mr. Castillero would have been sufficient, but as you seem to be doubtful on this point, we have spoken to him, he being now here, and his opinion is that if this grant is not tenable it will be better to go upon the three thousand varas of the Alcalde, granted at the time of giving possession of the mine, and approved of by the Mexican Government, which approval will be taken from the Mexican archives and sent on to you. Although by the Ordenanzas of Mineria, the Alcalde or Judge may not have had strictly a right to grant these three thousand varas, yet being approved by the Mexican Government, would make this valid as a *grant.* We hope, however, you will find the Berreyesas' lands not to include the hacienda, and consequently either the grant of the two sitios or the three thousand varas would be a sufficient title. Mr. Alexander Forbes will, however, write you more particularly on this subject.

In Alex. Forbes' letter of the 3d of February, he says (page 400):

Castillero has returned and is also here, so am I and William Forbes. This leaves out only the four California barras, and I think I may venture to act for you, if necessary, as you verbally told me I might. * * * * I have every reason to believe that the document you mention will be *found* in the City of Mexico.

I beg that the Court will notice that that man, James Alex. Forbes, forged a mark of italics under the word "found," thus altering the meaning of the letter—as is shown by the press copy of the original introduced by us.—See Transcript, 475:

And as Mr. Castillero will return there they will no doubt be procured, but we are at some loss to know what is exactly wanted, and I beg you will, by next steamer, give a sketch of the document you allude to, particularly a description of the limits of the grant. I think you must not have received the information sent you of the existence of the grant of the two sitios directly to Castillero, and registered in Monterey, nor am I sure if this will mend the matter.

He then speaks of attacking the Berreyesas' title as not valid.

Finally, he says (page 402):

We think at present that it may be the best plan to get an authenticated copy of the approval of the Mexican Government of the grant of three thousand varas, given by the Alcalde, on giving possession of the mine. As a doubt may be started as to whether the Alcalde, acting as the "*Juez de Mineria*," had a right to make this grant, yet, if approved of by the Government of Mexico before the possession of the country by the Americans, there could be no doubt upon the subject. This takes in our hacienda, and unless opposed by the Berreyesas, would, I should think, settle the question. Castillero says such approval was given, and that on his arrival in Mexico he will procure a judicial copy of it; this is the plan we shall adopt if we hear nothing from you to alter this resolution.

This is the state of the correspondence in February. On the 2d of March this man Forbes is turned out of the mine. Capt. Halleck is put in possession when the owners resume their property, and then Eustace Barron goes to Mexico. I now intend to follow him to Mexico, and see what he did there. It is in the record.

Look at pages 59, 1794 and 1797, if you wish to see what Eustace Barron did in Mexico, when he went there in 1850.

MR. JUSTICE HOFFMAN—In the letter of March 2d, 1850, addressed by Barron, Forbes & Co. to J. A. Forbes, they say

Eustace Barron is about to proceed to the City of Mexico, to attend to what he (Forbes) had recommended. Is that part of the temporizing policy?

MR. BENJAMIN—That was a part of their policy to keep James Alex. Forbes quiet until De la Torre could get up here and turn him out of the mine. Having previously stated what they were going to do, that is, to procure certified copies of the proceedings in Mexico in May, 1846, alluded to by Castillero; and they carried this out by what they actually did do, as I shall show the Court. They actually took steps in Mexico, and the first step taken the Court will find at page 59.

Forbes recommended a ratification and confirmation of the mining possession by the Supreme Government. The other recommendation was to procure a grant of the land.

Now, what did Eustace Barron do in Mexico on the 30th of June, 1850? He procured the following certificate (page 59):

I, the undersigned, Minister of Internal and External Relations, certify that, although in the communication which the Junta for the Encouragement and Administration of Mining directed on the 5th of May, 1846, to his Excellency the Minister of Justice, relative to the mine of cinnabar discovered in California by Don Andres Castillero, it was written that it was situated in Lower California, and the same is said in the answer from the Ministry, dated the 9th of the said month, of which documents copies were given on the 23d of April last, to Señor Eustaquio Barron. This is a mistake, for said mine and the lands granted to Castillero are in the territory of Upper California, which grant the Supreme Government afterwards approved and declared legitimate.

And I give this certificate for the purposes which it may serve, at the request of the said Señor Barron, in Mexico, on the 30th of June, 1850.

(Signed) LACUNZA.

MR. RANDOLPH—Where is the evidence in the record on that subject? Has there been a witness called to testify to this?

MR. BENJAMIN—That is one of the documents offered in the Land Commission.

MR. RANDOLPH—Not proved there.

MR. BENJAMIN—The certificate of Lafragua proves that exhibit in the Land Commission. I am going on to argue on it, if it is a forgery.

Now, your Honors have in your possession what Eustace Barron goes to do in the City of Mexico. Suppose he wanted to forge a grant. According to the gentleman (Randolph) is it not easier to forge these papers than to get a certificate from the Minister that these contain clerical errors? for Castillero to get a correct description copied, than to apply to the Minister of Relations to correct clerical errors? He has no difficulty in getting that certificate. That is fair, honest and upright.

Where are the clerical errors? If your Honors will look at page 1794, you will find that in the communication of the Junta of the 5th May, 1846, to the Minister of Justice, it is stated that the specimens of cinnabar came from the Mission of Santa Clara, in *Lower* California. On page 1797 you will find the same error.

Eustace Barron has gone to Mexico, to get the documents to send up here in proof of the title. What does he send up? A certificate that the errors are clerical errors; that the true place where the mine is situate, is *Upper* and not *Lower* California, as stated in the papers.

MR. RANDOLPH—When you first dwelt on the certificate of Lacunza, I asked you to point to anything that proves such a thing in evidence.

MR. BENJAMIN—I can't. I will state that it is; you prove the contrary, if you can. It was filed before the Land Commission at the time.

The application is made to Lacunza, Minister of Relations, to correct the errors; that is done by Eustace Barron. Is anything else done? Look at the letters here. Eustace Barron is stated in the letters to be busy in the public offices of the City of Mexico, searching. Searching for what? For proof that the Berreyesa title is not good, that according to the archives neither Micheltorrena nor Alvarado had the right to make the grant; and to seek up evidence of that fact to send up to beat off the disputes about the boundaries. That is what Eustace Barron is about. He is getting proof from the archives to beat off the neighbors who are going to swindle him and his partners out of the mine by extending their limits. He is correcting clerical errors.

These things show conclusively that what he did in Mexico was honest and correct; and it is before the Court.

Mr. Randolph, the other day, called the attention of the Court to a document "the most extraordinary," he says, "ever produced before a Court within his knowledge." Upon it he called the Court to stamp its reprobation, because the party subscribing it had presumed to say that the doubts of the Court in relation to the authenticity of these papers were offensive. That *is* an extraordinary document, and the most precious used in the case. It overthrows your whole case, little as you seem to think it. It proves, notwithstanding all you say of the power, the influence, the wealth of this house, and of its chief, that with the aid of the American Minister and the British Minister, Eustace Barron was unable to obtain from the Mexican Government the poor privilege of putting the great seal of State on his papers. They told him: "No, sir! our laws are against it. We will not do it." You (Randolph) say we could get them all to make up such papers as we pleased; put *espediente* after *espediente* in their archives; buy up nine different witnesses. All of those witnesses, it is said, are perjured; Castillo Lanzas is a perjurer; these documents do not exist in the original; were never issued. All this you say. We have procured the forgery of all this by means of our power in Mexico! And yet, all this vaunted power you ascribe to us could not procure the poor privilege of having the seal of State appended to these papers, because refused by law. Castillo Lanzas is the man who refused it, a year before he came up here. Instructed by the President, he told Mr. Barron: "We cannot violate the laws for you. I am instructed to tell you that, by the laws of the Mexican Republic, the great seal cannot be used for certifying such papers, and we will not go outside of those laws for your interests. The President instructs me to tell you that if the American Government will not respect that document, authenticated as we authenticate all such documents, it is not our fault. As regards the matter of national interference, that is a matter we will attend to at the proper time. As regards this seal, which you pray for, it cannot be granted."

They could not get it. All the British and American Ministers' influence could not get it. It cost nearly $200,000 to bring up witnesses, all of which might have been saved by that little seal. Our Government would not admit evidence taken there without it. All the money in Mexico could not get it. If disposed to bribe or forge, what was easier than to hand $50,000 or $100,000 to Castillo Lanzas, to put that seal on this paper. If he is a man who could be bribed to come up here and give false testimony, is it not much easier to bribe him to put the seal on our papers? All the power in Mexico could not do that; yet, your Honors are asked to believe that we could put a bundle of forgeries in the archives in Mexico, and then bring up all the Ministers to swear to it. All THAT we could do in Mexico, while all the power and influence of the British and American Ministers and of our house could not get those officials to swerve one hair from the letter of the law in relation to their archives. Those archives, it is said, we cannot offer in evidence, because the Government, which is claiming our land, and which ought to ask them itself, will not do so; therefore, it has a right to take our land!

This is all I have to say, may it please your Honors, in defense of Eustace Barron. After the existence of these facts came to his knowledge, he turned James Alex. Forbes out, put the mine in possession of another person, went to the City of Mexico, busied himself in getting such certificates as the laws of the country would allow, endeavored to get proper authentications, and was refused, and finally was driven by the persecutions of the Government to spend $200,000 to defend his character; the faith of the nation, bound by the Treaty, being violated by the acts of the Government.

This is what Eustace Barron did. This is ALL that he did And it is a labor of love for me to say so; because I knew him personally; because I became indebted to him in the City of Mexico for acts of kindness I will ever remember; because I loved his son Eustace, as noble-hearted a gentleman as I ever knew; because the other members of the family are my friends and acquaintances whom I also respect and honor. It is a labor of love for me to defend that family, because I believe them to be men of as high and pure character as any gentlemen on this floor. I believe it, and believing it, say it openly. Eustace Barron, senior, was a man of austere manners, upright, respected by everybody who knew him, with extended influence by reason of his character for probity and honesty, against whom no human being ever uttered a single word of reproach or calumny. Yet, Eustace Barron had his dying moments embittered by the statements spread broadcast throughout the country; and it was at his request that his friend Negrete came up here to give testimony to the truth of the documents brought forward, without fee or reward; testimony unimpeached and uncontradicted, and which the whole power of the Government has been defied repeatedly to deny or disprove. It *cannot* be impeached or contradicted.

Protected against the slanderous charges of the Government by such a character and by such testimony, even the lightning of Edmund Randolph's invective plays innocuous around the name of Eustace Barron.

THIRD DAY.

FRIDAY, Oct. 26th, 1860.

MR. BENJAMIN—In the course of my argument on the facts of this case, I omitted one reference to a communication not hitherto adverted to in the case, to which I now desire to address the attention of the Court for a moment only, before I proceed to discuss the questions of law. In Executive Document, No. 17, (House, 1st session, 31st Congress, 1849–50) previously referred to, is a communication from Col. Mason, then Governor of California, addressed to the Adjutant General at Washington, giving a report in relation to California mines, and a history of a visit made by him to this particular mine in the spring of the year 1848. As a part of the history of the country in relation to this mine, I desire to read a passage or two to the Court. (Page 534.):

Before leaving the subject of mines, I will mention that on my return from the Sacramento I touched at New Almaden—the quicksilver mine of Mr. Alexander Forbes, Consul of Her Britannic Majesty at Tepic. This mine is in a spur of mountains, 1,000 feet above the level of the Bay of San Francisco, and is distant in a southern direction from the Pueblo San José about twelve miles. The ore (cinnabar) occurs in a large vein dipping at a strong angle to the horizon. Mexican miners are employed in working it, by driving shafts and galleries about six feet by seven, following the vein.

The fragments of rock and ore are removed on the backs of Indians, in raw-hide sacks. The ore is then hauled in an ox-wagon from the mouth of the mine down to a valley well supplied with wood and water, in which the furnaces are situated. These furnaces are of the simplest construction, exactly like a common bake-oven, in the crown of which is inserted a whaler's frying kettle; another inverted kettle forms the lid. From a hole in the lid a small brick channel leads to an apartment or chamber, in the bottom of which is inserted a small iron kettle. This chamber has a chimney.

In the morning of each day the kettles are filled with mineral (broken in small pieces) mixed with lime; fire is then applied, and kept up all day. The mercury, volatilized, passes into the chamber, is condensed on the sides and bottom of the chamber, and flows into the pot prepared for it. No water is used to condense the mercury.

During a visit I made last Spring, four such ovens were in operation, and yielded in the two days I was there, 656 pounds of quicksilver, worth at Mazatlan $1.80 per pound. Mr. Walkinshaw, the gentleman now in charge of this mine, tells me that the vein is improving, and that he can afford to keep his people employed, even in these extraordinary times. This mine is very valuable of itself, and becomes the more so, as mercury is extensively used in obtaining gold. It is not at present used in California for that purpose, but will be at some future time. When I was at this mine last Spring, other parties were engaged in searching for veins; but none have been discovered that are worth following up, although the earth in that whole range of hills is highly discolored, indicating the presence of this ore. I send several beautiful specimens properly labeled. The amount of quicksilver in Mr. Forbes' vats on the 15th of July, was about 25,000 pounds.

At the close of my argument yesterday, my brother Randolph asked me on what authority I read the certificate of Lacunza, annexed to one of these espedientes. He will find it a part of Exhibit A, produced by Lafragua, offered in evidence in the Land Commistion, and received without objection by the United States.

I have now, may it please your Honors, closed my argument on the facts of this case. Imperfect and incomplete as it is, I trust sufficient has been shown to satisfy the Court to the fullest extent of the genuineness and truth of the papers brought before it as the titles of these claimants. Of the vast mass of testimony contained in these four or five octavo volumes, comprising 4,000 or 5,000 pages, necessarily a good deal must be omitted in the resumé of counsel.

But we rely on the dilligent care of your Honors, on your candid consideration of the evidence, both on the part of the claimants and on that of the United States, and on your conclusions, drawn from all these facts produced, that there is not one shadow of doubt remaining in relation to the truth of the titles presented before you. We rely on your conclusions therefrom, that the very documents and letters offered by the Government of the United States, for the purpose of establishing that at one time some antedated and fraudulent papers were contemplated by some of the parties to these transactions, referred to the documents, now offered by us in evidence, as being true and genuine—describing them with an exactness which makes their identity absolutely certain—and proposed the forgery of papers containing features, not one of which is to be found in one of the papers now brought into Court by us; which, on the contrary, possess all the defects which the forger, James Alex. Forbes, proposed to remedy by fabrication. We pointed out the defects in the title now produced before the Court; the statement of those defects being admitted by the Government counsel, and urged on the Court as ground on which this title is to be declared invalid, even if genuine. I leave, I say, the argument on the facts, and proceed to discuss the law.

LAW ARGUMENT.

And it appears to me that nothing could have struck your Honors with more surprise—certainly nothing struck *me* with more surprise—than to hear the counsel for the Government, after this case had been argued before the Land Commission, and there determined, brought into this Court, and petitions put into the record by the United States, asking the Courts to take cognizance of this mining possession; after years had been spent in an examination of the facts; and after the United States had filed a bill in Chancery, asking the interposition of the Court in Equity in aid of the rights of the Government, on the ground that this mining title was at issue—declaring that our petition before the Land Commission did not ask for a confirmation of this mining right.

I have read that petition with the light of the comments of counsel. It is so plain and so clear that I am unable to understand to what his objection points. The petition, as I read it, is this: It is a claim for land; it claims two square leagues of *land;* it declares that the owners and claimants have two titles; that one of those titles is a mining title—which of itself is a good title; and if that title be not a good one in law, we still have a title to the two leagues, because we have had a colonization grant in fee.

That was the petition presented to the Board of Land Commisioners. The demand was, that the two leagues should be confirmed to them; confirmed to them as comprised by them in their mining title and in their colonization title. The Land Commission took up that petition and determined, first, that the *colonization title* did not confer any interest or title in the land itself; and therefore it refused to confirm that title for the two leagues. But the Commission decided that the *mining title* did confer a vested interest in the land to the extent of three thousand varas all round the mouth of the mine; and therefore confirmed that portion of the two leagues which was embraced in the mining title.

The claim is a claim for land; for two leagues. The petition sets up the two titles. It closes with a prayer that this claim to two leagues of land be adjudged to be valid in law, as being embraced in both titles presented, on the faith of which the petitioners claim this confirmation.

The Commission examines the two titles—one of them for a part, the other for the whole; refuses the title for the whole, and confirms the title for the part.

Then, when the case is brought into this Court, what is the prayer of the United States? Is there any statement by the United States that this mining title is not in dispute; or that it is not a claim for land?

Let us see. On page 132 is the petition presented to this Court in behalf of the Government of the United States, on which your Honors are now acting:

(Page 132).—The United States, by their attorney, represent that this cause is for a review of the decision of the U. S. Land Commission, whereby the claim of the appellee was confirmed in part, as will appear by reference to the record in the case; that a transcript of said record was filed in this Court, February 26, 1856; that a notice of appeal was filed on the part of the United States, April 15, 1856; that the land claimed lies in this District, and that said claim is invalid.

Wherefore appellants pray that this Court reverse the decision of the said Commission, and decree the said claim invalid.

The Government is asking your Honors to decree our claim invalid; and at the same time says, no such claim is before you! Let this petition of the United States Attorney be compared with the position taken by the counsel for the United States in his brief, and your Honors will see how they conflict with each other—one saying the claim should be declared invalid; the other, saying that no claim at all is before the Court, that no demand at all to be declared valid or invalid was ever put in the petition. It is plain to me that the counsel for the Government never had read the petition put in before the Land Commission, when he made the point to which he has called the attention of the Court.

My brother Randolph also says that this Court has no jurisdiction of this claim; but although he argued it with great tenacity and vigor, expressed the desire that your Honors should determine the point against him. I will assist him in his wishes. I will satisfy the Court that it is perfectly consistent with law to grant his prayer; and as your Honors will thereby be doing a favor, both to him and to us, I trust you will lend a favorable ear to anything I may say on the subject. Both parties want that point decided in the same way. Let us then agree that your Honors so decide it.

No jurisdiction in this Court! What does the law say?

The petition is, that the "Hon. Board of Land Commissioners confirm to him (Castillero) the aforesaid tract of two square leauges of land, as embraced in his mining possession and grant, as aforesaid."

This prayer is preceded by a statement that he holds the mine under mining right: "That the rights required under said denouncement and juridical possession, have continued uninterrupted until the present time, and now are vested solely and wholly in the petitioner, Castillero, his agents, and those holding under him"; "that by the working of the mine from its denouncement as aforesaid, he acquired by the laws of Mexico a perfect title to the minerals of said mine, and a perfect right to use the said land for mining purposes, even if he had no title in fee, to the land itself; but he avers, and has shown, that such title in fee absolute was conveyed to him by the grant aforesaid on the 20th day of May, 1846."

Now what says the law under which your Honors are now acting?

The 8th section of that law (Act of 1851) says, that "each and every person claiming lands in California, by virtue of any right or title derived from the Spanish or Mexican Government, shall present the same to the said Commissioners when sitting as a board, together with such documentary evidence and testimony of witnesses as the said claimant relies upon in support of such claims; and it shall be the duty of the Commissioners, when the case is ready for hearing, to proceed promptly to examine the same upon such evidence, and upon the evidence produced in behalf of the United States, and to decide upon the validity of the said claim."

If we claim land by *any* right, the law says, "you shall decide upon the validity of that claim." There is nothing in the law in relation to our claiming a fee simple in the land. There is nothing in the law to distinguish the claim in fee simple from a claim under a lease for one hundred years. If we claim title or ownership in this land for one hundred years under a lease from the Government, you are told by the law to decide on the validity of that claim as we put it before you; not to decide whether we have a fee simple or not. That is your duty. It is a duty you cannot escape without violating the words of the law under which you are holding your sessions.

The 11th section repeats the same thing, "that the Commissioners herein provided for, and the District and Supreme Court in deciding on the validity of any claim brought before them under the provisions of this act, shall be governed by the Treaty of Guadalupe Hidalgo, the law of nations, the laws, usages and customs of the Government from which the claim is derived, the principles of equity, and the decisions of the Supreme Court of the United States, so far as they are applicable."

Your Honors are, then, not only to take into consideration *any* claim to land, by virtue of *any* right, but you are to determine upon the validity of that claim, and give a decision upon it, whatever the claim may be. And, in coming to a conclusion upon the validity of the claim, you are to be governed by the Treaty of Guadalupe Hidalgo, by the law of nations, by the laws, by the usages, and by the customs of the nation from which the title is derived, as well as by the common law of England, the special law of California, and the statute law of the United States. That is your duty, pointed out by the law.

How, then, and on what pretext, can it be said that you have no jurisdiction over this matter? Is it because our title does not carry the land? That does not prevent you from holding jurisdiction over our claim to it, and so determining. The question of jurisdiction is one thing; the question of the validity of the title, another. If we claim land, our claim may be a bad one, but all you have to determine for the examination of the case is, whether we claim land by a title; and, for the purpose of ascertaining that, you have only to look to the petition itself. Our title may be a bad one—that is a matter of ulterior consideration—but we claim the land. That gives you jurisdiction to examine into our claim, and upon examination you can determine whether or not our title is valid. We certainly claim the land.

Your Honors have jurisdiction, under general law, of claims for money over a certain amount. Did anybody ever pretend you had not jurisdiction, because on examination you came to the determination that less was due than was claimed? Is your jurisdiction determined on an examination as to whether we hold in fee simple? What is there in the law that restricts you to an examination of fee simple titles, as contended by the counsel for the Government? You may show me a distinction in the law itself, but I hold the point. All I can understand from the argument of the special counsel for the Government is this: That the law provides for the issuance of a patent after confirmation has been granted, and that only one kind of patent is known to the United States. Well, that may be a difficulty in executing your judgment, if you shall find we do not hold the land in fee simple. That is a matter for you to determine when you come to execute your judgment. And if the law has not provided for carrying that judgment into execution by the issuance of a patent, that is a subject for future legislation by Congress, or for the consideration of the Court under its general powers; but surely cannot curtail its general jurisdiction, which is granted in the broadest and most general terms.

I state this question, may it please your Honors, because I desire to get rid of all mere technicalities—not that I consider it can at all control the judgment of the Court in the case. If we have no title at all in fee simple to our mining possession, we have a better than in fee simple—an allodial title; a title held without the obligation which the feudal vassal owes to his lord; a title which contains both the *dominio directo* and the *dominio util* of the vassal; a legal title to the property and a legal title to its fruits, both held by no tenure, and subject to no conditions other than those by which lands were held under the Mexican Republic by everybody who had a grant of a tract of agricultural land. Those conditions were solely, that if they chose to abandon it, and to announce that abandonment by going away and taking away the workmen, then the nation had power to give it to another. The nation gave a title with conditions subsequent; when the party abandoned his land and went elsewhere to live, anybody desiring to settle on his land came and certified to the authorities that the previous grantee had abandoned the tract given him by the nation, and he wanted it, just as is the case with agricultural lands and colonization titles; so with our title to the mine—no more, no less. There were certain provisions in the law which were to be complied with, such as keeping a certain number of operators at work in the mine, and keeping the mine itself in a condition of safety to the workmen, etc.

MR. JUSTICE McALLISTER—Was there not a prohibition to alienate the property?

MR. BENJAMIN—On the contrary, in the plainest and broadest terms the grant is to the grantee and his heirs forever; with power to alienate, to sell, to grant, to devise by will, to pass by succession intestate, and to convey by every species of alienation by which other property could be conveyed.

MR. JUSTICE McALLISTER—I mean colonization grants.

MR. BENJAMIN—Oh yes, sir; but that prohibition did not exist in the mining grants: the mining grants are of a higher quality than those of colonization.

MR. JUSTICE McALLISTER—They could not be called "fee simple?"

MR. BENJAMIN—No; because "fee simple" is not known to the civil law. In the civil law it is known as "feud." If determined on the words "fee simple," then you have no jurisdiction over any mines or lands derived from the Spanish or Mexican Governments.

I have, for the purpose of aiding the Court in an examination of the authorities, and in order to relieve you from that tedious and irksome duty, prepared a small digest of all the mining laws of Spain and Mexico to which I have had access,—and I believe I have had access to more than can be found in any other library in the United States. And I trust I may say without disrespect for the Court, that neither your Honors nor the Judges, nor the bar of any part of the United States, have hitherto found it necessary to examine with any degree of care the mining laws of Spain and Mexico. There are but few litigated cases—and those cases of minor interest—and the question of the true tenure of property held in the mines, and the subject matter so held, has hitherto been without discussion in our Courts, and almost neglected by our lawyers; the subject matter of controversy not existing amongst us. For this case the duties of counsel have made it necessary that we should make a special study. Therefore I take pleasure in presenting to your Honors, as the result, this digest. [Handing pamphlet to the Bench.]

I take up, then, this digest of the principles of mining law, as established by Gamboa, under the Ordinances of 1584, with a note of the changes made by the Ordinances of 1783. I have given the references to the original Spanish—which I have here before me—and I have put within brackets the references to Heathfield's translation in two volumes. If your Honors cannot see the original, you can see the translation.

MR. RANDOLPH—Where are the references in Rockwell?

MR. BENJAMIN—I would have referred my brother to Rockwell, only Rockwell has only a part of Heathfield in it.

I will now quote from Mr. Randolph's brief, page 25. After quoting certain sections from Gamboa, as translated by Rockwell, he says: "It would be as presumptuous as useless to attempt to add to this elaborate discussion by Gamboa."

I am going to be so presumptuous: and I am going to do so, by adding what Gamboa adds just after what my brother has quoted, for his quotation stopped with the discussion at section 23. The principles resulting from that discussion are laid down in sections 24 and 25. I have stated those principles—and shall in an instant refer your Honors to the authority—in these three principles of Gamboa:

I. The title of the subject in his mine, is perfectly consistent with the King's right of *eminent domain* (*alto dominio*). The subject's title is a complete legal title.

II. It is not only a title to the beneficial interest, but is also a legal title to the property itself.

III. There passes to the subject the "*dominio directo ó de propiedad*," as well as the "*dominio util.*" I quote from the original Gamboa and refer to the chapters and the sections.

In section 24, chapter II, he says:

The rights of his Majesty in the mines of the Indies being then established; and those rights being perfectly consistent with the dominion and ownership (*el dominio y propiedad*) of the subject therein; it is incontrovertible, that this property, passing as it does pass to the subjects, so that they may dispose of mines as of their own things (*como suya cosa*) there are proved to exist in their favor, the consequences which result from ownership and dominion (*propiedad y dominio*); wherefore they may exchange them, sell them, lease them, and alienate them by contract, donation or inheritance, give them in dowry, charge them with rent (*censos*), and recover interest on the price for which they be sold, while it remains unpaid; as this kind of real property yields products (*como el fundo es fructifero*), as the Cardinal de Luca teaches, along with Garcia, Pedro Barbosa, Barbacia, Molina, Castillo and Gutierez, when speaking of those which are durable. And in the place cited, de Luca also deduces very sound rules as to the justice of reserving rents on a lease of mines. But all this must be understood with the necessary qualifications, that those to whom the property devolves by universal or particular succession, must conform to the ordinances, and fulfill the obligations which they impose, by virtue of the law, which thus directs.

Then in section 25, he says:

And there passes to the subject this *dominium directum*, or right of property (*ó propiedad*), and also the *dominium utile*, by virtue of the gift and concession of the sovereign, which we hesitate not to name, a gift with a charge (*una modal donacion*), bearing in mind the rules by which that species of gift is defined in the law.

And these rules, he continues—

Are reduced to these, that the gift is a free and complete act, which being consummated, a charge attaches to the donee from that time forth, although it is worded as if it were a condition.

And that upon the failure to comply with this definitive charge which the donor stipulated for in favor of himself or a third person, or of the kingdom, or of the republic, the gift expires, as may be seen by referring to many texts and doctors.

In section 26, he says:

Which rules [meaning the rules of law defining a *modal donacion*] apply justly to this second ordinance: in which his Majesty gives, "*and makes a grant* (*gracia y merced*) *to his subjects of the property, and possession of mines, discovered or to be discovered, and that they may dispose of them as of their own things*," this being the complete act of gift; since for the gift no price is paid, neither for registering nor for denouncing. "*But*," proceeds the ordinance, "*observing both in regard to what they have to pay us by way of duty, as in all other respects, the regulations and orders established by this edict, in the manner hereafter mentioned:* which is the charge, or qualification, looking to the payment of the fifth from that time forth, and to the observance of the ordinances in what relates to the mode of working the mines, the number of hands to be kept at work in them, their boundaries, and other matters required to be observed, upon the omission or non-performance of which the right of property (*el dominio*) is extinguished, and the mine becomes liable to be denounced by another."

Gamboa here defines the grant; it is a "*gracia y merced*" from his Majesty to his subjects; it is a gift of the property and possession of mines (*propiedad y posesion de las minas*); it is a gift of the *dominium directum ò propiedad*, and also of the *dominium utile:* but it is a gift subject to future charges. These charges, he says, referring to the words of the ordinance, are, to pay the duty and to observe the regulations and orders established by the edict. This, says Gamboa, is the charge which is imposed upon the gift, which attaches to the donee, although it is worded as if it were a condition.

Now, we have got something like legal precision of ideas. What is a miner's title? It is a sovereign grant of a legal title. It is a grant of the lord's title and the vassal's title combined; the *dominium directum* and the *dominium utile.* Both are given by the crown to the subject.

But this grant of a legal estate is not a grant that is not determinable. It is not a grant that is free from conditions. It is a grant that may be lost by the violation of conditions subsequent. No conditions are attached to the acquiring of the property; and there are no conditions which precede the acquisition. The title is conveyed by the registry as I shall show in a moment. Once conveyed it is the miner's forever unless he choses to forfeit it; it is his and his heirs forever. He may sell it, or constitute it as a dower; he may lease it, or convey it; he may subject it to usufruct, or use it in any other way in which all estate is used; it is his property; the King grants it and retains nothing but the right to demand a certain charge, on failure to satisfy which, the title is taken back.

Can your Honors distinguish this title in its nature from the feudal fee or the feudal tenure, in which the vassal takes the estate from the lord, as a gift upon the spot, with a title in himself and his heirs forever, which cannot be divested except upon his failure to perform the duties upon which the fee is granted? In such fees, there is a continuing obligation to render service and pay tribute; and as long as that service is rendered, or that tribute paid, the title remains in the tenant or vassal. "Yet," says Gamboa, "by virtue of the royal words the title of the miner is above that, because he has no feudal service to render. He pays a duty of one-fifth of the revenues of his mine, by way of a tax, to the sovereign; but that is what all other subjects pay when they are taxed. Therefore, the right the King reserves to himself, is not the right of the feudal lord, but it is the sovereign's right of eminent domain (*alto dominio*)." That is what the King of Spain reserved, and that alone, in his grants of mines.

Now, let me refer your Honors to some other passages which place this beyond the possibility of being controverted, in the opinion of Gamboa, and I accept his authority, which my brother Randolph eulogizes so highly.

In chapter XVII, section 18, Gamboa again says this:

> The age of pupilage, absence on public business, banishment and other causes privileged by the law, do not and ought not to serve for maintaining the ownership (*el dominio*) of mines without peopling them, because this is required by the law, and is a condition, without which the sovereign did not consent to grant the ownership and property of them (*el dominio y propiedad de ellas*).

Well now, my brother Randolph quotes one or two passages from Gamboa, in which he speaks of the sale or transfer of the sovereign's right to the subject, as being that of a partial interest only. That is true, and he explains exactly what he means. He says the King's right is one-fifth and the subject's right four-fifths, because the King reserved one-fifth, and in that way reserved the *alto dominio* and a part of the mine, being that part of the product which the subject pays as tribute, anually.

This point about "partial interest of the crown" does not come up in this case, because it was abandoned soon after these ordinances were issued, as I shall hereafter show to the Court. The whole of the mines now belong to the people in Spain and Mexico, and have for more than half a century. The entire mine now belongs to the subject. The partial interest was abandoned, and all the King retained, by ordinance of the Cortes of 1811, was a certain duty on the silver when it came to be coined in the mints. It is a *tax* on the *silver* that the king reserves, and the entire *ownership* of the *mine* is in the subject.

Now, the commentator (Gamboa) says, that this mine that is acquired by the subject is land. At page 103 of the original, he says:

In addition to the aforesaid purposes, the principal object is that there may be a record of all mines which have been and shall be discovered, as these ordinances express it, for the purpose, not only of certifying the ownership (*dominio*) of individuals in them, and of regulating landed properties (*fundos*) so important, but also that they may not escape the payment of the tenth or fifth of the silver which may be extracted.

Again, at page 377 of the original, Gamboa speaks of the nature of the title and the nature of the property, in language that no civilian can mistake. He says:—"Certain authors lay great stress upon another right conferred (*privilegio*), that is of acquiring nine parts of the metal, paying only the tenth to his Majesty in acknowledgment of his having conveyed to his vassals the *dominio util y directo* of such precious landed property (*fundos*)"—in acknowledgement of the sovereign having granted the subject both the direct title and ownership in said precious mines.

Again, at page 335 of the original, in a passage where he is speaking of the duties of executors, administrators of estates of deceased persons, and of the duties of tutors of minor heirs, with respect to forfeiture of mines for their being left unworked by the death of the owner (*por muerte de el dueño de la mina*), he says, they must sell them as they sell all other real estates, in thirty days : " Like other real property within thirty days (*como otros bienes raices*)."

Then, at page 447 of the original Gamboa calls them by another name. He says: "All kinds of interdicts and their subordinate varieties, whether formed for originally acquiring, retaining, or recovering possession of immovable or real property (*bienes immobles ó raices*), may apply in the subject matter of the possession of mines."

All species of remedies that the law grants in relation to any *immovable or landed property*, necessarily apply to the subject matter of the possession of mines. The absolute right in real estate, in *bienes raices,* known as *fundos* in law, applies to mines. The whole of the rules of law, relating to *land*, apply to mines, because mines are such precious lands (*tan preciosos fundos*).

Then he divides these actions, and instructs the Courts how to distinguish between a suit which is for the ores only, as they are extracted, and the mine itself.

My brother Randolph wants your Honors to suppose that under Spanish law the right of the mine was only to extract the ore. Gamboa distinguishes different actions; he says, one action is for the ores, another for the mine itself. That your Honors will find in Heathfield's Translation, on pages 255 and 256, 2d vol.

At page 453 of the original, he says the Court is first to ascertain whether the plaintiff is claiming, when he brings a suit, for the ores, or for the property itself; and in relation to that, it shall give different dispositions, according to the matter.

Now, may it please your Honors, the mistake that runs through all my brother Randolph's argument is, that he supposed that a grant of the mine is a grant of the ores under the surface only; whereas it is a grant of the *surface* and of the *depth* both. He says, in his printed brief, that he cannot understand what we mean by one party owning the land, and another party owning the depth under it; that he cannot gather into his mind this idea of a disembodied "tract or ghost of an acre."

Well, now, the mining laws do grant this "tract or ghost of an acre" to the miner; not only in its superficies, but also in its depth. They grant the land itself with reference to depth; in the language of the mining laws, "Even to the infernal region" (*el infierno*). Some thinking that the infernal regions would be reached before you get to the antipodes; others, that it would be reached before you get to the center. I do not know which is right, but—

MR. JOHNSON—And you don't want to find out by actual inspection, I suppose.

MR. BENJAMIN—Not yet. In this digest, which I have made of Gamboa, my fifth division is:

V. And the title to the mine is of the surface land as well as of all under the surface, even to the antipodes, or "*el infierno*" (and I won't translate that, from respect to the Court). "And the miner cannot go underground outside of the area of his surface measurement, but perpendicular lines are to be drawn from his surface lines toward the center of the earth, and these lines inclose his land."

The surface, and under the surface! If that is not land, I would like to know what is? Gamboa gives the rules for measuring the surface; gives instances in which miners underground, following their vein, went outside their surface lines, on the pretext, he says, that as they discovered the mine itself and worked the surface, when underground they had a right to go outside their *pertenencias*, if they did not intrude on anybody else's possession. But Gamboa says that is an error, and cites cases litigated on that subject, in Mexico. He says, in all those cases—many of which he argued himself—the principle was established beyond contradiction, that the miner's land is embraced between the surface lines and perpendicular lines drawn to the center of the earth. Those lines inclose his property, and he cannot go outside of them when undeground, because *that* is all the sovereign has given him.

Now, how is it, if the mine be found on private land. My brother Randolph thinks the surface belongs to the private owner; and underneath the surface, to the mine owner. Not so. The mine owner, in that case, owns the surface too; being compelled by law to pay to the private owner, if he asks it, *immediately*. If he does not immediately so ask it, he is considered to have abandoned the land. If a mine is found on private property, the sovereign will grant that mine, with the surface, to the discoverer, and the private owner of the land has only a right to be paid for that part of the surface which is taken as *pertenencias*, in addition to such damages as are suffered by reason of the using of any other portion of the surface for the purpose of carrying on the mining operations.

The Ordinances of 1783 leave no doubt on this point. At pages 236 to 339 of Halleck's compilation of mining laws, your Honors will find (Title 8, Art. 11 to 17 of the ordinances), all the rules for measuring out a mine. These rules are to be applied on the surface of the earth, and the lines are to be carried down to the center. These ordinances change the former ordinances of 1584, which allowed a miner underground to go upon his neighbor's *pertenencias*, if he was following out his vein, in good faith.

Art. 14 of the Ordinances of 1783 says, on this point:

> Whereas experience has shown, that the license or permission to enter into the *pertenencia* of another, working at a greater depth and within the vein, following its metal, and enjoying it until its owner can dig down to it, has been, and is, the most fruitful cause of the most bitter lawsuits, dissentions and disturbances among miners; and, moreover, that this intrusion more often occurs from fraud or accident than from the merit or industry of the intruder, resulting generally in nothing but the serious injury or total ruin of the two mines and the two neighboring miners, to the great prejudice of the public and of my Royal Treasury, I order and command that no miner shall enter upon the *pertenencia* of another, even though it be by going to a greater depth and by working his own vein, but that each one shall keep and observe the boundaries of his own, unless he amicably agrees and stipulates with his neighbor for the right to work in his *pertenencia*.

Again, after the *pertenencias* have been marked out, Art. 11 says:

The *pertenencias* being regulated in the manner prescribed, there shall be measured to the denouncer his *pertenencia* at the time of taking possession of the mine, he being required to set up in his boundaries, stakes (*estacas*), or land-marks (*mojones*), secure and easily distinguished, with the obligation of having them kept and preserved perpetually, without the power to change them, even though he allege that his vein has varied in course or inclination, (which are unusual things), but he must be content with the lot which Providence has assigned to him, enjoying it without disturbing his neighbors; if, however, he shall have no neighbors, or if he can, without injury to them, improve the location of his stakes (*mejora de estacas*), or change his boundaries, he may be permitted to do so for such causes, with the previous intervention, knowledge and authority of the Deputation of the district, which shall summon and hear the parties as to whether they may be and are legitimate.

There is no pretense in the case under consideration, that claimants have disturbed or entered upon anybody's stakes or *pertenencias*.

Now, how is this title acquired? How does a man acquire title to a mine, to this "precious real estate," as Gamboa terms it? How does he acquire "the direct and beneficial ownership," the title of the lord and vassal both combined? How does he acquire it? Gamboa repeats in some twenty places, the same expression: that the registry is the fundamental title, "*el titulo fundamental de las minas.*" Heathfield translates this "that the registry is the *bases* of the title," and Rockwell so copies it into his book. Now, the *basis* of title and *title itself* are two different things; and Gamboa repeats again and again that registry is *the title itself.* The translation is wrong. Gamboa's expression is, that the registry is, "*el titulo fundamental de las minas;*" repeated over and over again everywhere through his book. I have, in the digest, given your Honors some references to passages where he says so. The beginning of the 3d section of the 5th chapter of his commentary is in these words: "The registry is the fundamental title (*el titulo fundamental*) of mines, and the attributive cause of the *dominio* in favor of the vassals, under the charge of making which his Majesty granted them." It is *the fundamental title* of the mine, and the "*causa attributiva*" of ownership in favor of the subject!

At page 108, treating of another subject, Gamboa says again

that none are mines but those which are registered; because, as he says, registry is *the fundamental title* of a mine, and how can there be a mine without registry? The law did not recognize a mine in the bosom of the earth, as a mine, until it is registered; it did not exist as a mine until registered; and the reason is, because the law has made registry the fundamental title of a mine.

At page 124, he repeats—as he is constantly doing all through the book—the same thing. It is almost idle for me to quote to your Honors, for he uses the same phrase all through his commentaries, repeating it in innumerable places.

In chapter XI, towards the close of the section 2, speaking of another subject, he says:

> Because a mine without a registry is no mine, as we have shown at length in the proper place, for that is the fundamental title of the ownership (*el titulo fundamental de el dominio*).

We have now, at all events, this point established: that the title to a mine, according to Gamboa—whom my brother Randolph quotes as his authority, for which I am much obliged to him—is a full allodial title. It is not merely the title of the vassal held from the lord; it is the title of the vassal and the lord both combined, called *dominium plenum* in the civil law. By common law writers it is more commonly called "allodial." Combining both titles to property, it is called by writers the highest title the subject can possess; being a title to property which transcends all other right, except that of *eminent domain*, which is invested, in our country, in the people; in other countries, in Kings and Emperors. As representatives of society, States have always had that right of eminent domain—the right by which the property of the individual is taken from him, against his will, for public purposes; sometimes with, sometimes without payment, as the powers taking may choose. In our country, and under our laws, nothing can be taken without full and ample indemnity being paid.

Now, what is the registry which constitutes title? Is it an administrative act? or is it a judicial act? I have cited twenty or thirty passages of Gamboa, plainly and clearly stating that it is *always* a judicial act. The reason is obvious—the Ordinances of 1783 always give the reason for their requirements. It is because there may be a dispute about the discoverer; a question as to who first discovered the mine. One man may discover a mine, go to the office and register it; but a term of ninety days remains in which to question the registry, in case of a dispute between parties, as to who was the first discoverer. The law provides that this question shall be summarily determined, and the first discoverer allowed to register; but if the Judge is unable, from the evidence before him, to determine who did first make the discovery, then the first registerer shall be considered by him as the first discoverer, and shall have the preference. That is the provision of the Ordinances of 1783. Gamboa had laid it down in accordance with the Ordinances of 1584. It was incorporated in the Ordinances of 1783, and has become a principle of Mexican and Spanish law. It points out that the registry is a judicial act, and is to be made before a Judge. It might give rise to contention; therefore it belongs to the contentious jurisdiction of a Judge.

The next principle is, that a failure to register does not forfeit the rights of the discoverer, if he makes a registry before some one else denounces, even after the delays fixed by law. —Gamboa, page 108.

He is declared, by the ordinances, to be bound to register within ten days from his discovery; but Gamboa says, he may register it at any time, even after the time fixed by the law, except where the rights of third persons have intervened. Now, suppose a man fails to register; or suppose, having registered, he abandons his mine; he merely incurs the penalty of forfeiture provided by the ordinance.

My brother Randolph says in his printed brief, that the ordinance is *strictisimi juris*, and that he instantly loses his rights. Gamboa says just the contrary. Gamboa says, that in case of his failure to register, or in case of his abandonment, a third person may denounce the mine, and after judicial proceedings, pronouncing forfeiture, may register it, and thus acquire the title belonging to the discoverer. The judicial proceedings which are to take place in case of a mine being denounced for a forfeiture or for any cause set forth in ordinances, are described by Gamboa. A man never lost his mine under Spanish law;

and let me ask your Honors, how any case of forfeiture can appear or be carried into effect without investigation. Even where forfeiture was declared to take place *ipso facto*, can a man determine that fact without it be judicially decided. Both by the Common Law and by the Civil Law, abandonment must be proven judicially before the forfeiture takes place; and no other individual or authority has a right to take possession of a man's property as forfeited, or to impose a penalty upon him before judicial proceedings have shown the fact of accruing forfeiture. The whole of the law on this subject I have referred to in my list of authorities in Sec. XXIII of my printed brief.

The Ordinances of 1783 say that forfeitures are a penalty which must be lawfully established and proved. I refer your Honors to Title 6, Articles 8, 9, 10 and 11, and to Title 9, Articles 10, 11, 14 and 15, which establish proceedings for forfeiture.—Halleck's Col. pp. 225, 236, 227, 243, 244.

Now, Gamboa further says that proceedings for acquiring title by registry and denouncement differ only in form, not in substance; because a judgment for forfeiture against the discoverer, gives the denouncer no title unless he register anew. For registry in the *fundamental title*.

A man cannot own a mine if he has not registered it. He has no title except he registers it. If he finds a mine abandoned and goes on and works it according to the ordinances, but does not register it, he has no title. For registry is *the fundamental title*, without which no man has got a mine. So the denouncer must register his claim in order to have a title in himself.

But my brother Randolph says, "*Titles to Mines!*" Why, they are mere licenses to dig! How then when the King sells the mine? Now, Gamboa explains that by the ordinance of Peru, the King takes his share in this way: When a man discovers a mine he registers it, and stakes out one pertenencia for himself and one alongside for the King. He takes title to one pertenencia by his registry; the King takes title to the adjoining pertenencia by virtue of his royal prerogative. What is done with that? The Viceroy is ordered to sell it for the King. Who does it belong to, then? Does it belong to the King after he has sold it? How then are we told here that the subject cannot own a mine? The title of the discoverer to the mine marked out by him is precisely the title which the King conveys to another one of his subjects who purchases the adjoining pertenencia; and when you shall succeed in proving to me that when the sovereign sells property he retains it, after receiving the price, you will succeed in also proving to me that the subject by discovery acquires no title in a mine; then, and not till then.

Now, something has been said in this case, may it please your Honors, about the rights of discoverers and of partnership, and about the translation of certain Ordinances of 1783, of which my brother Randolph gives his own translation in direct contravention of that contained in some authors, and, in my humble judgment, in direct contravention of the language of the original.

Now, First: In the old ordinance the law was that the discoverer or ordinary miner could take only two mines by registry or denouncement. The old ordinances called a mining *pertenencia*, a *mine*, "*mina*" and *vice versa*. You cannot make any difference between them at all. Each *pertenencia* was called a "mine" in the old ordinances. In the new ordinance the *pertenencia* is distinct from the *mine*, the latter being composed of several *pertenencias*.

The ordinary miner, under the old ordinance, could only take two mines by registry and denouncement, and he must leave between them a space equal to three mines.

But, by the Ordinances of 1783, none but the discoverer could denounce two contiguous mines on the same vein. (T. vi, art. 17; Halleck, p. 229.) But, in considerable enterprises the miners might have allowed to them many *pertenencias* by grant of the Viceroy on report of the Tribunal General.

But *discoverers* are worthy of special rewards, and the discoverer may therefore take as many mines as he please in the vein which he has discovered, and *even the whole vein*.

That was the Ordinance of 1584 in relation to the discoverer; and Gamboa says there is nothing wrong in that. There is no danger in giving this privilege, because he is pledged to work this vein. He cannot abandon them, for if he does, they will be denounced by others. If he takes two mines, he must work each one of them.

But in the Ordinance of 1783 there was a change. Discoverers may take in an absolutely new hill, in which no mines or pits are opened, *three pertenencias*, continuous or interrupted, in the principal vein, and one in each of the other veins; but if they discover a new vein in a hill already worked, then they only take two *pertenencias.*—Tit. VI., arts. 1 and 2; Halleck's Col. p. 223.

Now, as to partnerships. What was the condition of the law in 1584 as to partnerships?

Partners were to take twice as many mines as there were partners in the partnership. The number extends to four, or six, or twelve. Gamboa recommended that each partner in a company should represent two mining *pertenencias;* but this proposition did not prevail. By the Ordinance of 1783, partners were restricted to four new *pertenencias*, without prejudice to what they might have a right to as discoverers, if they be such.

Now we come to the difference in translations. How is it to be determined? The Court is to determine this question, not by brother Randolph's assertions or my own opinion in the premises, but by a direct examination of the language of the original. The Court happens to have within its reach, competent translators, who have been employed in this very case. In my presence, a few days ago, Mr. Tobin was asked to sit down and give a translation of the disputed sentence. He did so. I will read it. It agrees in every essential particular with that given by Mr. Halleck on page 252 of his collection.

Mr. Randolph—Mr. Hopkins says that it is wrong.

Mr. Benjamin—If he does say so, I'll give it up. I know that neither Mr. Hopkins, nor any other Spanish scholar, will say that that translation is wrong. It is an exact and literal rendering of the original, which I will now read from Title XI, art. 2, and compare it word for word with Halleck's translation.

I will say nothing further about this translation. There is the language of the original; your Honors may translate it for yourselves; but I will say this: it is utterly stupid and without meaning if it does not mean what we say it does; for in no part of the Spanish mining law, under no circumstances, in no contingency, is anybody prohibited from discovering mines and registering them; and for an ordinance to say that because a mine is worked in partnership the partners shall not lose any right in discoveries which they may make in future times, would be extremely unnecessary and absurd. I think that this matter does not admit of an intelligent doubt.

But this was a change from the old law which provided the number of mines which might be owned in partnership. The old regulation was, twice as many mines as partners in the company; but the new law says, that if you are not discoverer you shall take only *four pertenencias*, and this is without prejudice to your rights in case you be discoverers.

Now, my brother Randolph says that "*en caso de que lo sean,*" "in case they be (or are) such," means "from becoming such," or "in case they *shall become such.*"

MR. RANDOLPH—No. I say that that was the opinion of a gentleman who translated it a hundred years before Mr. Halleck did.

MR. BENJAMIN—These gentlemen of the Inns of the Courts in London, having very little knowledge of the terminology of the civil law, are constantly making ludicrous mistakes. A distinguished member of the American bar, now deceased, once wrote an entire treatise on a civil law topic, when he did not know the meaning of the word used, as was evident from his language.

MR. RANDOLPH—May I inquire who you refer to?

MR. BENJAMIN—Judge Story.

MR. RANDOLPH—What was he writing upon?

MR. BENJAMIN—Upon Mandates, I believe. He had, or seemed to have, no idea of the real meaning of that word. He never suggests that it means *power of attorney* (its real meaning) in the whole of the treatise. That is its meaning; and in the civil law it means nothing else.

Now, as regards the *forfeiture* determined by the ordinances for those who, being in partnership, fail to announce their partners' names when they register or denounce, they are plain and distinct. The forfeiture against one who registers or denounces in his own name a mine owned in partnership, is only *in favor of his copartners*, and cannot be adjudged unless he has omitted the names *fraudulently*.

Now, in this case, Castillero denounced in his own name, but the articles of partnership were there in the archives of the Alcalde's office. There is no pretense in this case that this omission was *fraudulent*. That question does not enter into this case.

My brother Randolph has expended a great deal of time, and gone through a vast deal of cross-examining of witnesses, for the purpose of showing that this Pico act of possession of the mine was an idle and ridiculous ceremony; that it was accompanied by no act by which the taking of possession remained marked or fixed upon the soil. He says that such "possession" may fly over a mountain as one shadow does over another, each leaving as little track as their predecessors. That is in his "Discussion Reviewed." He has entertained your Honors with a view of a pic-nic party of neighbors going out upon their mining possession, and amusing themselves riding around it; and he asks if the law can be vain and stupid enough to give binding effect to such trifling and idle follies as this?

I propose to read a single passage from Heathfield's translation, p. 79 of the 2d volume.

After speaking of the denouncement of a mine for abandonment, he says:

3. Whence it follows, that the mere keeping possession of the mine, is not sufficient to authorize saying that it is kept at work, possession being distinct from keeping at work; the former depends on receiving and having the custody of the instrument conferring the title, on the delivery and receipt of the keys, or some corporeal act, such as the tearing off boughs, throwing aside stones, or walking upon any part of the estate, or on some other symbolical act, performed with the

intention of taking or maintaining possession of the whole, as may be seen in various texts and doctors. But the keeping a mine at work is the employing the labor of four persons, either within or without the mine, in some object tending to the working and habilitation of the mine. So that a mine may be in the possession of a party, though at the same time left insufficiently worked; it may have a guard and be under custody, and yet not be kept in sufficient work; it being an indispensable requisite to the latter object, that four laborers or workmen should be employed; and the performance of acts of possession, without observing the indispensable requisite of keeping the mine at work, is not sufficient to obviate the penalty.

If brother Randolph wants to consult the authorities on this subject, they are all in the note to Gamboa. So much for taking possession of this mine by walking about it. That is exactly what Gamboa cites as a usual ordinary formality for taking possession of a mine. That, however, has nothing to do, as he says, with the fact of keeping it at work. You may have somebody in possession in that way at any time. But if that possessor remains for four months without doing some work thereon, then this mine will be liable to be denounced; but during these four months that possession is as good as any other.

Now, when was possession to be given to an owner of a mine who registers it? Why, under the Ordinances of 1783 we are not permitted to doubt. If language means anything, we are not permitted to doubt. There can be no discussion on that subject. The ordinance gives the discoverer ninety days after his registry to make a pit. That is the time that is allowed him. If he has not done it within the ninety days, and cannot show good excusing reasons for such neglect, he loses his right in the premises. But he is not prohibited from making a pit on the next day. He is allowed to work the mine as soon as, and as fast as he pleases. But he is allowed ninety days in all, except for special causes which Gamboa points out. Now it says, "immediately as soon as"—not brother Randolph's "as soon as"—it says, "*luego que*,"—which I never heard translated before "instantaneously"; but, "*luego que esto que haya verificado*,"—as soon as he has done making his pit, the Justice is to examine it. If it is as deep as the law requires, he is instantly to give him possession. He is not obliged to wait the ninety days.

What are the ninety days for? Why, the next article explains it. Even after this act of possession has been given, it would not become conclusive against third persons till after the end of ninety days. The Justice may have given him possession after making registry, but at any time within the ninety days any person may come in and show to the Court that *he* was the true discoverer; that *he* was about to go in and get possession. The Judge shall take that into consideration, and determine who shall take possession of and work the mine. But, say the ordinances, after ninety days no third person shall be heard. All this was a mode of putting an end to controversies between private contestants, in which the sovereign had no interest. The sovereign cared not whether A or B possessed the mine, provided the mine was registered and worked. Therefore, so far as the sovereign was concerned, the pit might be dug and possession given the day after the discovery. But the ninety days were allowed in order that any person might come in during that period, if he so chose, and contest the discoverer's rights. Upon such contest being made, the Judge determined who had title in registry.

Now this is precisely what was done in Castillero's case. He registered this mine. After that registering, nobody ever came to denounce it, and the sovereign had no interest in the matter.

After discovery, Castillero made a pit. He went and told the Judge: "I have made my pit according to the mining laws"; and the Judge put him in possession, after forty or fifty days had elapsed, as indeed he might have done the day after the denouncement. His possession, which the law required to be given him *immediately*, was perfectly good against the sovereign. The sovereign knew that a mine would be worked after it was registered, and was certain to get his fifth. He had no interest in any controversy which might arise. But anybody else, during the ninety days, could come in and say, "Castillero is not the true discoverer; he has come in and stolen my discovery; I pray the Court to determine this matter summarily." Any person who should do that within ninety days, would have the right to have the question decided by the Judge. There is no pretense, however, that anybody else did so claim this mine.

So far as the Mexican Republic was concerned, the instant the pit was dug and possession given, Castillero's rights to the mine became perfect and complete.

I do not speak now of the *gracias* of 3000 varas. Your Honors will understand me to speak now only of the Alcalde's power to act *judicially* under the ordinances which govern that branch of the subject. The language on this point is so plain that I do not see how it was ever doubted. I read from Mr. Halleck's translation of Articles 4 and 5 of Title VI of the Ordinances of 1783 (Halleck's Col. pp. 224, 225):

Art. 4. Those mentioned in the preceding articles must appear with a written statement before the Deputation of Mining of that territory (*territorio*), or the one nearest, if there should be none there, stating in it their names, and those of their partners (*compañeros*) if they have any, the place of their birth, their residence, profession and employment, and the most particular and distinguishing features of the place (*sitio*), hill (*cerro*), or vein of which they ask the grant (*ajudicacion*): all of which circumstances, and the hour in which the discoverer presents himself, shall be noted in a book of registry which the Deputation and Notary (*Escribano*) of Mines, if there be one, shall keep; and, this being done, his written statement shall be returned to the discoverer for his due security, and notices (*carteles*) shall be affixed to the doors of the Church, Government-houses (*Casas Reales*), and other public places of the Town for due information. And I order that, within ninety days, he shall have made in the vein or veins of his registry a pit (*pozo*) of a vara and a half wide or in diameter at the mouth, and ten varas down or in depth; and that, as soon as this is done, one of the Deputies shall personally go, accompanied by the Notary (*Escribano*) if there be one, and if there be none, by two assisting witnesses, and a professional Mining Expert (*Perito facultativo de Mineria*) of that territory, to inspect the course and direction of the vein, its width, its inclination to the horizon, which is called *echado* or *recuesto*, its hardness or softness, the greater or less firmness of its sides, and the species or principal indications (*pintas*) of the mineral, taking an exact account of all this in order that it may be added to the corresponding part of its registry, with the evidence (*fé*) of possession which shall immediately be given in my Royal name, measuring to him his pertenencias, and causing him to fix stakes (*estacas*) in his boundaries, as will hereafter be mentioned; which being done, there will be delivered to him an attested copy of the proceedings, as a corresponding Title.

He gets a copy of the proceedings when they are closed, and gets a copy of the registry of possession the instant he digs the pit. The authentic evidence of the title given him cannot be perfect until a copy of all those proceedings is completed and handed over to him. There is no mistake about that. Now, although he may have denounced the mine, made this registry, and obtained this authentic evidence of possession, perchance he has done so fraudulently or in violation of the rights of others; and the next article goes on to provide for that:

Art. 5. If, during the said ninety days any one shall appear, pretending to have a right to said discovery, he shall have a brief hearing in court (*en justicia*), and it shall be adjudicated to the one who best proves his claim; but if he appear after that time, he shall not be heard.

I do not think that language can be plainer. I think it is a model of precision for the lawgiver. Well, now, how is it if the miner does not mark out his stakes. If he does not put the boundary marks or stakes separating these pertenencias, that he claims by virtue of his registry from the public domain, and from the neighborhood where everybody else has a right to search and ask for a mine, what then? Why, simply, anybody else who comes along and wants to register a mine, has a right to cite him before the Justice to compel him to set out his stakes. He does not lose anything. The law provides, that if he has failed to set out his stakes, and anybody else has an interest in the neighborhood, this third party may call upon him to mark out his boundaries, and the Judge is to give him ten days in which to do it. The new comer is the one who asks stakes (*pedir estacas*), the one who is previously in possession is the one who gives stakes (*dar estacas*). That was the position of two miners. One man went upon a mine and chose to remain for years without marking out his stakes, because the law always gave him the right to better his location—(*mejorar estacas*)—to swing his mine round and round until he got it in the most eligible position—unless some one else intervened. The moment some one else wanted to mine there, he applied to the Judge to compel the discoverer to mark out his boundaries.

John Smith discovers a mine, and the law entitles him to three pertenencias. But John Smith has not marked out his lines. I ask him to mark them out, as I wish to take up one adjoining. As discoverer he was required to mark out his boundaries and set his stakes. Does he lose his mine if he does not so mark it out? The ordinance says that in that case the Judge shall do it for him, and that he shall be bound by what the Judge does. He shall not have his choice, but the choice of the Judge; that is all. The Judge does what he thinks proper, and the discoverer can no longer complain if the second discoverer comes and takes his mine along side the boundaries of his original mine. A plain, simple, equitable series of rules, in strict accordance with the simplicity of that patriarchal people; for they are a patriarchal people, particularly in mining districts. They are now what they were one hundred years ago, to a very great extent.

Of all the causes that the ordinances point out for forfeiture of a man's property in a mine, there was one to which special importance was attached. The public did not care much about forms. The King had no interest about particular measures being carried out in relation to the ownership of mines beyond this. They must be kept safe. The pillars must be in proper order to protect the lives of his subjects; and the mines must not be abandoned but worked, for in that the Royal Treasury had an interest. Gamboa says, the King has specially prohibited his Viceroys from extending the term, when a mine has been abandoned. No informant can go, and through favor or interest, ask the Viceroy to let him abandon the mine for more than four months. The King says that the Viceroy shall not grant that favor. Yet even in that case, Gamboa says, the ordinances contain a penalty which must be most strictly construed. Therefore, he says the failure to work must extend to four *continuous* months. If you work two or three days, then stop three months, and so on, the mine cannot be considered forfeited. True, you are evading the ordinances, contravening public policy, and not doing what is required; but that is the fault of the ordinances, and the Judge cannot pronounce forfeiture, except under the terms of the law. And, therefore, unless proven to have abandoned your mine for four full consecutive months, your title remains untouched. That is one of the difficulties in the ordinances pointed out by Gamboa. Gamboa suggested a change in that respect. So the Ordinance of 1783 changed it, and changed it to this effect: That if a man during a year failed to work a mine four months—if all the intervals of failure put together amounted to that time—then he should lose his mine. He could not work a few days and then not work for months, but if all his intervals of failure put together amounted to four months in a year, he lost his mine. Under Gamboa's advice, that was established as a cause for judgment of forfeiture.

I state these things, not because they are in the case at present before your Honors, but as is the case with many points to which I have alluded—having been introduced by brother Randolph in his propositions, I have seen fit to make this in answer. Brother Randolph states that these ordinances were *strictisimi juris*, and that the miner was treated by the mining laws of Spain with the utmost rigor. Just as much as Judge Hoffman treats seamen in the Admiralty Court with the utmost rigor of the law. The miner was the special favorite of the law. He was considered in the character almost of a ward to the King. The ordinances are full of provisions giving him privileges, promising him nobility, directing the different Governors and Viceroys how they are to favor him in all his business, and prohibiting them from allowing any monopolies of provisions necessary to him. The ordinances also prohibit their paying the miner with provisions and with clothing instead of money. The ordinances thus take care—first, of the mine owners, then of the mining workmen—with the same liberality as the admiralty law treats sailors.

If your Honor (Judge Hoffman) treats sailors with the utmost rigor of the law under the admiralty jurisdiction, you illustrate just that rigor with which the mining laws of Spain treated the miner.

Well, now, we come to this case; all that I have said hitherto being in relation to the mines on public lands granted to miners both in their surface and depth. How was it with mines found on the lands of private individuals. Was the same surface granted to the discoverer, or was it a mere easement upon that land that the miner took? The miner took the property in the land, the title to the land, the title to the area, the title to the soil, neither more nor less. Gamboa says it over and over again. Why quote the clause? The mining ordinances say it in so many words. I quote the ordinances, Title 6, Art. 14, Halleck's Collection, p. 228:

Art. 14 Any one may discover and denounce a vein or mine, not only in common land, but also in the private lands of any individual, provided he pays for the land of which he occupies the surface, and the damage which immediately ensues therefrom, according to the valuation of the Experts (*Peritos*) appointed by both parties, and a third in case of disagreement: the same being understood with respect to him who denounces a place (*sitio*) or waters for establishing works and moving the machines necessary for the reduction of ores, which are called *Haciendas*, provided they do not include more land nor use more water than may be necessary.

This is not an easement, but a purchase of the surface of the land; for the proprietor is to be *paid* for the land of which the miner occupies the surface, and also any damages which may immediately ensue to him from the acquisition and working of the mine which has been discovered and denounced in his land. The miner acquires his vein or mine by discovery or denouncement, but if this vein or mine be in private land he acquires the surface of that land by paying for it according to the valuation of experts, if such valuation be immediately asked for. That is what is stated in the ordinance just quoted, and this is what Gamboa says. At page 94 of the original, in speaking of mines upon private lands, he says: "According to the 15th of the old ordinances, the owner of the land (*fundo*), besides being paid the damages, was entitled to be paid one per cent. before deducting the payment to the treasury; and this payment of one per cent. was likewise established by the ordinances of Peru. But by this 16th ordinance of the new code, nothing is to be paid to the owner of the inheritance or land (*predio*) of the silver or gold which is extracted. And there is to be paid only to the treasury the fifth, or the tenth, of whatever may be imposed."

Then in Sec. 5 of the same Chapter IV, he continues: "In consequence of this liberty and absolute concession in favor of all to search for mines, it is not to be understood that the Prince grants them to the private owners of the lands (*fundos*), unless special mention is made of them, in virtue of a special grant, or by time immemorial. But, except in these cases, no person, of whatsoever condition, rank or dignity he may be, can prevent their being searched for in his estates, pastures, inheritances, or other places, for the Royal power has made them common everywhere in favor of those vassals who shall first discover or occupy them."

And now in Sec. 6, he speaks of the private owner of the surface of the land under pastural, agricultural or other title, not being a concession of mines:

Sec. 6. But this liberty is to be understood without prejudice to a third party in his same pastures or inheritances; because it is understood that the damage which results is to be appraised by two experts, which the court must appoint, and not the parties, and by a third in case of disagreement; and that which they shall assess, under oath, may, in default of payment, be collected by means of execution.

He then proceeds to describe the manner of appointing additional experts in case those appointed cannot agree, and says that they must take into consideration all the circumstances of the damage which results from the establishment of mines, and other arrangements required for working them—the digging and excavation of the soil, the building of houses, erection of smelting furnaces, etc. I ask the Court to compare the remarks of Gamboa with the Ordinances of 1783, on page 228 of Halleck's Collection, and with Lares' Commentaries on "Derecho Administrativo," p. 93.

Your Honors will see from these authorities that the discoverer or other grantee of a mine has the right, as incident to his ownership of the surface of the land included within his *mining pertenencias*; if the land belongs to the national domain he pays nothing for this surface; but if it be private land, he can be forced to pay to the owner the appraised value of this surface. Moreover he has, in virtue of his mining concession, certain easements, servitudes and usufructs in other lands not within his *mining pertenencias*, as a right of way, of water, of wood, of pasture, of drainage, etc., but if this land be of private ownership, he must pay the owner the appraised damages, as provided in the ordinances. In the one case the private land-owner loses his right to the surface, by what is called in Spanish law, forced alienation (*enagenacion forcado*), and the mine-owner acquires the right to that surface by what is called in Spanish law, an onerous title (*titulo oneroso*). It becomes his property. But not so with the surface upon which the miner, in virtue of his mining concession, acquires certain easements, as the right of way, wood, water, etc. The title to that surface remains in the private land-owner, and the mine-owner must pay the appraised amount of the damages which result from the exercise of these easements.

Your Honors will also see from these authorities that the payment which the mine-owner is required to make to the private land-owner, is not for the mineral substances, or the portion of the earth's substance which contains them, or with which they are incorporated, but for the *surface* of the land included within the *mining pertenencias*, and for the damage which may immediately result to the *surface* of the land upon which the mine-owner exercises his easement, or right of wood, water, drainage, etc.

If I understand Gamboa's reasoning on this subject, he says that when a man discovered a mine on the ground of another person, he registered it. If he was the discoverer, he took three pertenencias. But the surface of the land belongs to a private individual. Then the law says that the miner shall pay that private individual for the piece of surface, and also for damages that accrue from his working the mine and going on his land. Besides that, the miner was entitled to denounce other pertenencias for putting up houses, works, fabrics, etc. In other words, he had to buy the surface of the land and pay for it. He obtained the title of the owner of the private property, by payment of the purchase money. But the payment was only required, and only necessary, if demanded immediately. Otherwise the land could be taken without it.

And again: Gamboa says that the judges must be especially careful to avoid appointing receivers in cases of miners, for they almost always receive the fruits and defraud the owners.

The XXII point of my digest is, that in New Spain the special mining judges do not exist, nor is the ordinance in force there on that subject. In their stead the ordinary judges must act—*alcaldes* and corregidores.

Now, in no point of view is the change between the ordinances of the time of Gamboa and those of 1783, more striking than on this subject of forfeiture. In the ordinances upon which Gamboa commented, there were fifteen cases of forfeiture of mines, which I have cited in my printed digest, and which I will here read:

1. Art. 18. For failing to register a new mine already owned and worked.—Halleck, p. 77.
2. Art. 20. For registering a mine not really belonging to him who registers it—p. 78.
3. Art. 21. For registry without declaring names of partners—p. 78.
4. Art. 22. For failure to dig pits and fix stakes when pertenencias are defined—p. 82.
5. Art. 32. For taking a mine for another when the person taking it has no power of attorney, and is not the hired servant of that other—p. 85.
6. Art. 35. For failing, within 90 days after registry, to sink a well three *estados*, or seven varas deep, (unless excused for causes stated in Art. 36)—p. 86.
7. Art. 37. For failing to work the mine for four consecutive months—p. 87.
8. Art. 38. For failure on part of denouncer to deepen former works three *estados*—p. 89.
9. Art. 42. For purchasing a mine which has not been sunk to the depth of three *estados*—p. 92.
10. Art. 42. For not registering purchase, or other change of ownership—p. 93.
11. Art. 53. Of mine and metal; for smelting in another's furnace without permission of the Administrador of Partido. —p. 100.
12. Art. 59. Of mine and silver; for reducing ores otherwise than with quicksilver, after obtaining license to reduce them in that mode—p. 105.
13. Art. 67. Of mines; by certain officers who are forbidden to hold them, and by those who participate in the offense of these officers—p. 112.
14. Art. 68. By certain salaried persons; for becoming owners of mines within two leagues of the Partidos where they are employed—p. 112.

15. Art. 73. Of all property, and ten years' labor in the galleys for purchasing or dealing in unstamped metal—p. 114.

All this was intended to secure the King's revenue. Gamboa advised one or two other cases of forfeiture; but when the new ordinances were made, they were made by a body of miners in the City of Mexico, who framed and sent them to the King, begging him to accept them. The King and Council announced them authoritatively, and they became the Ordinances of 1783. Instead of *fifteen* cases of forfeiture they left only *five*. I will read them from page 10 of my digest:

1. For abandonment.—Title 9, Art. 13; Halleck, p. 243.
2. For failure by secular ecclesiastics to sell mines belonging to them, after a certain day.—Title 7, Art. 2; Halleck, p. 231.
3. For removing or impairing the strength of pillars and arches of a mine.—Title 9, Art. 7; Halleck, p. 241.
4. For repeated violations of rules relative to the security and preservation of mines.—Title 9, Art. 10; Halleck, p. 142.
5. In cases of partnership, for omitting the partner's name in the denouncement, or refusing to contribute for four months toward expenses; in which cases the forfeiture inures to benefit of copartners.—Title 7, Art. 6, and Title 11, Art. 8; Halleck, pp. 233, 254.

But by these Ordinances of 1783, there is no forfeiture for irregularities of any kind; there is a careful avoidance of forfeitures for such cases from the beginning to the end of these ordinances. So far from forfeiture, as brother Randolph says, taking place for failure to comply with *any* condition of the ordinances, they expressly say that it is a failure to comply with the conditions *which impose that penalty*.

And now we come to brother Randolph's own individual translation, of which *he* assumes the responsibility. On pages 26 and 27 of his printed brief, he says that the condition of the miners' property is, "that they take their grants under two conditions: the first, that they shall contribute to my royal treasury the designated part of the metals; and the second, that they shall work and enjoy the mines in compliance with the provisions of these ordinances, in such manner that they should be understood as lost so soon as there should be any default in the fulfillment of those ordinances *which regulate that matter*." I don't understand this. What does he mean by "that they are to be understood as lost so soon as there should be any default in the fulfillment of those ordinances *which regulate that matter*"? What matter? Regulate the matter of the forfeiture or losing of mines? Then we agree. If that be the meaning of his translation, then we agree as to the meaning of the original; and, although it is not a correct translation, I make no change in it. But if he means anything else, then I want your Honors to look at the original.

The mines are to be "lost so soon as there shall be any default in the fulfillment of those ordinances which regulate *that matter*." Does he mean *the matter of title?* Is that what the ordinances say,—that the mines shall be lost and forfeited for a violation of *any* of these ordinances, if proved, or that they are to be forfeited for a failure to comply with *any* of the conditions of these ordinances?

Now, there are one hundred and odd of these ordinances, and but four or five of them pronounce forfeiture. I admit that a violation of *those ordinances which pronounce forfeiture* for failure to comply with *their* conditions works forfeiture; provided, always, the man is brought before the Court and the fact proven on him, for which judgment is pronounced.

Now there is not a single thing in this title attacked, upon which your Honors are called to declare it illegal and invalid, for which the ordinances pronounce it forfeited. If by the grossest informalities and defects in form—not on the part of the person discovering, but on the part of the Judge—a man's title was lost *ipso facto*, then if there was the slightest error in drawing up the paper, or if the Judge forgot to do something which the law required, the property would be forfeited. Now that proposition is startling. The Judge is the agent of the sovereign, not of the discoverer. The Judge is directed by his sovereign, the law, to do certain things; hence he performs certain ministerial duties. The right of the subject cannot depend on the Judge's failure to comply with the forms prescribed, because the right of the subject is acquired as soon as the mine is registered. You cannot take that away from him; his registry does not become void because the Judge did not draw up the act of possession.

Mr. Johnson—Suppose he refuses to put him in possession?

Mr. Benjamin—Mr. Johnson suggests, suppose the Judge would not put him in possession? Place another Judge in his stead. A man cannot so lose his property.

Is there any system of law where the defect of failure of an executive officer to perform certain ministerial duties directed by law destroys the title of a third person? If there is any such example, I should like a citation of it. There may be such, but none suggests itself to my memory. Where do these ordinances so state?

This article, which brother Randolph translates in the manner which I have pointed out to the Court, is translated by Mr. Halleck with a felicitousness that is remarkable, for he translates every word literally, and gives precisely and exactly the meaning of the original. I read his translation of Articles 2 and 3 of Title V, from pages 222 and 223 of his collection:

Art. 2. Without separating them from my Royal Patrimony, I grant them to my subjects in property (*en propriedad*) and possession, in such manner that they may sell them, exchange them, rent them, donate them, pass them by will, either in the way of inheritance or legacy, or in any other manner alienate the right which in the mines belongs to them on the same terms on which they themselves possess it, and to persons capable of acquiring it.

Art. 3. Let this grant be understood to be upon two conditions: first, that they shall contribute to my Royal Treasury the prescribed portion of metals; and second, that they shall work and enjoy the mines complying with what is prescribed in these ordinances, so that they shall be considered forfeited (*perdidas*) whenever a failure shall occur in complying with those (ordinances) in which it is so provided, and that they may be granted to any person who for that cause may denounce them.

Now this is a perfectly literal translation, and there cannot be the slightest doubt of its correctness. The language of the original Spanish is not susceptible of any other meaning; it is impossible to force any other meaning from that language. The ordinance is as plain as it is possible for language to make it. I defy any one at all acquainted with the Spanish language to make anything else out of it. You may change the words of the translation, but you cannot change the meaning unless you change and falsify the meaning of the original.

Let us now refer to Article 11 of Title VI of these ordinances. It is translated on page 227 of Halleck's collection:

Art. 11. If any one denounce a mine as forfeited (*perdida*) on account of the non-observance of any of the ordinances which carry this penalty imposed, it shall be granted to him, provided that any of the prescribed causes shall be lawfully established and proved.

Denounce a mine as forfeited! Forfeited for what? For the non-observance of any of the ordinances? No! "for the non-observance of any of the ordinances *which carry* (*or have*) *this penalty imposed.*" What penalty? The penalty of forfeiture. And this is not all that is required: "Provided that any of the prescribed causes shall be lawfully established." Any of what causes? Any of the causes of forfeiture prescribed in the ordinance. To denounce a mine as forfeited, it must be on account of the non-observance of some one of the ordinances *which prescribe forfeiture* as the penalty for such non-observance, and this *fact* must be established and proved lawfully, that is, in the manner prescribed in the laws.

But my brother Randolph insists, the mining ordinances and commentators to the contrary notwithstanding, that the mine-owner forfeits his mine by the non-observance of *any* one of the ordinances, no matter whether or not such ordinance imposes that penalty for its non-observance; no matter whether the neglect is his own, or that of a ministerial officer of the Government over whom he has no control; no matter how trivial or unimportant may be the provision not observed; no matter whether such non-observance be proved or not. Nay, more, he forfeits and loses his mine unless *he* prove and establish affirmatively, that all the formalities and conditions prescribed and imposed both upon the officers of the Government and upon him, have been strictly and literally observed and complied with.

Why, may it please your Honors, so far as these Ordinances of 1783 are concerned, a great many of their provisions were out of use in Mexico long before Castillero discovered this mine. In proof of this, I refer your Honors to the notes to titles I, II and III of the Mexican editions of these Ordinances, and to the Mexican editions of the learned commentaries of Sala and Febrero. In the former you will perceive numerous modifications in the organization of the Mining Courts provided by these ordinances, and you will learn from the two last named authorities that the special Mining Courts provided in these ordinances, with all other special tribunals, were abolished in Mexico soon after the adoption of the Federal Constitution of 1824, as being incompatible with the principles established by that instrument. Sala Mexicano, Tomo 4, p. 60, says distinctly that the *Consulados* or Mercantile Courts, and the "Mining Tribunals," were abolished in Mexico after the adoption of that Constitution, and that their jurisdiction fell to the ordinary Judges (*jueces ordinarios*). The learned La Pascua in 1834, published his celebrated work, now in the hands of every Mexican lawyer, called "Febrero Mexicano." This work was founded upon the Spanish Febrero of Tapia, but a large portion of it was either re-written by La Pascua, or so changed as to make it conform to the then existing laws and practice in Mexico. That part relating to mines is entirely re-written, and most essentially changed from the language of the old Spanish Febreros, which were based upon the mining ordinances as they were originally enacted, or as they had been modified in *Spain*. But La Pascua was preparing his book for the use of *Mexican* lawyers, and he wished to give the law as it then existed, and was understood in *Mexico*. He therefore gives the substance of articles 4, 5, 7 and 8, of Title V of the Ordinances of 1783, with important modifications. Instead of saying, as in article 4, that the discoverer shall present his writing of discovery for registry to the mining deputation (*ante la deputacion de mineria*), he says it shall be presented to the Justice of the place (*al juez del lugar*); and instead of requiring one of the deputies to personally inspect the mine and give the possession (*pasara personalomente uno de los diputados*, etc.), as in the original text, he alters it so as to read that the *Justice* shall do all this, (*pasara personalomente el juez*, etc.) And so throughout, wherever the word *diputacion* is used in the Ordinances of 1783, La Pascua, in his "Febrero Mexicano," substitutes the word *juez*. Vide lib. 2, tit. 1, cap. 2, secs. 20, 21 and 22.

This proves what is said by other writers, that in 1834, when La Pascua published this book, the jurisdiction conferred by the ordinances upon the Deputations of Mining, with respect to the registration and the giving possession of mines, were performed by the ordinary judicial officer of the place—*el juez del lugar*.

In another place (vide Appendix to lib. 3, tit. 2, cap. 2), he gives the organization of the judiciary of Mexico; but no mining courts are mentioned in his list of *tribunales speciales*, thus confirming what is stated in the Sala Mexicana by Peña y Peña, and by other Mexican law writers, that there were no special mining tribunals in the Republic of Mexico.

Now, if it be true, as brother Randolph says, that the old ordinances were in force in their entirety when this mine was registered; that they were *strictisimi juris;* and that nullity or forfeiture resulted from the non-observance of *any one* of these ordinances, in its word and letter, how could any person acquire a title to a mine in Mexico by discovery or denouncement? All such titles were absolute nullities, because there were no "Mining Deputations" to registry discoveries and denouncements, nor were there any mining *diputados* to examine the pit and give the possession.

Again, I call the attention of your Honors, in this connection, to what La Pascua says respecting the *time* when this possession is to be given. He says distinctly, that as soon as the miner has dug his pit the *juez* is to personally inspect his working and to put him in possession, without waiting for the expiration of the ninety days; with the proviso, however, that during the ninety days anybody else may come in and dispute the possession of the mine.

This exactly corresponds with our interpretation of Articles 4 and 5 of Title VI, and is in direct conflict with the meaning which brother Randolph attempts to force out of the plain and unmistakable language of the original.

RECESS.

After recess, MR. BENJAMIN continued: May it please your Honors, I have thus far discussed the general principles of mining law, as applicable to all mines—gold, silver and other mines, even those of baser metals,—rather as a subject of curious judicial inquiry than as directly pertinent to the present case; because there was a special legislation of Spain for quicksilver mines. That legislation comes down to a much later period than the general ordinances in relation to the other mines of the kingdom. In the Ordinances of 1783, quicksilver mines are included; but certain special provisions are made in regard to them.

These are found in Art. 22 of Title VI, and at page 230 of Halleck's Collection. They are in these words:

> I declare that, although the free discovery and denouncement of mines of quicksilver is permitted, it must be on the express condition of giving an account of them to the Viceroy and to the sub-delegate of quicksilver in Mexico, in order that it may be considered and determined, whether such mine or mines shall be worked and carried on at the expense of the particular subject who discovered and denounced them, he punctually delivering the quicksilver extracted from them into the royal store-houses on the terms and price stipulated; or whether it shall be done at the expense of my Royal Treasury, indemnifying the party for it by some equitable reward, according to the circumstances of the said discovery and denouncement—the whole of this important subject being regulated according to my sovereign intentions recently declared on that subject.

Under this new legislation in relation to the quicksilver mines, it was found that the mining interests of Spain were much prejudiced; the discovery and working of quicksilver mines being so impeded by reason of the uncertainty of the rights acquired by discovery, that no person went in search of them. People searched for gold, silver, lead and copper mines, for all other mines rather than those of quicksilver. They were deterred from searching for quicksilver mines by the consideration, that if they found such, the mine would not belong to them, but would be subject to be taken from them by the King's officers, to be worked at the expense and to the profit of the Royal Treasury.

The King wanted quicksilver. The mining interests of the colonies required quicksilver. Quicksilver was the very breath of life to the mining industry of New Spain. The mines consisted of vast bodies of ores, not very rich, but of very ordinary quality, which could be worked when quicksilver could be afforded at a certain price; but the moment quicksilver rose above that price the mines were abandoned; the Royal Treasury suffered; the interests of the whole kingdom were prejudiced. It became necessary, therefore, to change this special legislation. In the year 1811, whilst Ferdinand was prisoner at Bayonne, the Cortes assembled at the Island of Leon; a Council of Regency assumed the control of the affairs of Spain; an extraordinary Cortes was invoked, and deputies from all the provinces of Spain, from Mexico, the colonies, and all parts of the Spanish dominions, were present.

One of the first things done by the Cortes, was to declare the prejudice which had resulted to the interests of the kingdom by reason of the restriction to the discovery and working of quicksilver mines. And the Cortes passed a special law, declaring that it was time the jurisprudence of Spain should be changed on that subject; that thereafter no Viceroy should manage the quicksilver mines discovered; but that discoverers should own the mines, and not only own the mines, but own them *en propiedad y usufructo*, *propiedad y utilidad*, in that full title which concentrated within itself both the legal title and the beneficial interest; and that legal and full ownership, it was declared, should be in them and their heirs forever. And, further, it was declared, that they should not be obliged to sell them, even to the crown; that the right of retaining them should be absolute in the mine-owners, and that the King should not force them to convey to the crown, for a price even, unless they chose. This was to be as an incentive to persons to search for quicksilver mines.

This ordinance of the Cortes was sent to the Mining Tribunal of Mexico, with a recommendation from the Council of Regency to offer large rewards out of the Quicksilver Mining Fund to everybody who would search for and find a mine; and a very large reward to the one who should find the best quicksilver mine. An assurance was also given, that it would belong to him; that he might sell it, or devise it to whom he pleased; that his heirs, and those to come after him, should enjoy the fruits of the discovery of their ancestor. That was to endure forever.

A little later, the Cortes informed the Mining Tribunal of Mexico—and those laws are published in the Mining Laws of Mexico—that further rewards were offered; that the Council of Regency would ennoble with honorable distinctions those who invest capital in quicksilver mines. And the Council of Regency, on the 2d of February, 1811, informed the Tribunal of Mexico, that the Cortes had provided for giving rewards to discoverers of quicksilver.

In 1818, quicksilver was restricted from all duties whatsoever, by express orders to the Viceroy.

In 1822, when Mexico acquired independence, the first thing done was to decree a revenue on gold and silver; but under Art. 13, pure quicksilver was to remain as previously, absolutely free from any duty.

Nothing was to be paid. There was no obligation to convey to the State, even for a price, the *propiedad y usufructo* conferred. All this was done to incite search for quicksilver mines.

That is the quicksilver legislation of Mexico. It is to that we are to look for the purpose of ascertaining the rights of Castillero in his discovery.

But they did not stop there. Again and again Mexican authorities incited search for quicksilver mines, by promises of every possible reward which ingenuity could suggest. At one time they offered a reward of $25,000 each, to any four quicksilver miners who should produce a certain quantity within a year. Then they offered a premium of $5 per quintal of quicksilver; exclusively to owners of mines in the Republic. The importance of this interest—and its importance can be understood by anybody who has ever traveled throughout that country and seen the mines, and the way in which they are developed by the aid of quicksilver—was such, that the Mexican legislators have never been able, apparently, to sufficiently reward its furtherance. No longer are any duties paid; no longer does the State reserve to itself the right to purchase, or to take for its own use these mines, by indemnifying the owner, for the owner shall not be bound to sell, even for a price, to the Government; but he shall have the mine for himself, without obligation to sell, and without obligation to bring the quicksilver to the royal or state store-house. He shall sell it for the best price he can get; and producing proof to the Government that a certain quantity has been extracted within the year, shall receive $25,000. He shall also have his reward of $5 premium per quintal produced within the Republic.

That was the condition of Mexican legislation at the time Castillero discovered this mine. Well, then, might he say, that if the mine was what he supposed it to be, his fortune was made! Well, then, might he refuse with disdain $30,000 or $40,000 offered him for this mine! He would not listen; turned a cold ear to offers made for his discovery. He was going home to Mexico; he knew the public feeling in his country; knew what the President would do for him. He had previously made a grant to him of an island on the coast. He knew what the Mining Junta would say when there should be displayed before their astonished gaze ores of such richness, as according to his own report, threw the ores of the mine of Old Almaden in the shade. That was the condition of Mexico when Castillero made his discovery.

Yet we are told here, that all he obtained was a license to dig; no property; no right to his mine, but a *mere easement* upon the surface of the proprietor of the soil. What says the ordinance of the Cortes of Spain? See what is his title under that law:

Sovereign Resolutions of the general and extraordinary Cortes, and of the Supreme Council of Regency, conceding the full dominion (*pleno dominio*) and acquisition of Mines of quicksilver, free trade in their products (*frutos*), and exemption from all kinds of duties; and offering rewards to their discoverers and to those who invest their capital in them; communicated to the Royal Tribunal of the important Body of Mining of New Spain.

A full dominion! *pleno dominio!*

I will take the translation in Halleck's compilation, which I have compared and find correct. I pledge my little knowledge of the Spanish language that it is perfectly accurate. (*Vide* pages 138, *et seq.*) What is the communication of the Council of Regency to the Mining Body of New Spain. After the title of these resolutions which I have just read to your Honors, it says:

Under this date I notify the Viceroy of that Kingdom, that the prerogative of Seigniory, which, from remote times, the Exchequer (*Fisco*) has reserved to itself, with respect to mines of quicksilver, when it considered them advantageous, after having paid to their owners their just value, has been annulled by general and extraordinary Cortes, in consequence of the resolution and manifestation of the Council of Regency, enacting, at the same time, that the said mines shall be worked under the same rules and ordinances as those of gold, silver and other metals, and that their possessors shall preserve their proprietorship (*propriedad*) and usufruct, and that in no case shall they be obliged to transfer them (*enagenarlas*) to the State; giving them permission, moreover, to sell their products (*sus frutos*) to any one who will pay the highest price for them. This measure (*providencia*) affirms, in a manner inviolable, the ownership (*propiedad*) and profits (*utilidad*) of this class of real estate (*tales fincas*), and dispels the reasonable fears which prevented individuals from taking them under their care.

The zeal of Your Honor (V. S.), and your love of the public wellfare, must cause you to take interest in promoting, among those miners, the search for and examination of (*cateo*) mines of cinnabar, to which there can be no more powerful stimulant than the offer of a large pecuniary reward, to be paid from the funds of that Body, to the person who shall discover, and fully prove that he has discovered, a rich and abundant mine of quicksilver; promising also, that the Council of Regency will recompense and ennoble, by honorable distinctions, those individuals who shall invest their capital with evident advantage to the said works, and especially those who shall excel in it by extraordinary progress. To serve as an example to all of the necessity of application to this kind of industry, it will be of advantage for Your Honor to undertake methodically the working of a mine of quicksilver which, of all those of that Kingdom, offers the best prospect, after careful and repeated examinations, and exact and well verified reports establishing a very simple and well regulated Administration, so that the amounts which shall be dedicated to the enterprise may be actually expended in its advancement, and not in unnecessary works and buildings, nor in large salaries, which, without any profit, are commonly charged to such speculations; the accounts of which shall be presented, for examination and approval, at each one of the general Juntas held, at the usual time, by that Body, which shall thereupon order what may be most expedient for the management of the affair (*negociacion*).

So that the Mining Tribunal would not only inform the Mining Body of Mexico, that for the future they would own the mines themselves, and be allowed to dispose of them as they pleased, with the prospect of large pecuniary rewards, but the Council of Regency requested the Mining Tribunal to establish a model quicksilver mine, that others might profit by the working of such a mine. That was what was done in 1811. These ordinances are still in force in Mexico. These decrees are now established and published in the Mexican ordinances of Mineria, as part of the mining laws of Mexico. They have never been changed. So far from having been changed, they have been added to by farther rewards and farther incentives for the discovery and ownership of these mines.

In 1823, the law of mines was modified in Mexico, except as to quicksilver, which is again expressly exempted from all contributions to the State (Halleck, p. 404).

In 1842, foreigners were allowed to acquire in property (*en propiedad*) mines (including quicksilver) which they might discover (Halleck, p. 427). Foreigners were invited to come into the country and discover and work quicksilver mines.

In December, 1842, the Junta de Fomento was established; and its special attention was directed to making regulations, fixing the mode in which the working of quicksilver mines is to be supplied, rewarded, stimulated and protected (Halleck, p. 437). The whole legislation of the nation tended to that one thing.

On the 24th of May, 1843, rewards of $25,000 and $5 per quintal, were offered to quicksilver miners (Halleck, p. 453).

In July, 1843, the Junta was empowered to work, to supply and to protect quicksilver; to search for such mines, to examine those discovered, and to take partial measures for efficaciously encouraging the production of quicksilver (Halleck, p. 455).

Again, in September 1843, a further decree was made to

encourage and assist the working of quicksilver; and researches were ordered to be made through the whole Republic. Commissioners were accordingly appointed by the Mining Junta, and sent all through Mexico in search of mines; and some $30,000 had been expended in a vain effort to discover a mine in Mexico, when in 1845 Castillero arrived at the City of Mexico, and said to the Mining Body: "What you have so long desired, promises now to be speedily accomplished. All that Humboldt predicted would one day happen, has occurred. I have found the mine, which for a generation all Mexico has been seeking for. I have found the mine which all your Commissioners who have been searching for a year past in all the Departments of Mexico, have been unable to discover. *They* have found for you simply fractional deposits, which yield but 2 and 2½ per cent. *I* have found a vein, of which the ores excel in richness those of the ancient royal mine of Almaden of Old Spain—the royal mine which supplies the gold and silver mines of the world. I want you now to carry into effect those powers invested in you by law. I want you to carry them out, in accordance with the design for which you were established. What is that? Why, I want you to protect my mining discovery. I want you to assist me in working it. I do not want much. I want you to advance me $5,000 to start my mine. I will pay you back, during the year, in quicksilver, delivered at Mazatlan, at one dollar per pound. I want you to let me have some iron retorts, and some empty flasks, which you have in the *negociacion* of *Tasco*, which are not used. I will pay you very fully in quicksilver. I will make a bargain hereafter by which you may enter into the business. Now I am in a hurry to get back. There is no time for *that* now; but I will now make a provisional contract. Give me this money, these flasks and retorts, and I will pay for them in quicksilver. I will give you the preference of all the quicksilver that comes out of the mine for a certain time, at a certain fixed price. And then, you must, on your part, ask the President to ratify the favor given me by the local Alcalde, by extending my possession in the mine to 3000 varas in all directions [which of course he (the Alcalde) had no authority to do, because an Alcalde could not make a *gracia* of 3000 varas of land]. You must recommend to the President that *he* approve that, and also beg him to give me a grant of two leagues of land, as a colonist, that I may cut wood for my burnings in my reduction of the ores to quicksilver. That is all I ask for my discovery."

The Junta snapped at these propositions on the subject. Here was something all Mexico had been ardently desiring for generations. The demands were but for a simple advance of money, to be returned in a single year, in the thing which they most wanted—quicksilver; it was to be returned at the moderate price of $1 per pound, when it was then worth $1.80 in many parts of the Republic. In consideration of being returned at that low price, no interest was to be charged. The Junta said: "Although, under our general powers, we have a right to make all these general contracts in relation to quicksilver, yet we will present your propositions to the President of the Republic, instead of acceding to them ourselves; and we agree to recommend your proposal in all its parts, with a single exception—that is, the grant of two leagues as a colonist. That does not concern the Mining Board. That concerns other Departments of the Government of the Republic, and in relation to that, we will submit your petition to his Excellency, without any recommendation, in order that he may do what he thinks proper on that subject."

Then they write a communication to the President of the Republic, explaining the value of this discovery. They give Castillero's propositions in detail; they speak of the ratification which he wants for his mining possession; they commit some errors in detailing the actual nature of those propositions; they state, for instance, that three thousand varas is equal to fifteen pertenencias, and that is made a great affair of by the special counsel for the Government, as if it were not a very common thing for men in making calculations to commit an error! It is not such a very remarkable thing. They forget that the terms were "three thousand varas in *all directions from the mouth* of the mine," which would make a square of six thousand varas on each side. But that is now a matter of little consequence.

Where was the mine? In the distant north, hundreds of

leagues away off in California; so far that their Commissioners did not penetrate to it. It was in "Santa Clara." They did not know where Santa Clara was; some called it a "Mission," some called it a "Presidio;" the clerks copying the terms spoke of it sometimes as in Upper, sometimes as in Lower California; it was in a remote and distant section of the Republic; and if he had asked for three thousand miles, he would, in all probability, have got it. That is all there is about that.

How about the colonization grant? The Junta said to the President: "You know how to act in relation to that subject; that concerns another Department."

What does the President do in relation to that? He tells the Minister of Justice, in whose office the espediente was: "Issue orders to the Minister of Foreign Relations to make this grant; I agree to it; and let him send out the necessary orders for possession to the proper Departments." That is what the President says.

The Minister of Relations so understands it, and he writes to the Governor of California that the President of the Republic had acceded to the petition for two leagues in colonization, and directs the Governor to put the petitioner in possession.

We are told to-day that all this was done, but that it is a nullity. We are told to-day, that the Alcalde did not mark out certain stakes. We are told to-day, that this mine, thus granted, thus possessed, thus vaunted of, by its discoverer, as his property of great value, was proclaimed to the Mexican authorities or Mexican people, and by them recognized as Castillero's, even before the declaration of war; but that notwithstanding all that, it does not belong to him, but now belongs to the Government of the United States, by virtue of a treaty, in which that Government most solemnly pledged its word in the face of mankind, that every right of property belonging to a Mexican citizen who did not live in California, would be respected in the same manner as if he was a citizen of the United States. Not "fee simples," not "perfect titles," but "*property of every kind.*" See Art. 8 of the Treaty of Guadalupe Hidalgo.

After first providing for the rights of Mexicans who remained in California, provision was made for those who live in Mexico.

"In the ceded territories, property of every kind"—(and mines are property of some kind—does anybody doubt that? I should be sorry to believe the contrary. I have taken some small interest in mines myself since I came to California, and if I have got no property for what I have paid my money, I shall be badly off. It may happen that the Government will turn me out some day. But I have digressed from my quotation from the treaty)—"In the ceded territories, property of every kind, now belonging to Mexicans, not established there, shall be inviolably respected."

But, how the Government of the United States has inviolably respected this private property of Mexican citizens! How the faith of the nation has been duly respected and preserved by the Attorney General of the United States!

He has come here, and through the agency of the law singled out the only mine owners who owned a mine in California prior to the cession, and called upon the Court to turn them out of their property; and alleged as a ground for so doing that they were *foreigners!* It has been thundered in the ears of the Court, it has been proclaimed to the American people—the Attorney General of the United States has not been ashamed to proclaim it—that *because* these mines belong to *foreigners*—I say, that in the administration of public justice, the man whose office imposes on him the special duty of seeing that the laws are observed—and what higher or holier law than the treaties that our Government has made with foreign nations—this man has not been ashamed to proclaim that he will endeavor to oust these claimants, *because they are foreigners!* Where may he find a higher or holier law than this Treaty, as regards the duties of his office? This Treaty is the supreme law of the land. If made in accordance with the provisions of the Constitution—and none here dispute that—it is the supreme law of the land. And this supreme law of the land says that these men's rights of property shall be inviolably respected; whereupon they are turned out. "The present owners, the heirs of these, and all Mexicans who may hereafter acquire said property by contract, shall enjoy with respect to it guaranties as ample as if the same belonged to citizens of the United States." Then why denounce the foreigners? Does the law denounce them? If the law does not, why does the minister of the law denounce them? His rights and duties are commensurate with the law which he is placed in power to observe and enforce.

Now, we have this valuable property under the Mexican law. We have it by virtue of Mexican laws, usages and customs. Your Honors are told by the law-giver, that if we have this property under Mexican laws, usages and customs, and make claim to it before you, proving these laws, usages and customs, and that our title is in conformity therewith, then you shall declare the validity of that title, whatever the nature of the right may be. You may say, if you please,—I think if you say so, you will err,—that this mining title is a mere license to dig. But if that is so, we then say, confirm that.

You may say, that he (Castillero) is a mere tenant at will of the Government; holding the mine on condition that he shall perform certain services in it, but liable to forfeiture if he fail so to do. If that be his condition, say so, and confirm that; but don't turn him out of Court, and say that he has got nothing. Whatever his claim be, tell us what it is; and when you have so told us, say it is confirmed; that it is valid; for whatever you say it is; because you are to judge any right of property in land, and you are to judge in conformity with Mexican laws, usages and customs.

Will the special counsel for the United States stand up in this Court and say that any Mexican miner has no property at all? Entire districts of the Republic may be found, in which there is no other kind of property; where the entire land is covered by a mining population, all holding titles under mining laws. What have *they* got? Can your Honors be induced for one moment to sanction the conclusion, that all that vast, that great, that predominant interest of the Mexican Republic, rests upon no title whatever? That, whereas twenty or thirty millions of dollars are annually exported from Mexico, in silver, at the present time, those who extract this silver from those mines have no title at all? You have either to determine that, or, you have to determine what that title is, and say that our title is the same, and confirm it as such. I say, and in this I believe I do not err, that that title is the *allodium* of the Common Law, the *dominium directum et utile,* or the *dominium plenum* of the Civil Law; the highest right of property the citizen can have; only subject or liable to be taken from the individual by the right of eminent domain—*dominio alto*—for public service, after the full value has been rendered.

Now, under the Civil Law, the King might own property. When he owned it, he combined in his person the *dominium eminens,* or eminent domain; that is, the right of sovereignty, the *domininum directum;* that is, the right of the lord; and the *dominium utile,* or beneficial interest, which was the nature of the interest of the vassal. All these were combined in the King when he owned property. But the King could sell his property, or the King could give his property away, whether belonging to him by purchase, or by title derived from his prerogative as sovereign. He never could divest himself of the *dominium eminens,* because that he held in trust for society. He might divest himself of the *dominio directo,* and the *dominio util,* one or both, or give the *dominio util* alone. That rested in his sovereign pleasure. If he required feudal service, vassalage or homage, as a condition of granting the property belonging to him, he retained the *dominio directo* as well as the *dominio alto* or *inmanente,* and conveyed to the vassal only the beneficial interest, or *dominio util.* But, if he did not choose to reserve any feudal or vassal service, he conveyed to the subject both the *dominio directo* and the *dominio util;* and then the subject possessed the *pleno dominio,* the highest right of property known to the law in any organized society. That I understand to be the law on the subject.

On the definition of these civil law terms, I will read pages 13, 14, and 15 of my printed brief:

DEFINITION OF THE CIVIL LAW TERMS "DOMINIUM EMINENS," "DOMINIUM PLENUM," "DOMINIUM DIRECTUM," "DOMINIUM UTILE."

I. *Dominium eminens,* or *altum,* as *Gamboa* calls it.

"All separate interests of individuals in property are held of the Government under this tacit agreement or implied reservation. Notwithstanding grants to individuals, the *eminent domain—the highest and most exact idea of property*—remains in the

Government, or in the aggregate body of the people in their sovereign capacity; and they have the right to resume the possession of the property in the manner directed by the Constitution and the laws of State." Chancellor Walworth, in Beekman *v.* Saratoga and Schenectady R. R. Co., 3d Page Ch. R. 72.

Bowyer calls it "a part of the sovereign anthority, and one of the *jura majestatis.*" Bowyer's Universal Pub. Law. pp. 227, 372.

Vattel, "the right which belongs to society or the sovereign of disposing in case of necessity and for the public safety of all the wealth contained in the State."—Book 1, Ch. 20, Sec. 244.

See also Cooper's Justinian, p. 456.

II. *Dominium plenum*, termed *allodium* in the Common Law writers, is subordinate only to the *dominium eminens;* it is the highest right of property in land that an individual can have. Its two parts are the "*dominium directum*" and the "*dominium utile.*" These two parts are united in the allodial proprietor.

"Allodial lands are enjoyed by the owner, independent of any superior or without any feudal homage. * * * No lands are allodial, unless the grantor who makes them over expressly exempt the grantee from the obligation of homage and fealty, and renounce the rights of superiority which were formerly competent to him against the vassal in the subject of the grant. By the usage of Scotland, no lands are allodial except, first, those of the King's own property; second, *the superiorities, which the sovereign, as the fountain of feudal rights, reserves to himself in the property-lands of his subjects;* and, third, churches, churchyards, etc."

Erskine's Institute, B. II, T. 3, No. 8.

III. But in the feud or fee, these two parts are separate. The *dominium directum* is retained by the lord, and only the *dominium utile* is conveyed to the vassal, and the vassal's estate takes the name of feud or fee.

"*Dominio directo.*"—The right of superiority in any real estate, without the right of the beneficial ownership (propriedad util). Such is the ownership of the proprietor of land alienated in fee (*feudo*).—Escriche.

"*Dominio util.*"—The right of receiving all the fruits of a thing under service, or tribute paid to him who reserves the "dominio directo;" such is that of the vassal in the land which he holds in fee.—Escriche.

"*Dominio pleno.*"—The right which one has in a thing, to alienate it *without dependence on another person*, to enjoy *all* its fruits, and to exclude others from its use.—Escriche.

See also Escriche—*Verbo* "*Feudo,*"—a reciprocal convention between the lord and the vassal, by which the former transfers to the latter the "*dominio util*" of anything, and the latter acknowledges the former as the direct owner (*como dueño directo*) and promises him fidelity, military or other personal service, and sometimes the payment of some tribute (*derecho*)."

Erskine, in the Institute of the Law of Scotland, says:

"The grantor of the feudal right is called the *superior*, because he stands in a higher rank than the grantee, who is styled the *vassal*. * * * But our writers, particularly Craig, have, after the example of old feudists, expressed the interest retained by the superior by "*dominium directum*," or, as we translate it, the superiority; and that conferred on the vassal by "*dominium utile*," or the property. The word *fee* is often used promiscuously for both.—Book II, Title 3, No. 10. See also Title 6, No. 1.

Dumoulin, as quoted by Pothier, defines the "Fief" as being "*benevola, libera et perpetua concessio rei immobilis vel æquipollentis, cum translatione utilis dominii, proprietate retenta sub fidelitate et exhibitione servitiorum*," and as translated by Pothier, "a gratuitous grant made in perpetuity to another of a thing immovable, or reputed immovable, on condition of the grantee's rendering faith and homage and military service, and with reservation of the direct lordship (*seigneurie directe.*)"—Pothier; Coutumes d'Orléans, sec. 1.

The *dominium directum* is purely a title of honor, thus described by Pothier: "The direct lordship (*la seigneurie directe*) which the lord retains in the land held by the vassal in fee, is purely one of honor, consisting in the right of calling himself the lord, and of having himself recognized as such by the vassal who owns the property."—Pothier, sec. 2, No. 8.

Chancellor Walworth, in the case above referred to, defines this *dominium eminens* in our country, as the right reserved by the people of the United States, or the people of the different States, to take private property in accordance with their constitutions and laws, whenever the public interest requires it. He says: "All separate interests of individuals in property are held by the Government," and "notwithstanding the grant to individuals, the *eminent domain*, the highest and most exact idea of property, remains in the Government, or in the aggregate body of the people in their sovereign capacity; and they have a right to resume the possession of the property in the manner directed by the constitution and laws of the State, whenever the public interest requires it. This right of resumption may be exercised, not only where the safety, but also where the interest, or even the expediency of the State is concerned; as where the land of the individual is wanted for a road, canal, or other public improvement."

That power of resumption in every Goverument like ours —the right of resumption by the State of the property of the individual—is checked by constitutions. The right of an American citizen to the land that he owns, cannot be divested under American constitutions by any power in the State, without adequate indemnity previously paid. But the constitutions might direct otherwise. The people of a State might declare in their constitution, that every man held his land as property, subject to confiscation without payment whenever the interests of the State required the interests of the individual to be sacrificed. Such, fortunately for us, has never been the principle adopted in the constitutions of a free people. In despotisms, these things have been known; and yet history is full of interesting examples of the most mighty despots ever known shrinking from this invasion of the rights of private property, by authority of their title to the eminent domain. Don't we all know the story of Frederick the Great and the miller? how the King coveted the miller's possession and could not get it? how he respected his right? how the miller told the King, "You can't have my property, because there are Judges in your kingdom." So, fellow despots have shrunk from insisting on this right of eminent domain, without making indemnification for despoiling the subject of his property.

But, after that, comes the *dominium plenum.* I do not know as I have ever seen anywhere a better definition of it than in Erskine's Institute. I will not detain your Honors by reading the whole passage, but here and there a sentence, leaving you to inspect the authority on which we rely.

[Mr. Benjamin here read extracts from Erskine's Institute of the Law of Scotland, Book 2, Title 3, secs. 8, 9 and 10; and then continued.]

According to this author, both the *dominium directum* and *util* are in the civil law termed the "*fee*"; "the word fee is often used promiscuously for both."

And Pothier, who, I think, may with propriety be styled the greatest civilian of modern times—I know no man who can dispute the title with him—gives us in the Coutumes d'Orléans, Sec. 1, his definition of "Fief." He says that it is "a gratuitous grant made in perpetuity to another of a thing immovable or reputed immovable, on condition of the grantee's rendering faith and homage and military service, and with reservation of the direct lordship (*seigneurie directe*)." I think I have translated it word for word.

Well, now, what is the *dominium directum* that the lord reserves when he transfers the *dominium utile* to the vassal? Brother Randolph, in his printed brief, has translated this *dominium utile*, or *dominio util*, by the word *usufruct*, and *dominio directo* by the word *property*. But that is not a correct rendering, for the words of the translation do not convey the same meaning as the words translated.

Pothier says, (Sec. 2, No. 8), in inquiring into the nature of this *dominium directum*, that the lord retains when he transfers to the vassal the *dominium utile*: "The direct lordship which the lord retains in the land held by the vassal in fee, is purely one of honor, consisting in the right of calling himself the lord, and of having himself recognized as such by the vassal who owns the property." He has no right of *property* in it—only the right to be called, my lord. That may be valuable in monarchical governments—in Old Spain, England, or France—but when Mexico declared her independence of her sovereign and established herself into a Democratic Republic, all these lordships disappeared, and the ownership direct and ownership beneficial became centered in the same individual, who held the *plenum dominium*, or the *allodial* title of the common law,—a title higher than the old fee, because the discoverer of a quicksilver mine owed allegiance to none; higher than titles to mines of gold and silver—for it was free from taxation; higher than any other kind of property—for the State could not force you to sell it for a price; higher than anything else in the Republic of Mexico—for the State gave you a premium for owning and working it, and paid you for taking possession of it as owner. Such was your *property* under Mexican laws, in mines of quicksilver. And yet it is that peculiarly favored species of property, which, we are told, is held at the pleasure of the sovereign, and confiscated for any and every failure to comply with the slightest prescription of the ordinances!

Now, may it please your Honors, the description which I have given of the miner's title to his mine, I believe to be the true exposé of the mining laws of Mexico. If I have overpainted it, I have done so unconsciously. It is the first duty of every advocate, in the Court in which he practices, to aid the Judges in understanding what he believes to be the law. I have done so candidly. If I am mistaken in my views, there sits the man (pointing to Mr. Randolph) who will put me right. Anything said which is open to cavil or correction will be reviewed by him; let him put me straight, and let your Honors decide between us, as to who has most correctly stated the law, after you shall have examined the authorities.

Now, we are told—taking some of the minor points in this case—that our titles are worthless, because the Alcalde had no power to receive the registry of this mine. I, in deference to my brother Randolph, treat some of these points, although I do not consider them of the slightest consequence in this case.

I do not consider it of the slightest importance whether anything was done, or what was done in California. Holding, as I do, that the President of the Republic, the then Constitutional President of the Republic, had full and plenary power to dispose of mines and lands in California, under a general administrative authority, I think it is a matter of little consequence here how regular or irregular his (the Alcalde's) proceedings may have been. But I hold them to have been regular, legal and valid.

Now, what were the powers of an Alcalde in California? Your Honors have had that question before you often enough. I believe it has been universally admitted, that they had jurisdiction in the first instance of almost all civil cases. I understand that your Honors hold, that they had the same jurisdiction as Judges of First Instance; I know the Supreme Court of

California has held that they had such original jurisdiction. The books are full of decided cases on the subject.

But, in relation to the denouncement and registry of mines, I mean to inquire not what the written law was, but rather what were the usages and customs of Mexico, because these latter are to govern the Court in this case. They are to govern the Court by virtue of the express text of the law. They are to govern the Court by virtue of the decision of his Honor Judge McAllister, in the case of Adams v. De Cook *et al.*, confirmed at the last session of the Supreme Court of the United States, and reported in 23 Howard—which number may be here to-night in the Uncle Sam. We are looking for it on every steamer.

What were the laws and the usages of Mexico? what its customs in relation to the denouncement of mines? What have we in the record? We have the depositions of Danglada, Brodie, De Leon, Villagas and others.

Then we have four denouncements, produced here by the Government itself, all made before an Alcalde. They are put in here as evidence by the Government. We have four denouncements of mines, made before the nearest Alcalde two years afterward, offered in evidence by us. We have the certificate of the Chief of the Mining Tribunal of Oajaca, to be found in the record on page 3046:

The citizen Licentiate Francisco S. de Enciso, matriculated Miner and President of the Territorial Mining Deputation of the State of Oajaca ——

MR. RANDOLPH—If the Court please, I object to this. There is no evidence that this man ever said any such thing. It is no evidence, if he did, that it is true, because it is an unsworn *ex parte* statement of a thing. It is utterly irrelevant to the issue.

MR. BENJAMIN—I suppose that if Enciso had written a book, then we could have quoted it as authority. Of course, if Enciso had put that in, it would be authority. The gentleman quotes Alaman's History as authority. But we have better authority than this. The Government has furnished us with it. The Government has furnished us out of its own archives, by the communications of its own officers to it, at a date prior to the conquest, the proof that such was the custom of Mexico.

Pages 2670–71 of the Transcript—Larkin's Letter to Mr. Buchanan, then Secretary of State. On the 4th of May, 1846, before the war broke out, Mr. Larkin communicated to our Government the fact that such was the custom of Mexico:

> By the laws and customs of Mexico respecting mining, every person or company, foreign or native, can present themselves to the nearest authorities and denounce any unworked mine; the authorities will then, after the proper formalities, put the denouncer in possession of a certain part of it, or all, which is, I believe, according to its extent; the possessor must thereafter occupy and work his mine, etc.

Then, the custom of Mexico, as known to our Consul, and certified to our Government before the declaration of war, authorized any man to denounce a mine before the nearest authority; and that authority put him in possession as owner. Larkin reports that, in another letter, at page 2674 of the Transcript.

But, may it please your Honors, what better or higher proof do we want than the action of the Government, so declaring in California, through the Governor of California, during the war?

The Governor of California, appointed by the President of the United States, in his capacity as chief of the army of occupation, exercising full powers over the conquered territory, acted legally and constitutionally in his official position, according to decisions of the Supreme Court of the United States. Gov. Mason issued his proclamation recognizing that custom, and on the 12th of February, 1848, declared that thenceforward he abrogated it. See Ex. Doc. No. 17, page 476:

> From and after this date, the Mexican laws and customs now prevailing in California, relative to the "denouncement" of mines, are hereby abolished.

Now, if anybody can find me any other denouncement of a mine in California than a denouncement before an Alcalde, I will enter into the question as to what was the jurisdiction of Alcaldes in regard to mines.

But your Honors have judicial cognizance of this matter; and I am perhaps not hazarding too much in saying that no person in California has any knowledge of a mine in California denounced before any other authority. There are none others in this case; and there are a large number presented here.

Governor Mason continues:

The legality of the denouncements which have taken place, and the possession obtained under them since the occupation of the country by the United States forces, are questions which will be disposed of by the American Government after a definitive treaty of peace.

And those that were before the occupation—what is to be done with them?

MR. JUSTICE HOFFMAN—How many were there before that?

MR. BENJAMIN—The Government has introduced four in the immediate neighborhood of this mine.

As soon as Castillero discovered his mine, the whole country round about there was covered by people engaged in denouncing mines. The Government has brought in four of those. Those which we brought in were made after the war, in 1848, but before Governor Mason's proclamation, which was dated at Monterey on the 12th day of February, A. D. 1848.

At page 492 (Ex. Doc. No. 17), in a contest between parties for the possession of a mine, he (Mason) states to the Alcalde of San José, that a person who actually and fully denounced a mine before the 12th of February, is entitled to it; but the donouncement must have been full, complete and in good faith before that date. He orders the Alcalde to give possession to a man who made a fair denouncement of a mine before the 12th of February, 1848. This letter is on the 9th of March, 1848.

So we find the Government of the United States, by its military authority in possession of the conquered country, exercising legal powers under the laws of the United States—so determined by the Supreme Court of the United States—not only recognizing the California laws and customs relative to mines and their denouncement prior to the occupation, but informing the inhabitants that up to the date of the proclamation by the Governor (February 12th, 1848), every denouncement made in good faith should entitle the claimant to his mine. From that date forward he abrogated the preëxisting law—a law which continued to exist as a matter of course, under the law of nations, until abrogated by the conqueror.

And now, what question remains in regard to all this matter about the powers of an Alcalde?

Are you going to search through the mining laws of 1783, in order to ascertain whether they specially invest the Alcalde with power to receive and register discoveries and denouncements of mines? Was not that attempted in the case of the *Will of Grimes* tried before your Honors? (Adams *v.* De Cook, 1 McAllister, pp. 253 *et seq.*)

Did not I try in the Supreme Court of the United States, to set aside the will on the ground that it was contrary to the old laws of Spain? Did not my brother Johnson make himself disagreeable enough to beat me on that point? He is always doing it. I have complained to him again and again, and repeatedly about it.

Mr. Johnson—I shan't beat you in this case.

Mr. Benjamin—I proved by the written law of Spain that the will was bad. He (Johnson) proved that by the usages and customs of California it was good. The Judges of the Supreme Court said, with his Honor the Circuit Court Judge (McAllister), that those usages and customs were good; and I lost a contingent fee and my case, and everything else. What my brother Johnson got, I don't know. I don't think he got enough for gaining that case, for it was a very hard one to gain.

Still, there is a law, as decided by the highest judicial authority. Now what have we got to do with the mining laws of Spain of 1783, when the record is full of California usage, and that usage recognized by the officers of the California Government, down to the 12th of February, 1848? So much for the question as to the power of an Alcalde to receive and register a discovory or denouncement. It is a question not so much depending on written law. If it depended on that, I think I could satisfy your Honors; but it is unnecessary, for on the authority of Adams *v.* De Cook, this question must be taken as decided (1 McAllister's Reports, page 260).

But, may it please your Honors, why, after all, should we be troubling ourselves in relation to the local proceedings in California? California was not a *State.* California had no independent sovereignty. California depended on the Central Government of Mexico. The President of Mexico was vested by the laws of Mexico and the Constitution of Mexico with the right of providing as he pleased for the colonization of the territories of the Republic.

One of the ordinary exercises of the administration of power by the ordinary Constitutional Presidents of Mexico, was to make grants in colonization in territories. They permitted, and even directed, the Governors of territories to make small grants; but they specially ordered them to make no large grants, no *empresario* grants, even with the advice of the Legislative Assembly, without sending them on to the President for his approval. He held the supreme power on that subject under the laws and the Constitution of Mexico. He made partial delegation of his power to the local authorities—the power which delegates can of course withdraw the delegation. The source of power can divert the stream. The President has the right, either himself in person, or through other agencies, to exercise the powers conferred on him by the laws, and especially by the Colonization Law of 1824, which, under the Regulations of 1828, he determined should be exercised to a certain extent by the Governors of California and the local Legislative Assembly; and in support of this proposition, we have cited to your Honors a long series of decrees by Mexican Presidents, Constitutional Presidents, and Extra-Constitutional Presidents, Revolutionary Presidents, Military Presidents, and Civil Presidents, Judges of the Supreme Court, in the exercise of executive power, and Chiefs who have for the moment taken the Executive Chair—every possible variation of executive power that the many-phased history of Mexico presents. Every one of these Executives has exercised, unquestioned, the power to grant lands in the territories, and the power to give mining concessions, even in the States. We read to your Honors the other day mining concessions made by the President of Mexico to the States of the Confederacy. Has anybody ever denied in Mexico—was any instance to be found—or can one be cited—where the acts of the Constitutional President of Mexico under such circumstances have been questioned or disputed?

But if you leave aside that constitutional power, and inquire into the power exercised by those men, who, sword in hand, fell on the authorities for the time being, wrested from them the keys of the Republic, and assumed the Executive chair by strength of their own will and arms,—where do you find a limitation on them? Is no grant in Mexico good unless made by the Legislature? Then there can be no good grant in Mexico, and there has been none there for years and years gone by: for no Congress has sat in Mexico since Comonfort's time. Prior to that, year after year President succeeded President and Government succeeded Government, almost like the shifting of kaleidoscope. Has nobody ever heard of grants being made by the Executive for the time being?

The Executive of Mexico, with the powers of a dictator, brother Randolph says, had no power to make a grant of two leagues of land in California! A man of extraordinary powers, who acquired the Presidential Chair by revolution, and held it by the sword, with the army at his back, had no power to confirm mining possessions made by an Alcalde in a remote province of the Republic! Well, now, there is something that suggests itself instantly to one's mind on this matter. What title does the Government of the United States hold to the lands and mines of Arizona? Does it lie in the mouth of the Government of the United States to say that the President *pro tem.* of Mexico had no power to grant away the mines and the soil? I never heard that we had any title to Arizona, except the one that came from General Santa Anna. Yet, I have never heard a question raised in this Republic as to the validity of our title to the land comprised in that territory. But yet that was no grant of a little mine; no grant of two square leagues on the summit of the top of a mountain. It never has been ratified by any Mexican Congress. It never has been passed upon by any other power than the will of the Dictator who consummated the act by which his country was dismembered. And yet, this very Government of the United States, that accepts the transfer of entire districts of territory, with all their ungranted mines—and they are included—from a Mexican Dictator, stands up here, and by its special counsel denies his (the Dictator's) power to give two leagues of land in one of his territories; denies his power to extend special favors under the mining ordinances of his Government!

And, observe you, these mining ordinances provide expressly that the Executive *shall* give special favors in cases of extraordinary merit—even when the law is in force; those extraordinary favors, being grants given in such cases, were given under the law, not in violation of it. The law says that the Alcalde may put a person in possession of a certain quantity of land. The Alcalde could not go beyond the law. The very same law said, that under those provisions the Alcalde could not go beyond a certain number of *pertenencias;* but it also said, that the Chief Executive might give *any number of pertenencias* to a person who undertook any considerable enterprise involving a large outlay. And the law further provided, that the Executive *should* give special rewards to such miners.

Yet we are told that the King of Spain had a right to do in the provinces of New Spain what the Chief Executive of the Nation had not a right to do after its independence had been acquired. Now, something more than mere assertions is required to establish this. I have yet to hear any good authority on that subject. I heard my brother Randolph say that this Dictator (Paredes) had no such power. The only argument I have heard on the subject, is this—I trust that I shall not misrepresent it. It is, that every revolutionary President in Mexico, in order to get into power, promises and proclaims to the Nation all that he is and is not going to do. This, he calls his plan of government; and after he gets into power, this *plan* serves as a Constitution which he cannot violate!

According to that, Napoleon had no power in France, and all his acts were void, because he had positively sworn before the nation to preserve the Republic. Now, at the moment he became Emperor, in violation of that oath, all his decrees were null and void, because he had published a plan by which he proclaimed that he was going to be President of a Republic, and nothing else. Is a usurper and dictator limited by previous promises in regard to the extent of his power? We are not now speaking of right. We are speaking of the exercise of power. And to talk to us of a usurper, or a military chieftain, not being able to do anything beyond what he promised to do when in search of power, is really something which I do not understand—when brought here in the shape of an argument in this case. Yet I am willing to give my brother Randolph the full benefit of that principle, whatever it may be. And I tell him, that by the plan of Paredes, it was specially provided that none of his acts should ever be called in question. That was *his* plan. His Ministers might be held responsible, but the acts of his Government "*were at no time to be revisable.*"

MR. JUSTICE MCALLISTER—Not by any *human* tribunal.

MR. BENJAMIN—Perhaps your Honor is right. Perhaps he did admit that they would be revisable by some tribunal *not human.* But unless your Honors are not a *human* tribunal, these powers exercised by Paredes are not to be revised by you!

Now, so far as these powers of Paredes are concerned, let me call your Honors' attention to the Mexican laws and decrees. I do not mean to read them; but you will find them in this volume of "Leyes y Decretos, 1844–1847," from pages 316 to 394.

The following are some of the laws established by Paredes: He established a general electoral law, by virtue of his own authority; he assumed the authority to govern the Department; violated the constitutional law, and put in his own Governors; extended the elective franchise; organized Courts; appointed Ministers; levied taxes; made laws, etc., etc.

Here are eight or ten decrees evincing the exercise of a most sweeping and absolute power over the interests of the Republic, unquestioned to the present hour. They may have been changed and modified since, but all that was done under them has never been questioned to the present hour. The only thing that he (Paredes) could *not* do, it seems, was to give Castillero two leagues of land for his mining possession! *That* he could not do! *That* was such a stretch of power that no human being can imagine that he could do it!

Now, if your Honors please, I have closed what I intended to say in this case, with the exception of calling the attention of the Court to the decision, in another claim of Castillero, by the United States Supreme Court, on these powers; being the decision already read to your Honors by brother Peachy.

In this case the Court, after holding that the disposition made by the Central Government in Mexico, authorizing the Governor and Departmental Assembly of California to dispose of islands on the coast, could only be exercised by them conjointly, came to consider the claim of Castillero to the island of Santa Cruz, under a grant from the Governor alone, which grant was based on another special communication from the Government of Mexico to the Governor and Departmental Assembly of California, simply recommending to them that Castillero be allowed to select for himself one of the islands on the coast before they disposed of the rest in accordance with the previous instructions.

The grant in this case, says the Supreme Court, was not made under the general authority conferred by the first dispatch exhibited in Osio's case, but under another dispatch bearing date at the City of Mexico, on the 20th of July, 1838, signed by the Minister of the Interior, and addressed to the Governor of California. The Court then goes on to state the contents of that dispatch, as follows:

By the terms of the communication, the Governor is informed that the President, regarding the services rendered by the claimant to the nation, and to the department, as worthy of great consideration and full recompense, has directed the Minister to recommend strongly to the Governor and to the Departmental Assembly, that one of the islands, such as the claimant might select, near where he ought to reside with the troops under his command, be assigned to him, before they proceed to grant and distribute such lands, under the general authority conferred by the previous dispatch. Beyond question, the legal effect of that second communication was to withdraw such one of the islands as should thus be selected by the claimant, from the operation of the previous order, and to direct that it be assigned to this claimant.

To accomplish that purpose (the purpose of this special dispatch) and carry into effect the command of the President, two things only were necessary to be done; one was to be performed by the claimant, and the other was a mere ministerial act. It was the claimant who was to make the selection; and if it was a proper one, near the place where he was stationed with his troops, nothing remained to be done but to make the assignment, as described in the dispatch. Emanating, as the dispatch did, from the supreme power of the nation, it operated of itself to adjudicate the title to the claimant, leaving no discretion to be exercised by the authorities of the Department.

Here we have a despatch from the Minister of the Interior to the Governor of California, saying to him that the President has directed him, the Minister, to recommend to the Governor and Assembly, that Castillero, on account of services rendered, have assigned to him the one of the islands on the coast which he might select; and the Supreme Court says that this dispatch constituted a perfect title, requiring only that the claimant should make his selection, and that the Governor should perform the ministerial act of making the assignment, which act could not be neglected or refused by the Governor without disobeying the orders of his superiors in Mexico.

Mr. Justice Hoffman—Who rendered the decision?

Mr. Benjamin—Mr. Justice Clifford, who was Minister of the United States to Mexico, and is familiar with the history of that country, and the laws, usages and customs of its people. He says: "Emanating, as the dispatch did, from the supreme power of the nation, *it operated of itself to adjudicate title to the claimant*, leaving no discretion to be exercised by the authorities of the Department," that is, of California.

Now let us look at this New Almaden grant of two leagues. Here also is a special dispatch, dated at the City of Mexico, signed by the Minister of the Interior, and addressed to the Governor of California. *It* also communicates the will of the President with respect to a grant of land to Castillero. It does not merely recommend to the Governor and Assembly to assign to him such two leagues of land as he may select, for he had already made his selection. In his petition he had described the two leagues which he wanted—he wanted them on his mining possession, which the President confirmed to him, of 3000 varas in every direction from the mouth of his mine. He did not, I say, direct the Governor to perform the ministerial act of making the grant, or of issuing the title papers, for the grant was already made by the President, and the dispatch itself would serve as the documentary evidence of that fact. He merely required the Governor to put Castillero into the possession of what he, the President, had already granted to him. That was all the Governor had to do. He had no authority to do anything else, and he had no discretion to refuse to do that.

If the dispatch of Minister Pesado left "no discretion to be exercised by the authorities of the department" of California, what discretion was left by the dispatch of the Minister Castillo Lanzas to be exercised by the Governor, when it directed him to put Castillero in possession of the two leagues granted to him on his mining possession? Was he, the Governor, ordered to exercise any judgment or any discretion in this matter? Was he allowed to say, "I had rather give him possession in some other place; I prefer that he select some other land, and then I will put him in possession?" No! "Put him in possession," says the dispatch, "of the land which the President has granted him, and which is here described. If any prior proprietor has rights to a part of the land here decribed, his rights will not be affected by this grant and possession. Give him the possession as here directed, and any conflict of rights and boundaries between him and the neighboring land-owners, will be settled at the proper time and by the proper authorities. I give you no discretion in this matter; you have only to obey my orders."

The Court continues:

Neither the Governor nor the Assembly, nor both combined, could withhold the grant after a proper selection, without disobeying the express command of the Supreme Government. Nothing therefore remained to be done after the selection by the claimant, but to issue the title papers, and that was the proper duty of the Governor as the executive organ of the department.

Now here in this New Almaden case, we had not even a selection to be made. Our selection was already made—made in Mexico. There was nothing here absolutely to be done, but to take juridical possession. That juridical possession was interrupted by the war. The title issued before the war; but the juridical possession was interfered with by the war.

Now, does the interruption of the juridical possession by the war, forfeit the grant? If it does not, what is there wanting to our title to these two leagues of land, issued by the Supreme Government; the land is designated in such a manner as to be easily segregated from the national domain, the Ordinanzas de Tierras y Aguas furnishing the rules for laying it out; there was nothing to be done but to run lines and put stakes on the boundaries described.

Am I at this day and hour to argue to your Honors, that a man has a good title to his land, even if he has not been put into formal possession, provided the granting power has ordered possession? More especially, is his a good title to land, if there has been a lawful impediment to his taking possession.

This man (Castillero) was an officer—an officer of the army—an "officer of the permanent cavalry," as he is sneeringly styled by the special counsel for the Government. He was bound after the war broke out to defend his Government.

Could his Government, if California had remained a province of Mexico, have urged against him as a reason for forfeiting his mine, that he had served his country in the camp instead of in person taking possession of these two sitios? But, if that grant was binding on the conscience of Mexico, tenfold is it binding on the good faith and honor of the American people. If the obligation Mexico was called upon to respect, implied a promise, we have pledged its fulfillment here in the face of the civilized world. We have published it to Mexico; we have published it to the world. We have bound ourselves, hand and foot, that every right of property shall be respected; and the Court is ordered here, by the law, to declare any right of property valid when held in accordance with Mexican laws, usages and customs.

What then remains for me to say in this case?

Is there any man in California, not having an interest against these claimants, that has any doubt that every paper brought here is genuine? Not one.

Is there a man in California, or is there a man to be found anywhere, prepared to deny in the face of the evidence in this case, the possession and ownership of this property at the date of the acquisition of this territory?

And if this possession and ownership was published, claimed in the face of the authorities, and by them admitted, and the party felicitated upon the fact of his possession and ownership, and the nation congratulated that this man *did* own this property, because the public interest would be subserved by this private proprietorship acquired by virtue of his discovery—if all this existed before the war was declared, as it *did* so exist, with what face does the Government of the United States now pretend to take this property away from us, and confiscate it for national purposes?

Take the pretext offered!

I do not now speak of what is believed, but of the ostensible pretext—the United States taking the property for their own use. I speak of the public professed object of these proceedings. With what face, in this state of facts, will the Government of this mighty empire violate its word, forfeit its honor, and commit a most atrocious spoliation of the rights of foreigners, after telling the world that it was precisely as foreigners that their rights were to be respected; to be just as faithfully and inviolably preserved as those of any citizen of the United States.

Your Honors are told by the decisions of the Supreme Court of the United States, that it is not legal and perfect titles alone that you are to confirm; for, as regards such, no treaty was necessary. The most shamelessly rapacious nation on the face of the earth would shrink with horror from any assertion of right to *private* property by virtue of treaty or conquest. The law of nations recognizes in modern days no such right in the sovereign. You are told by the decision of the Supreme Court, and by the express language of the law, that "equity is to be your guide and the pole star of your decisions."

Equity! equity! to attack these people in possession of a property upon which nearly a million had been expended, and turn them out of that possession and the enjoyment of its fruits, on the ground that they were "wasting" it. I appeal to the common sense and judgment of every man within the hearing of my voice, or any man who ever saw a mine anywhere—is not that mine as it now stands, with its improvements and its works, even deducting what has been taken from it, is it not worth more now, to-day, than it would be worth if all the ore were put back whence it was taken, and the mine restored to its original state, as a desert piece of land in the mountains and in the wilderness? We have "wasted" this mine! and there is not a man in California who would not give a million of dollars more for it now than when we took possession of it. We have "wasted" the Government property! We have spent a fortune on it and made it invaluable—more valuable than it ever could have been if we had done nothing. All this is "waste," technical "waste," common law "waste." But is it waste in the judgment of the American Congress and of the American people, according to the manner in which they administer mineral lands in California?

What gave Mansfield his great judicial reputation? He knew that where reason ceased, the law ceased too; and that gave a fitness to innovations made by him in the affairs of commerce. When these old precedents and legal technicalities were brought before him, he said: "We have changed all that. Commerce cannot bear that kind of restriction any more. That was the old common law. I established new principles, because the reason of that old common law has ceased; and with it ceased the law."

Cannot we apply a little of Mansfield's sense here in California? Cannot we forget these old common law restrictions, and the rules which governed *waste* between citizen and citizen, and cannot we see that the general policy of the United States, as indicated by the administration of the Government and the laws, is not in accordance with that monopolizing and proscriptive spirit under which it is claimed that the mines in all public lands belong to the sovereign; that no person is permitted to work them; that whoever does so is a trespasser, and may be driven away at the pleasure and by the fiat of an Attorney General?

When California was first discovered to contain gold, and a new system of laws and principles was to be applied here, public officers, high in power, members of Gen. Taylor's Cabinet, advised the Congress of the United States to place restrictions upon the mining operations in California. To make the minerals there a source of revenue to the Government: they advised leasing them out—not selling them. Thomas Ewing recommended this. But the Congress of the United States was wiser than Thomas Ewing. It would do nothing of the kind. It declared that the mineral lands of California should not even be surveyed, that they should not be subject to settlement and pre-emption.

The Government of the United States has left those lands as the common heritage of the people of the United States; subject to the peculiar regulations of the miners in their respective localities—each community of inhabitants or settlers in a particular district, taking such a share of the lands as by their common agreement is deemed equitable and expedient.

What has been the result? Look at your State! Look at this Pacific coast! An empire grown up under the liberal administration of the Government. All this would have been lost, had the old law counsels of Thomas Ewing been followed What would have become of your mines, had his advice been carried out? Where would have been your fifty, sixty or seventy millions per annum, exported from San Francisco, nourishing the streams of commerce in all parts of the world, and promoting the general prosperity of the Republic? What would have become of all that under a system, such as Great Britain pursues, of licensing and staking out the rights of the people in different localities where they choose to mine?

I have been looking at the mines in your State. I have seen in your mountains long veins of gold-bearing quartz projecting from barren ridges, and out-cropping for miles and miles together. Thousands of hardy miners are digging and delving in these gulches and mountains for the gold and silver hidden beneath the earth's surface, and millions of money are invested in fabrics *on* that surface, constructed for the extraction of these precious metals. Why are not these miners enjoined from prosecuting their mining operations, and turned out of their possessions, all over the State? *They* are on the Government lands. *They* are committing technical "waste" on the public domain. Nobody disputes that. Why are *they* not turned out? Is it because *their* possessions are protected by no solemn treaty? by no pledge of the nation's faith, proclaimed in the face of the world? Is it because Congress has passed no law recognizing *their* rights of property, and directing the Courts to confirm and protect those rights? Is this the reason why the Attorney General has not sought to arm himself with the powers of the Federal Courts to stop this technical "waste"? Is this the reason why these untold millions, daily, weekly and monthly dug out from federal lands, have hitherto escaped the prying gaze and eager clutches of that faithful law officer of the Government?

There is no treaty of the Government of the United States pledging the faith of a great nation in favor of one single quartz-miner in this State of California; but there *is* a published treaty, pledging the faith of a great nation in favor of a Mexican mine-owner, whose Government ceded this territory to the United States; ceding to us nought but what it held itself. We acquired from Mexico nothing but what Mexico owned; and she did *not* own this mine when she ceded this territory to the United States. It belonged to one of her citizens. She did not claim it—did not pretend it was hers. Whence then the right of the Government of the United States?

May it please your Honors, I do not know but that the technical rules of Equity or Chancery Jurisprudence may have made it imperative on the Court, when requested by the law officers of the Government, to drive these people from their possession, and absolutely stay their hands from wasting and despoiling what is claimed by these officers to be government property. I am bound to receive with respect the decision of your Honors on that point.

But, I put it to you, is there equity, *now* that you know the truth, in binding their hands and depriving them of their just posssssions? I put it to you *now*, whether you do not feel in your souls and in your consciences, that, although done conscientiously, you have done an injury to the owners of this property?

I know not what your Honors may think of the past, but I think I can share what you feel now. I think I can share the satisfaction, the glow of honest pride, with which you will be actuated when you are enabled to say, "*Now*, *now* we are able to undo what we have done, because our consciences are informed. We know *now* the rights of the parties; and no longer shall the equity powers of this Court be made the instruments of spoliation and oppression."

[At the conclusion of Mr. Benjamin's argument, the demonstrations of applause in the Court-room were loud and prolonged. The Court called on the Marshal to preserve order.]

As soon as silence was restored, Mr. Justice McAllister leaned over his desk and said: The "spoliation and oppression," of which this Court has been guilty, according to the suggestion of the counsel—

Mr. Benjamin (interrupting)—Your Honor mistakes me. I do not say that the *Court* has done this knowingly or wrongfully. I say that the *power* of the Court has been used for that purpose *by the officers of the Government.*

Mr. Justice McAllister—What has been done by this Court heretofore in connection with this case, has been done without any regard to the alleged frauds in this case. Whatever has been done by the Court heretofore, has been done in accordance with what the Court believed to be the eternal principles of Justice, Law, and Equity.

FOURTH DAY.

MONDAY, November 5th, 1860.

After Mr. Johnson's argument, and Mr. Randolph's reply to Mr. Benjamin and Mr. Johnson—

MR. BENJAMIN said:—I certainly did not anticipate uttering another word upon the merits of this case. I had supposed that my task would be confined to a simple congratulation of your Honors that your weary task was ended. I am sure that that was my own feeling on the subject, and I came here with no other purpose.

But my brother Randolph is a brave man. He argues this case with a tenacity and a courage utterly unparalleled. It is in vain that the chimeras which his imagination conjures up are destroyed to-day; they reappear to-morrow. And there is scarcely an assertion made this morning in relation to the facts in this case which has not already been met and refuted by the clearest and the most incontrovertible evidence. Yet, all appears again; and, finally, when all other means of attack are exhausted, when everything else fails, the Legislature of California is attacked by a citizen of California, its members are accused of having been approached and influenced by bribery; and I, who come from a distant land, am called upon to vindicate them from the charge.

MR. RANDOLPH—I did not say that they were bribed. I said that they were influenced.

MR. BENJAMIN—I know not what was the precise language used by my brother in reference to this matter. I am sure, however, that there was but one impression left on the mind of a single individual who heard it.

But your Honors received the compliment of a warning against extraneous influences. Their "unbounded wealth" does not suffice alone to corrupt Legislatures, but it is about to approach the Bench.

Mr. Randolph—I did not say that either.

Mr. Benjamin—"Public sympathy" usually follows the poor and needy, and the unfortunate. You are told, that by a strange introversion of the order of things in California, it is wealth that commands sympathy.

And thus we have been going on from day to day with the strangest assertions; thus have we had repeated as facts in this case assertions which are disproved by the evidence on the record, and against which there is no controverting record. And thus have we had repeated assertions of law directly contradicted by the views of the writers. And how are these contradictions maintained?

I happened to say the other day—I am very sorry I mentioned it—I happened to say in the heat of forensic discussion that an eminent deceased Judge had written a treatise on a subject of the Civil Law, and that from the beginning to the end of the treatise it was apparent that he did not know the meaning in the civil law of the word about which he was writing a treatise. My brother Randolph asked me the name of the Judge and the name of the treatise. I told him that the name of the Judge was Story; that the name of the treatise was "A Treatise on Mandate." Brother Randolph comes here this morning to correct me by citing from Story's Treatise on Bailment. I think that if my brother Randolph had looked, he would have found other places where Judge Story treats of mandate without reference to its civil law signification.

Mr. Randolph—Give me the reference.

Mr. Benjamin—Really, I did not quote it as authority—merely as an incident in the course of my argument. Certainly it was not worth being answered this morning; surely, it is not worth spending the breath I have in defending and verifying it.

I come now to what touches the merits of the case. A strange spectacle is here presented. My brother Randolph urgently calls upon your Honors to say, what? To say that this mine is situated on private land. If so, what is he doing here for the Government? Either the land is private or public. If it is private, why all this parade of Government interference, and large sums paid for special counsel to defend the interests of the United States.

Mr. Randolph—(*Soto voce.*) Not so much as there ought to be. I wish they'd increase the amount.

Mr. Benjamin—If the land is not public, then the argument is worth nothing.

But, in point of fact, as your Honors well know, under the statute by virtue of whose provisions you are here sitting, you are forbidden from inquiring into the merits of private claims to the property in question, and are confined simply to the investigation of the fact as to whether the lands are public as against the United States, whether there is any outstanding claim of the Government to the land; and if there be none, and the private claimant's title be genuine, the Government gives him a quit-claim to the property. We will settle all private quarrels with other private claimants, that may be opposed to us, without the interference of the Government.

And that of which we must complain here is, that we are not allowed to do this, but that the Government interferes for the purpose of asserting one private claim in opposition to the claim of another individual.

But, may it please your Honors, the whole argument upon this subject of private property, and the necessity of an adequate indemnity previously made, is based upon an entire mistake of the mining law. It has no affiliation whatever with the case before the Court. From the beginning to the end of my brother Randolph's argument upon this subject, he assumes that there has been here a confiscation of private property for public uses, which is to be preceded by an adequate indemnity, ascertained and paid. That is a misapprehension of the whole theory of the law. The theory of the law is, that in the grant of agricultural lands, whatever may be necessary to be excluded therefrom by virtue of the mining ordinance, does not pass with the grant. The theory of the mining ordinance is, that if there be mines discovered upon lands granted by the sovereign, the discoverer, by virtue of the registry, will not only take the mine but will take the surface soil; and that by virtue of the primitive grant to the agriculturist. The mining ordinances provide the indemnity for his damages, in case any portion of his land is taken, and in case he immediately claims indemnity—no more, no less.

My brother Randolph, however, has a happy faculty of ignoring whatever is recent in legislation which is opposed to the principles he is desirous your Honors should assert, jumps back to the ordinances of 1584, passes over in convenient silence the express language of the ordinances of 1783, which declare that the surface owner is to be paid for his land within the limits of the pertenencias. After he has been so paid, does the same remain his? Brother Randolph says, yes. Well, that would be rather anomalous legislation, and until he has proved it to the satisfaction of your Honors, I shall stand by my interpretation of the law as the correct interpretation. Until then my brother must excuse my not answering that assertion of his.

Plowden's case is next and again referred to. Well, really, I did not suppose that that case would ever again be mentioned. I did not speak of it before, because I could not think that brother Randolph saw its bearing on the controversy before the Court. Why, brother Randolph draws the most extraordinary conclusions from it, which I had been content to leave to the judgment of your Honors. He takes up the case in Plowden, and he succeeds in establishing to his entire satisfaction that a mine of copper is a "mine," but a mine of gold is a "prerogative"; that a mine of lead is a "mine," but a mine of silver is not located in the ground, but in the Queen's crown. That I draw from this argument—that inference I draw. I have read his argument thoroughly. He gives us a lecture on the impropriety of confounding in our minds the two subjects under discussion—land, and estates in land—which he says are two entirely distinct things. Estate in land is not the land, nor is the land the estate. After having carefully guarded us against this error of logic, he proceeds to demonstrate that a mine of gold is a "prerogative," and I suppose that a "prerogative" is a mine of gold—sometimes he tells us that a mine of silver is a prerogative, not in the soil, but in the crown, just as the sturgeon or seal or the whale is. So that Queen Victoria, when she presents herself before her subjects on occasions of state, has symbolically on her crown all the whales and sturgeons of the English waters, all the shipwrecks on her coast, and all the gold and silver in the land. All this is symbolical, undoubtedly; but if carried in actual painting, it appears to me it would be somewhat anomalous and strange, certainly. It may be true that a mine of silver is not of the same character as a mine of copper—that a mine of lead is land and a mine of silver not; that a mine of copper is land and a mine of gold is not; but I have yet to hear of the first lawyer who has ever so stated at the bar, before brother Randolph; and I have yet to see the first authority in support of so anomalous a proposition.

Mr. Randolph—Doesn't Plowden say so?

Mr. Benjamin—You read it so; but you read everything so singularly. I cannot understand it as you did.

Mr. Randolph—That is to say, you are a better lawyer than I am.

Mr. Benjamin—Not at all. I give the language a different construction. I have heard you state over and over again certain matters asserted by you to be facts in this case, which on careful examination I could not find in the record. It is utterly impossible for me to see anything of the kind, after the most careful reading, of what you have said.

It certainly cannot be expected that I shall go again into the facts of this case. If answer is to be made every time an assertion is made, I admit that my brother Randolph can assert faster than we can disprove. But undoubtedly the mere assertions made in this case will not influence the judgment of the Court, but your Honors will for yourselves examine the record you have before you, and give its just weight to each piece of evidence according to its intrinsic value.

There is one thing in the argument last made, to which I wish to recall the attention of the Court. It was a suggestion that the Castillo Lanzas decree now produced was not the original decree issued in the year 1846; that it was not the decree spoken of in the letter of James Alex. Forbes in May, 1847, or December, 1849; but that it is a forgery.

MR. RANDOLPH—I always argued that; I argued that in my first speech.

MR. BENJAMIN—I understand that perfectly well. My brother Randolph says that he has always argued it. I knew very well that he had always argued it; but what I did not understand was that he still insisted upon it after my reply to him; that he had the gallantry and courage still to argue it after the evidence produced in a replying argument. It is therefore in relation to that matter that I desire to call the attention of the Court.

What is the proof that this is the genuine original document of May 23d, 1846? That is the proposition.

Now, first: The exact duplicate is found in the Mexican archives. This is not disputed any longer that in the Mexican archives in the City of Mexico it was seen by Mr. Forsyth, examined, compared; and we have the certificate of the Minister of the United States, that in the City of Mexico we found this grant there in the original archives of the Mexican Government, and that he saw no reason whatever to doubt the fact of its being a genuine paper.

But to this your Honors will, of course, pay no attention. It is very easy for you to say, when you come to take up that decree, that Mr. Forsyth must have been an ass not to have seen that it was a forgery, and a perjurer not to so have declared it after having seen it. It does not stand in the way of your decree at all.

Next, it was in existenc in December, 1846. This is a positive certainty, for it is proven by Mr. Negrete, who identifies the paper as the one that he himself forwarded to Alexander Forbes, of Tepic; which was handed to him in the City of Mexico, in the month of December, 1846. It was handed to him by Castillero, by Castillero who was the Commissioner of the Mexican Government; and who had expected when the document was handed to him, to return home to California with the answer to the messages which had been sent to the City of

Mexico, and for which purpose the Don Quixote was detained first in Acapulco, and afterwards sent to Mazatlan to meet him. Now Negrete swears that this is the original document that he had in his hand, and forwarded this identical document.

But does not the counsel for the Government say that Negrete comes from Mexico? It is true, he is not a Mexican. If he was, he would be utterly unworthy of credit; for are not Mexicans about the same as Chinese and Hindoos?

MR. RANDOLPH—A little better.

MR. BENJAMIN—A little better than Chinese and Hindoos.

But this is a Castilian gentleman, a native of Spain, a Spanish merchant, for a series of years residing in the Republic of Mexico, who attended to this business for a friend when by chance he happened to be in the City of Mexico.

I suppose, however, that does not mend the matter. Your Honors will only have to declare that Mexicans are little better than Chinese and Hindoos, and Spaniards very little better than Mexicans. That being so, you will at once discard the testimony of Negrete. That being established, you will also take care to say that you discard the testimony of this gentleman as being a perjurer, for he proposed, and proceeded after he came here, to forge some letters which he had written some fifteen years before, when he addressed his friend at the date when he transacted his business.

You will also pass an eulogium on the skill of the Spaniards in forging, and refer to these letters as evidence; notwithstanding there is not the slightest proof in the record that these are forgeries.

You will not stop here, however.

As you will ascertain by the testimony of Forsyth, that this document was in existence in the archives of Mexico, in July, 1858, you will proceed further to declare that the entire archives of Mexico are fraudulent, antedated, forged, and are worthy of no credit whatever. You will say that you are perfectly satisfied that it was possible to forge this paper and put it in the archives of a Notary who died years before; and that, in the absence of any testimony as to how it came there, it is your duty to call it forged and fraudulent.

Having disposed of this part of the testimony in an eminently satisfactory manner, you will next have to proceed to take up the testimony of Francisco Villalon, one of the Notaries who certified to the certificate of Nazario Fuentes at the time the copies were originally issued in December, 1846, or February, 1847. He came up here and swore to the signature of the Notary Nazario Fuentes, and the genuineness of the notarial seal of the College of Notaries in the City of Mexico. But what is all that worth? Don't the Government say that this is forged? How then can your Honors for one instant pay the slightest attention to such evidence. You certainly cannot pay any attention to such testimony as that, but you will proceed to guide your judgment according to the assertion of the United States special counsel.

Having also disposed of these inconvenient witnesses, whose names are scattered throughout the record, and who give perfectly consistent testimony, which no unprejudiced man can read without having conviction forced upon him, you will next inquire—it being fully proven in this argument made by the special counsel for the United States, that all these witnesses are perjurers, and all the documents they produce and swear to, are forgeries—you will next proceed to go back to the month of May of the same year, and ascertain what proof you have of the existence of any such papers at that date?

The first thing that you will find will be the testimony of Velasco, the clerk who wrote the grant, and who came up here from Mexico: in whose handwriting is the original, and who swears to his own handwriting put upon the papers at the time it bears date. But he is a Mexican—but little better than a Hindostanee or Chinaman. Of course you will at once dispatch such a witness from your consideration.

You next come to the Secretary of State of Mexico, who signed the grant, and who swears that he signed it at the time it bears date. But don't your Honors see—has it not been satisfactorily proven this morning—that he was either an impostor or an idiot? All that you have got to say is, that you are satisfied that the witness is an impostor, and that the United States, by the testimony of their special counsel, given in gratuitous assertions, have proven this paper a forgery, and this man a perjurer.

Or, you will go on to say, that although this individual did personate Castillo Lanzas, did present himself as Castillo Lanzas, and you are forced by the testimony to believe him as such, he was plainly an idiot; it being the custom of all nations to choose idiots for their Foreign Ambassadors and Secretaries of State. That being a well-known fact in history, you will at once set it down in your judgment.

You will next proceed to discover that the existence of a certain book in the archives was not known until the last two years; and you will find that brother Billings went down to the City of Mexico for the express purpose of bringing up evidence to prove the truth of whatever facts may have occurred there, and for the purpose of producing witnesses who would prove the authenticity of these archives. And you will find that in his search in the City of Mexico, he discovered on the margin of one paper in the Department of State certain hieroglyphic marks which he was unable himself to decipher. That the clerk of the office was called and asked, What are these letters on the margin of this paper? Why, these letters are "L. G.;" that means, Libro General—the "General Book;" and F. O.; that means "folio;" and this number means the number of the folio; and V. A. means Vuelta, "turn the leaf." This is because Mexican leaves in the archives are numbered only on one side of the leaf. They do not number every page as we do, but the first side of a leaf is numbered as a page, and the second side is considered as the same leaf—the same page, turned over.

You will find that the Clerk, when called on, gives that explanation as to what these hieroglyphics mean. And thereupon, amongst the dusty tomes that had been laid away for years, is found an enormous ledger, and it turns out to be—what? An abstract of all the important resolutions which had passed through the Department of State for a long series of years. And there, right upon the face of the paper, at the very spot pointed out by these hieroglyphics, is found the entry of this grant by the Mexican Government.

Mr. Randolph—Who found it?

Mr. Benjmin—Billings, I think.

Mr. Randolph—How do you know he found it?

Mr. Benjamin—Yrisarri tells you that. My impression is, he says it was brother Billings. He says he was asked what the hieroglyphics meant. He says that brother Billings was in the office with others, at the time, and that he was asked by those who went there what these marks meant. He also says, that he gave the answer I have stated; that the book was searched for; and he relates what occurred. On the very page on which this man found this entry made by himself, as he swears, at the date which it bears, on that very page he found,—and the fact which instantly brings the whole circumstance to his recollection, he was married at that time, and under the Mexican laws the clerks and persons employed in the public offices forfeited the right which would otherwise inure to their wives and children a certain annuity (*monte pio*) if they contract matrimony without the permission of the President—and on that very page, on the very day the contract is recorded by this witness in his own handwriting, on the same leaf, after the entry of the grant to Castillero, is noted the President's permission to him to marry, and the name of the lady whom he was to espouse.

Mr. Randolph—What authorities prove Mr. Billings' connection with that discovery?

Mr. Benjamin—I am not certain whether the witness said brother Billings was present and made the discovery; but the discovery was made, and brother Billings was there.

Mr. Billings—I found the book myself. I might have so testified.

Mr. Benjamin—My memory is not very precise on that point. It is said that brother Billings found the book, and that the book contained this entry.

Well now, your Honors will come to the further conclusion

that we had previously sent there and had his book of entries, running through some ten or twelve years, containing the entire record of the administration of the office of Foreign Affairs of the Republic, forged, or rather had their entry forged and placed in it by the clerk who made the original entry in 1846, and then carefully covered over with dust, so that it could be conveniently found when Mr. Billings went to Mexico.

Mr. Randolph—That book only contains thirty-six hundred recorded words in a year, I wish your Honors to take notice.

Mr. Benjamin—I will admit it. The book, if your Honors please, contains thirty-six hundred words in a year. Your Honors' attention is called to that. We concede it; but if that is proof of fraud, I am unable to see the connection. I concede the fact that the entries made in this book were at the rate of thirty-six hundred words per annum; the precise bearing of that fact I am utterly unable to define.

Mr. Randolph—It is simply this: that nothing can be drawn from the fact that that is a public volume, a dusty tome, where the whole business was set down. It only requires a nice calculation: thirty-six hundred words a year were written in it, which might be written in an hour—or at least, a day.

Mr. Benjamin—it is perfectly easy to write it in a day; but it is not easy to write an index to a long series of volumes, after they have been made up.

This is a book answering precisely to the Jimeno Index, on which your Honors and the Supreme Court constantly rely for the purpose of deciding the validity of a grant, and to which the Supreme Court expressly refers in the case called to your attention by Mr. Johnson on Saturday.

But, may it please the Court, these parties were not satisfied with forging a simple schedule; Mr. Billings went into it with all the *goût* of an amateur trying his first experiment. *He* was not satisfied with bringing up to the Court the minutes of the Mining Board, containing a statement, made day by day, of the entire transactions of that tribunal, which had in its charge the whole mining interests of Mexico,—but actually went to the trouble of taking a traced copy of the book, which being there, of course they must have forged also and put there. After they took the trouble to forge the original minutes, they went on to forge and trace a copy. Even that did not satisfy their gormandizing appetite for forgery. Brother Billings immediately proceeded to have prepared a rough draft of the minutes, as they were taken down by the clerk or secretary, at the meeting of the Junta, written with a dozen different pens, with all the erasures, interlineations and corrections made as the proceedings went on from hour to hour; all that was forged, with all the blots, all the erasures, and all the interlineations of a series of months, in order that they might have the pleasure of putting before your Honors duplicate forgeries, when one would have sufficed.

Mr. Randolph—It was only necessary to forge so much as relates to this business.

Mr. Benjamin—Certainly; it was only necessary to forge so much as relates to this business. I thank my brother for the concession. Then, I pass the book to your Honors. I ask you to glance your eye over it, and see if there is a possibility of the entries relating to this mine being forged, unless the whole book was forged.

Mr. Randolph—It might be rewritten.

Mr. Benjamin—Why, yes; the whole volume of the proceedings was rewritten! The whole volume of the proceedings of the Mining Tribunal of Mexico for nine years was rewritten, for the purpose of introducing these minutes, which can by no possibility be put into the middle of the book without its being rewritten, for they are in the middle of concessions constantly being made, and other business constantly in charge of the Mining Junta. The whole of these books were rewritten! and after being so copied as rewritten, then the whole of the rough drafts—which had disappeared and were found after careful search alone, being no longer records—were also hunted up for the purpose of being rewritten and reforged from these minutes.

Mr. Randolph—I don't see how you make out that was necessary.

Mr. Benjamin—No; it is very difficult to make you see what is plainly in the case; the evidence would satisfy any other person on the face of the earth; but I doubt very much if even an emissary from Heaven, or anywhere else, would ever make you believe contrary to what you have asserted of this case.

What conclusions do I derive from all that? I will not announce it; my respect for my brother is too high.

Mr. Randolph—Don't hesitate.

Mr. Benjamin—Then, if I must say it, I believe your mind is diseased on the subject. I believe on this particular subject your mind is diseased. I can't account in any other way for the persistency with which you have refused to open your eyes to evidence which is as plain and clear as the sun shining above us to everybody else I have ever spoken with on the subject in California.

Now, may it please your Honors, let us look for a moment at this letter (written by James Alex. Forbes, as late as Dec. 20th, 1849) which refers to the genuine grant, which complains of it, and which suggests the forgeries that he desires to have made. If this grant, which is now attacked by the special counsel of the Government as a forgery, corresponds with the recommendations of the man who wanted the forgery committed, you will find ground, and just ground, for suspicion; but if the grant corresponds precisely with the instrument which he describes as genuine, and which he states in his judgment to be insufficient to convey title, and therefore recommends to have changed and substituted by a forgery—then the conclusion is irresistible to any other man but my brother Randolph, that this is an original, and that no forgery was ever perpetrated. Let us, then, examine the description of the forgery recommended, and the description of the genuine instrument believed by James Alex. Forbes not to be a good and valid title.

Your Honors will find it at page 846. The proposition of my

brother Randolph now is, that the title is forged. Why forged? Why was any title required to be forged; under what circumstances, and upon what provocation? Because, as he says, James Alex. Forbes, on the 20th December, 1849, proposed to us to commit a forgery, and we did not repel the proposition. Let us read the forgery which he proposed to us to commit, and see whether that forgery thus proposed and designed, corresponds with the instrument produced in Court. What is the difficulty? (Page 846):

I shall now state why this document [speaking of the grant] is not sufficient to secure this property from the insidious attacks of Mr. Walkinshaw and his associates, or other villains, who are awaiting an opportunity of annulling the title of the same.

(Page 847.) The grant, after making allusion to the proceedings of Castillero and the local authorities, without any explicit statement goes on to state that the two sitios of land are conceded to him "como colono" upon the mining possession of the quicksilver mine he discovered in the *Mission of Santa Clara.*

Now, at page 1807, you have the document produced by us, as the true title, and you find it precisely corresponding with the description here given by James Alex. Forbes of the true title. For, in this title, the mine is granted to him as a colonist; it is granted upon the land of his mining possession, and it is stated that the quicksilver mine is discovered in the Mission of Santa Clara. That is exactly what our deed now says, (1807): "The petition of Señor Don Andres Castillero, for the encouragement of the quicksilver mine which he has discovered at the Mission of Santa Clara."

Now, what is the objection to that? That James Alex. Forbes goes on to say: "In your description of your mine, you have stated that it was on the land of Berreyesa. In your grant of land, it is stated that the mine is at the Mission of Santa Clara. Now, there are five leagues of private property intervening between the lands of Berreyesa and the Mission of Santa Clara. Therefore, your document is not a good title. There is an error in description. You must get that corrected."

Here is the statement (Page 847): "No judicial possession was given of the two sitios of land which constitutes in reality that occupied by the Hacienda, but which the document granted to Castillero declares to be in the Mission of Santa Clara, between the lands of which and this mine there is a space of five leagues of land belonging to private individuals."

So, then, the genuine title, of which James Alex. Forbes was complaining, was a title which described the mine to be in the Mission of Santa Clara, five leagues from its true location. That is exactly the document we have produced.

MR. RANDOLPH—I do not admit that to be a genuine title.

MR. BENJAMIN—I have not stated you do. I will prove it is, however. I know you admit nothing. I know you deny everything.

MR. RANDOLPH—I do not want you to mistake my proposition.

MR. BENJAMIN—I am stating my own propositions. I give up attempting to make you believe anything.

MR. RANDOLPH—But you are misstating my proposition. The paper which you are arguing about now, is one sent up in November, 1849; and is one of those papers which I contend in my argument is different from the one seen on the 5th of May, 1847.

MR. BENJAMIN—I understand that my brother Randolph maintains all those propositions, and does not prove one of them. You have at different times given such different dates to this matter, that I have become bewildered.

MR. RANDOLPH—You don't understand my propositions.

MR. BENJAMIN—I don't understand one of them.

Now, in 1849, there was a proposition to forge a grant. It is stated that we were parties to that forgery; that we were interested in that proposition, and that we carried it out.

What was the proposition made to us? The proposition made to us was to change the grant which Jas. Alex. Forbes evidently treats as a genuine grant—which he believed to be a genuine grant. The proposition made to us, in effect, is to change it; first, because it describes the location of the land in the wrong place—as being in the Mission of Santa Clara, and upon the lands of Berreyesa. The grant which we produce describes the mine as being at the Mission of Santa Clara—and not in its true location, on the lands of Berreyesa, as the mining petition describes it.

The next proposition is: "Your grant is not good, for the reason that you have no boundaries expressed between the mine and the land of the neighboring proprietors. I want a title forged which shall state the boundaries"; naming the boundaries he requires to be so stated. The grant which we produce has no boundaries in it; does not give any description such as James Alexander Forbes proposed should be forged. In a word, the grant in every respect corresponds with that which James Alexander Forbes is describing as being in his opinion a genuine but not a sufficient grant. It fails in every attribute which James Alexander Forbes was recommending to those parties in Tepic, to attach to the grant when it should be forged.

Mr. Randolph—Except the ratification.

Mr. Benjamin—I am speaking of one thing at a time. I speak now of the grant of two leagues. The two-league grant is one thing; the ratification another. The whole idle story about the ratification has been refuted again and again, until I am satiated with the subject. I cannot repeat as often as you do. I give *that* up.

I will not repeat my argument to your Honors. If I have not been sufficiently clear, it is my misfortune. But I will not repeat things over and over. I never have argued in that way, and never will.

I do not know, may it please your Honors, if I shall be justified in detaining the Court a minute longer in this case. I feel that we have engrossed the time and attention of the Court long enough. For myself, I am thoroughly weary, tired and sick of it. I have never before had my attention engaged in a case so long, and—in my judgment—so unprofitably. When a few months shall have passed away, when your Honors shall have done justice to all this charge of fraud and forgery, and the history of the charge shall be amongst the forgotten curiosities of the past, we will look back on this whole trial, amazed that six or eight intelligent men could have engaged for a month in the discussion of the question whether the titles in this case were forged or genuine. It will not be believed to the credit of any of us, that it could have required a discussion of this record. For myself, I feel ashamed of the time, for which I have engaged the attention of the Court, and shall say nothing further on the subject.

I will say, before closing, that we appeal to your Honors for as early a judgment in the case as, consistently with your other duties, it shall be in your power to give us. I know, in making this appeal, that I re-echo the desires of my friend on the other side. If we be indeed entitled to this mine, we have been long enough kept out of its possession without any person to look to for the damage suffered by being deprived of our property.

In the administration of the law, an appeal was made to the powers invested in your Honors, as Chancellors. I referred to an exercise of that power once before, and regret to say, that in the observations made on that occasion I was misunderstood by the Court. It is not for me to say, that not now, for the first time for nearly a generation spent in the practice of my profession, would I have voluntarily incurred the rebuke of the Bench for disrespect to the Court. It never had occurred to me before, and I trust never will again. I should have supposed, nay, would have felt confident, that nothing would have drawn the remarks of the Bench, but a misapprehension of what I had said. But brother Randolph was kind enough the following morning, with exceeding courtesy, to renew the observations made by the Bench, and to say that he concurred with those convictions of the Court. I was therefore satisfied, can but be satisfied, that I expressed myself very illy. I must therefore thank him for recalling to the Court the want of courtesy of which I was guilty; and now desire to say, that if anything fell from my lips on that occasion that could give umbrage to the most scrupulous or sensitive, or lead them to suppose I entertained other than feelings of the highest respect for the Court, the error is due to some misapprehension of my true sentiments, and certainly is not what I wished to have conveyed by my language. I thank your Honors for your attention.

In the United States District Court,

NORTHERN DISTRICT OF CALIFORNIA.

THE UNITED STATES
vs.
ANDRES CASTILLERO
No. 420.

"NEW ALMADEN."

OPINION

OF HIS HONOR

M. HALL McALLISTER,

Circuit Judge,

DELIVERED, JANUARY 16, 1861.

SAN FRANCISCO:
COMMERCIAL STEAM BOOK AND JOB PRINTING ESTABLISHMENT.
1861.

OPINION

OF

HIS HONOR M. HALL M^CALLISTER,

CIRCUIT JUDGE,

DELIVERED, JANUARY 16, 1861.

THE UNITED STATES vs. ANDRES CASTILLERO,	"NEW ALMADEN."

The jurisdiction of this Court in this case is a preliminary question, and must therefore be first discussed.

It is urged by the Government, that if the nature and character of a mine under the laws of Spain and Mexico are considered, it resembles an easement at common law, is a usufructuary interest, and therefore is a "property" which cannot be protected by this Court under the Act of Congress of the 3d March, 1851.

It is urged again, that such Act limits the jurisdiction of this Court to claims of land held in fee simple only; and lastly, it is asserted that no test of the jurisdiction of this Court is found, in inquiring into the question whether a mine owned by a person constitutes him the holder of real estate, or not.

Now, as to the idea that the jurisdiction is limited to the determination of only fee simple estates in land, such construction would render the carrying out the Act of 3d March, 1851, almost impracticable. Neither the owner of land under a colonization grant, nor the holder of a mine under titles from the Mexican Government, could be deemed tenant in fee simple. The former held under a grant fettered in many instances with stringent conditions; among them even one against alienation of the estate. Such was the fact in the Fremont case, and we have met with it in others. The owner of a mine held a peculiar estate under the laws of Mexico, and subject to the provisions of the Mining Ordinances. Neither can be said properly to have held an absolute fee simple estate.

To adopt the construction contended for, that fee simple estates in land are alone protected by the Statute of 3d March, 1851, would impute to Congress an intention in framing the Act to forfeit, or leave to their fate, all estates in land less than fee simple; for a clause in the 13th Section of the Act declares, that *all* lands, the claims to which shall not have been presented to the said Commissioners within two years after the date of the Act, shall be deemed, held and considered as part of the public domain of the United States. 9 Statutes U. S. at large, 633.

The word estate is derived from the latin *status*, it signifying the condition or circumstance in which the owner stands with regard to his property. 2 Blk. 103.

Under the Mexican law, two separate parties might be the owners of different interests in the land, and each recognized as the holder of a distinct estate.

In this case the claimant presents his claim for two different estates in land; the one under a title from the Mexican Government in the form of a grant for land *eo nomine*, the other under a title from the same Government in the form of a mine. In discussing this question, as the objection to the jurisdiction is limited to such portion of the substance of the land only as has been converted to mining purposes, the inquiry will be confined to *it;* nor will the Court consider the proposition urged by a portion of the argument of claimant's counsel, that the mining right being "property," such as is protected by the Treaty of Guadalupe Hidalgo, is entitled to the protection of this Court, under the Act of Congress of 3d March, 1851.

In the view this Court takes of this case, a decision of that question is unnecessary to fix a construction upon the Statute of 1851, upon which the jurisdiction of this Court solely rests.

The Treaty may well be borne in mind when the Court attempts to fix a construction upon that statute from its face, and when the intention of the law makers is to be ascertained, whether a literal and stringent, or a fair and more liberal construction, will best carry out the statute. Such regard to the Treaty, and to that extent on this question, we think salutary.

The object proclaimed by the Act of 3d March, 1851, and in its first section, is the ascertaining the "*private land claims*" in the State of California. The eighth section of the Act requires that each and every person claiming *lands* in California by virtue of *any right or title* derived from, etc. The subject to be adjudicated then is a "*claim to land.*" The *nature* of such *claim*, the *estate* and *interest*, are wholly *unlimited.* It would seem, therefore, from the very words of the statute, that a claim to *land* by *any right* or *title* derived from the Mexican Government, and whether acquired for the purposes of extracting the minerals from it, or for any other purposes to which it could be applied, every such claim would be within the Act. So it would appear from the language of the statute, that a claim under a title or right from the Spanish or Mexican Government to any *estate* or *interest in land*, whether a conditional or absolute estate, whether for term of years, for life, or in fee, is within the statute.

For all the purposes of this case, it is only necessary for this Court to decide, that the fair interpretation to be applied to the Act of 3d March, 1851, is to include all species of property claimed under titles from the Spanish or Mexican Government, which are considered and deemed by the laws of either of those governments, and by the law of our own country, as belonging to that class of property whose "*nomen generalissimum*" is expressly mentioned in the statute.

If *land* be designated in the statute in relation to title, as in this case, every portion which formed its original substance is deemed in the eye of the law a part of the land, the law not recognizing a change of substance by a change of use or name.

A simple license to dig gold or quicksilver is a mere incorporeal hereditament, and would not come within the rule of interpretation; but the *owner* of a mine, under the laws of Spain or Mexico, has a *jus in re*, and not a mere *jus ad rem.* Under that interpretation, he who claims a mine as *owner* under these laws, is, in legal contemplation, claiming land under a title from Mexico, as well as he who claims under a colonization grant.

We think that it has been correctly stated, that "Although the things belonging to each species (land and mine) possess all the attributes of land, being immovable corporeal hereditaments, the name *land* is therefore generic, and, in strict legal acceptation, applicable to both species; yet, in common parlance, and frequently even in legal language, this generic name is used to designate one of the two species, while the other, to designate its species, is called a mine, or by some characteristic name."

"The nomenclature is not, perhaps, perfect, but sufficiently so for all practical purposes, for, while the name *land* is usually applied to one species, from the greater frequency of its use in that respect, its definition (in a legal sense) is so plain that no one can fail to perceive that it includes both, when it becomes necessary to predicate of either the *generic* attributes of land."

But the three thousand varas of surface land, as well as a vein of cinnabar included in this mining claim, are certainly to be considered land.

But, leaving mere verbal speculation, let us ascertain the solution of this question by reference to authority. What, then, are the nature and character of the property held in a mine by its owner under the Spanish and Mexican laws?

We know no better authority to refer to than the celebrated commentator on the mining laws of New Spain. The references are to Heathfield's Translation, 2 vols.

Speaking of a mine, and the indispensable necessity of keeping it at work, Gamboa says, that this being required by law, and being a condition the Sovereign has thought proper to annex "*in granting the right of property*," it must be performed. (2 Gamboa, Heath. Trans., 92, Sec. 18.) Again, he states, in considering the privileges awarded to the miners by the authors he was at the time combatting: "Another circumstance treated by the authorities (authors) as a privilege, is the per mission given to the miner to appropriate nine parts of the produce, paying to the crown a tenth only, as an acknowledgment for giving its subjects a *beneficial* and *direct interest* in this *valuable class* of property." Both Gamboa and his opponents seem to have considered the miner's interest in his property both "beneficial and direct." *Ibid*, p. 157, Sec. 15.

Again: Gamboa makes reference, in a case he is discussing, to the laws of Peru, which directed that upon the death of the proprietor of a mine, his executors shall, if his heirs be in Spain, sell it, "*like other landed property*, within thirty days." *Ibid*, p. 95, Sec. 22.

Under the laws of Mexico, all remedies, whether framed for originally acquiring, maintaining or recovering the possession of immovable or landed property, were held applicable to mines. 2 Gamboa, 258, Sec. 6.

Gamboa, in the 14th section of his Commentaries, vol. 1, p. 20, discusses the doubts which had arisen whether the mines in the kingdom of the Indies were to be regarded as the peculiar right of the crown, or whether they are to be considered as the absolute property of the subject. This he does with his usual ability, and comes to the conclusion that the mines of the Indies "are a right of the Crown, *and that as this right is quite consistent with the property granted to the subject therein, it must follow, beyond dispute, as a consequence of their being made over to the latter; with the power to dispose of them as of anything of his own, that all the incidents* of property must attach in favor of the proprietor, and that, therefore, they (the mines) may be exchanged, sold, leased or alienated by contract, donation, or inheritance; may be given in marriage or charged with a rent, and that interest may be demanded for the purchase-money while remaining unpaid. Ibid, Sec. 25, p. 28.

"But all the above qualities," continues this eminent commentator, "are to be understood as governed by this essential condition: Those to whom the property devolves, by universal or particular succession, must conform to the ordinances and fulfill the obligations thereby imposed, being the law." 1 Gamboa, Sec. 24, p. 27.

He further tells us: "The grant of the Sovereign, therefore, conveys to his subjects a direct and beneficial right of property." This last clause has been translated more literally, using the technical terms employed by Gamboa, thus: "And there passes to the subject this *dominiun directum*, or right of property (á propiedad), and also the *dominiun utile*, by virtue of the gift and concession of the Sovereign, which we hesitate not to name *una modal donacion.*"

The learned writer then proceeds to say, that such qualified gift will appear upon considering the rules by which that species of gift is defined by law, that is to say, that it be a free and complete act, which being perfected, a charge attaches on the donee from that time forth (and the being worded as a condition makes no difference), and that upon the failure of the modification limited by the donor in his own favor, or in that of a third person, or the kingdom or republic, the gift determines, as will be seen by reference to various texts and doctors," a reference to which is made in a note. *Ibid*, Sec. 25, p. 28.

These rules Gamboa considers precisely applicable to the second ordinance he is considering, for he states it thereby gives "and makes a grant to his subjects of the property and possession of the mines discovered, or to be discovered, with power to dispose of them as of anything of their own," which amounts to a complete act of gift, no price being paid for the grant, nor for the registry or denouncement of the mine. But the ordinance proceeeds, "observing, both in regard to what they have to pay us by way of duty, and in all other respects, the regulations and arrangements established by this edict in the manner hereinafter mentioned," which is the charge or qualification, and which refers to the payment of the fifth from that time forth, and to the observance of the ordinances which regúlate the mode of working the mines, the number of hands to be kept at work in them, their boundaries, and the other matters required to be observed, upon the omission or non-performance of which the gift determines, and the mine becomes liable to be denounced by any one." 1 Gamboa (Heathfield), p. 29, sec. 26.

If any title has been acquired by the claimant—a question hereafter to be discussed—and the inquiry shall arise as to his forfeiture of his title, *it* will be more appropriate to the discussion of such forfeiture than to the question of jurisdiction, as to the matters alleged as ground of forfeiture. We have cited thus fully from Gamboa, although his Commentaries, which have given him so much celebrity, were made upon the Ordinances of 1584, known as the New Code, and are alluded to by a majority of the Board of Land Commissioners in this case as "the celebrated Commentaries of Gamboa on the Mining Laws of Spain, which, although published before the Ordinance of 1783 was adopted, is a work of inestimable value at the present time." Transcript, p. 85. The dissenting Commissioner in the case, also, does not place his dissent in the matter in which he disagrees from the majority of his associates to any want of authority in Gamboa's Commentaries, but upon their alleged misconception of certain passages in that work. *Ibid*, 108. The same favorable testimony is borne to them by the counsel of both parties in this case who cite from and rely on them.

But we rely more on them by reason of the statement, verified by examination, that the "Mining Ordinances of 1783," which were promulgated about twenty years after the publication of Gamboa's Commentaries, adopted his views both as regards the rights of the Crown and the rights of the subject in the mines, and made the law as he said it was:—"a comparison of the twenty-fourth, twenty-fifth, and twenty-sixth paragraphs of the second chapter of Gamboa's Commentaries with the fiftieth of the Mining Ordinances of 1783 will illustrate this."

The caption to this title is, "Concerning the fundamental ownership (*dominio radical*) of mines, of their concession to individuals, and of the duties for which they are to be paid." Mining Laws of New Spain of 1783; (Halleck), p. 222.

The first article declares the right of the Crown in mines. It describes that right by the term *dominio radical*, which conveys a similar meaning with the *dominium altum* of which Gamboa speaks. This we infer from the second article. This is in the words: "Without separating them from my royal patrimony, I grant them in property (*en propiedad*) and *possession* in such manner that they may sell them, exchange them, rent them, donate them, pass them by will, either in the way of inheritance or legacy, or in any other manner alienate the right which in the mines belongs to them on the same terms on which they themselves possess it, and to persons capable of acquiring it."

This article certainly did not intend to separate the mines from the royal patrimony. The *dominio radical* stated in the grant, the *dominium altum* spoken of by Gamboa, and "the rights of sovereignty in the mines," as they are designated by the Supreme Court in the case of Fremont, it was intended not to separate from the Crown; but in the same instrument in which this reservation is made, the royal donor granted the mines to his subjects in property and possession (en propiedad y posecion) in such manner that they may sell them, etc. There can be no doubt that this grants the mines to individuals in property and possession.

The framers of these Ordinances of 1783 undoubtedly adopted the views and opinion of Gamboa, in the sections cited at large from that commentator by this Court especially. (1 Gamboa, ch. 2, Sec. 24, p.27.)

The mine is land in the Spanish, as "fundo" and "bienes raices" and "bienes immobles;" the one translated, is land; the second, *real estate;* the third, *immovables.*

We deem the foregoing authorities, without citing others, sufficient to establish:

1. That the transfer of a mine under Spanish and Mexican law was the granting of "a direct and beneficial right in this valuable class of property."

2. That such property was deemed by the Mexican laws of conveyance, and those as to the remedies to be applied to the recovery of the possession of it, as property in land.

3. That while the *dominio radical* or the *dominium altum* remained in the Crown, the *dominium directum* or right of property (ó propiedad) and also the *dominium utile* were in the subject.

4. That there was no inconsistency between this right of the subject and the reservation to the Crown of the *dominium eminens.*

Having ascertained the nature and character of the property held by the owner of a mine under the Spanish and Mexican laws, the inquiry is: "What view does the common law which obtains in our country take of the property of a mine as one in land?"

This is an important investigation in discussing the question of jurisdiction under the statute. If property in a mine is considered, under the laws of Mexico, real estate because it is land, and if it is so viewed by the common law, these facts will illustrate the propriety of the interpretation this Court has placed upon the statute. Such illustration will have been derived from the laws and acts of the two parties to the Treaty of Guadalupe Hidalgo, to carry out which such statute was enacted.

The venerable father of the common law (Lord Coke) lays down the rule: "By the name of *minera* or *fodina plumbi*, the *land itself* shall pass in a grant, if 'Livery of Seizin' be made, and also be recovered in assize *et sic de similibus*." 1 Co. 6, A.

We learn from any text-writer of modern times on the subject, that an action of ejectment cannot be brought for *incorporeal* hereditaments that lie in grant, except for tithes, and that by statute in England of 32 Henry VIII., Ch. 7.

That ejectment can only be maintained for the possession of a corporeal hereditament, such as land or a mine.

Notwithstanding the difference which exists between the absolute ownership of the gold and silver in the King of England, and the distribution of the property in mines which subsisted in Spain and Mexico between the Crown and its subjects, it is held at the common law that land does not cease *to be* because converted to mining purposes, and grants for copper, lead and other mines have been held to pass estates in lands and be recoverable, as it always had been, by ejectment. Ejectment lies for a coal mine (or any other) upon the principle that it is not to be considered as a bare profit *apprender*, but as comprehending the ground or soil itself which may be delivered in execution. Adams on Eject. 19.

In the case of Stoughton *vs.* Leigh (1 Taunton, 402), an application was made for dower, out of *several mines and strata of lead out of the lands* of the husband, to a Court of Equity; a case was directed by the high Court of Chancery to the Court of Common Pleas of England for its opinion on the case. The counsel for the doweress admitted rather reluctantly in open Court, "that where mines have been actually *wrought* as part of the estate of the husband, they may perhaps be collaterally subject to dower, together with the *rest of his real property*." *Ibid*, pp. 404, 405.

In that case a second argument was prayed on behalf of the heir, which the Court refused, thinking the case *sufficiently clear.*

The Court of Common Pleas certified to the High Court of Chancery in the above case their opinion, arising out of the first and second statements in the case, submitted as follows. Their response to the first statement applicable to this question will be only given. The remaining portion of the case refers to the mode in which the Sheriff should act in the service of an execution against mines.

To the first statement of the case presented, the Court of Common Pleas certified: "That the widow of John Harbury (the deceased) was dowable of all his mines of lead and coal, as well those which were in his own landed estates as also the mines of lead, or lead ore and coal, in the lands of other persons, which had been in fact open and wrought before his death, and wherein he had an estate of inheritance during the coverture; and that her right to be endowed of them had no dependence upon the subsequent continuance or discontinuance of working them, either by the husband in his lifetime, or by those claiming under him since his death." *Ibid*, 409.

Such decision, which solemnly enunciates the principle that the property in a mine is *real estate*, and constitutes a part of the estate to which a widow is entitled at common law, as tenant in dower, affords a conclusive proof that *land* converted to mining purposes remains *land* in substance, and the law therefore considers it real estate.

A few observations on the above case of Stoughton *vs.* Leigh will illustrate to what extent a mine is deemed by the common law *land.* As early as the time of Littleton, a widow was entitled, as tenant in dower, only to a portion of the *lands* and *tenements* of which the husband was seized during the coverture; and such has been the law ever since in every country where the common law has obtained.

The Courts in that case (both the Chancery and the Common Pleas) could not have cut out the heir and awarded dower to the widow out of any property other than *lands and tenements* of which the husband had been seized. Now, it cannot be urged with propriety that a mine is to be deemed a tenement.

The Chief Justice said that "the words (lands and tenements) must receive the same exposition." The Court, though, evidently placed their decision upon the ground that the mines being landed estates were real estate. Apart from any legislation like that of Spain and Mexico, which creates a different ownership of the surface of the soil and of the soil beneath between individuals, land extends downwards to an indefinite extent, and, by the common law, beneath the soil is a part of the land, and belongs to the owner of the surface of the earth above.

The change of the ownership of the intermediate soil to another owner does not change it from its original substance of land, even in Mexico, where the change is made, as we have seen by the review of her laws. Blackstone, discussing the general character and attributes of land at common law, tells us that land has indefinite extent downwards as well as upwards * * * * * * * so that the word *land* includes not only the face of the earth, but everything under it * * * * * * and therefore, if a man grants all his lands, he grants thereby all his mines of metals and other fossils. * * * not but the particular names are equally sufficient to pass them. But *the capital distinction is this*, that by the particular name nothing else will pass except what falls with the utmost propriety under the name used, "but by the name of land, which is *nomen generalissimum*, everything terrestrial will pass." 2 Black. pp. 18, 19.

In the case of Townley vs. Gibson (2 Term Rep. 701), the construction of an Act of Parliament was before the Court; one of the Judges, delivering his opinion, says: "Whether by this Act of Parliament the mines passed to the tenants? That is the question here. The soil undoubtedly passed; now what are the mines but part of the soil?" When a review of the common law teaches us, that by the name of *mine* the land itself will pass; that an action of ejectment is an appropriate remedy to recover possession of a mine as well as land; that mines can be recovered by a widow as part of her dower, although she is entitled to a portion only of the lands and tenements of which her husband had been seized during his coverture; that mines pass by an Act of Parliament, by which the soil passes as part of it; that in a deed, where mines are mentioned *eo nomine*, that nothing but what properly and strictly comes within that term—such as his metals—will pass, but by the name of land which is *nomen generalissimum*, the grant passes thereby all his mines and fossils. When such review gives such results in our country, which harmonize with our view of the Spanish and Mexican law on this point, must not an American tribunal consider, that when Congress used the words "claiming *lands*," in the Statute of 3d March, 1851, they were used in the sense they are understood and interpreted by the laws of this country and of Mexico—the two parties to the Treaty of Guadalupe Hidalgo?

The case of Fremont vs. The United States (17 How. 542), has been cited as a controlling authority on this question of jurisdiction. It cannot be deemed so, without a violation of the rules of judicial construction. In Carroll vs. Carroll's Lessee (16 How. 287), the Court say: "If the construction put by the Court of a State upon one of its statutes (and the proposition is applicable to any Court) is not matter in judgment, if it might be decided in either way without affecting the right brought in question, then, according to the principles of the common law, an opinion on such a question is not a decision. To make it so, there must have been an application of the judicial mind to the *precise* question necessary to be determined, to fix the rights of the parties, and to decide to whom the property in contestation belongs."

Now, the Supreme Court in the Fremont case could have decided either way upon the title of the colonization grant, under which Fremont claimed, without affecting the question before us; and this is what they actually did do, and they in so many words tell us: "The only question before the Court is the validity of the title." What title? It could have been none other than that of the colonization grant, the only title presented to the Court. If the only question that was before them was the validity of that title, how can an authoritative decision upon a totally different question be imputed to the Court? All that was suggested in the argument of counsel for the Government in relation to mines, was referred to by the Court in these words: "And whether there be any mines on this land, and if there be any, what are 'the rights of the sovereignty in them,' are questions which must be decided in another form of proceeding, and are not subjected to the jurisdiction of the Commissioners or the Court by the Act of 1851." (565.) In a word, no claim to landed property known as such under the laws of Mexico and this country as a mine, was before them, and it therefore was not decided upon by them.

The Court does refer, as it seems to us, to the *dominio radical* found in the Title V. of the Ordinances of 1783, the *dominium altum* spoken of by Gamboa, and designates the property reserved by the Crown as "the rights of the sovereignty in the mines." But no decision was made which should control this Court in its action on the claim of an individual to a right in a mine which he alleges he has derived from the Government of Mexico.

The only case in which such question has come before a Court in this country, is that of Delassus vs. The United States (9 Peters, 117). The suit was instituted in the District Court of the United States for the District of Missouri, and carried on appeal to the Supreme Court.

In the language of Chief Justice Marshall (p. 131): "The suit was instituted under the Act of the 26th of May, 1824, enabling the *claimants to lands* within the limits of the State of Missouri and Territory of Arkansas, to institute proceedings to try the validity of their claims."

This case has been cited by counsel for claimant as authority. We do not consider the question was so directly adjudicated as to make it a decision which should control this Court.

It is true, that substantially a lead mine was recovered, and the decree of the Court was in favor of the petitioner's claim as a tract of land; and lastly, Chief Justice Marshall says (p. 142): "The lead mine has been mentioned, but the Act of Congress makes no reservation of lead mines." This leaves this implication, that the term "land" included "lead mines," and that it required an *express* reservation in the Act to exclude it; but the question arose in the case like that in the Fremont case, from the suggestion of counsel, and was not so raised so as to constitute a *res adjudicata.*

The Ordinances of 1783, under which we have heretofore discussed the question of jurisdiction, continued in force throughout New Spain to the time of the captivity of Ferdinand VII, when great changes were made by the general Cortes between 1811 and 1814. The Cortes passed the Act of January 26, 1811. This Act is to be found in Galvan's "Collection of the decrees and orders of the Cortes of Spain, which are *actually in force in the Republic of the United Mexican States*," is translated by Rockwell from the "Printed volumes, published by Authority," and by Halleck from "Note to Article 22, Title VI, Mexican '*Ordenanzas de Mineria*.'" This decree abolished the monopoly of quicksilver reserved by Law 1, Title 23, Lib. 8 of the *Recopilacion de las Indias*, and the right reserved by Article 22, Title 6, of said Ordinances of 1783, of taking *mines* of that metal from the *discoverer*, and working them on account of the *Royal Teasury*, which in the language of the decree, "leaving uncertain the interest of the owner, and taking it out of trade, necessarily restrain people from engaging in the useful and expensive undertaking of discovering and working mines of quicksilver." It consequently modified considerably the tenure by which the quicksilver mines had been previously held. Though this Act made no alteration in the mode of acquiring title, or in the principles of the mining laws regulating mines which previously existed, it certainly enlarged the tenure of the holder of a quicksilver mine, and rendered his right of property more secure and certain.

It is true that this decree of the Cortes, with all their other acts, were annulled by Ferdinand VII, on his restoration in 1814, but the troubles which ensued in Mexico constrained him to re-establish the Constitution on the 9th day of March, 1820.—Galvan's Decretos del Rey Don de Ferdinand VII, p. 284. Subsequently, by decree of April 15, 1820, he declared the "Decrees" of the said general and extraordinary Cortes in full force *through America*.—Galvan's Decretos de Ferdinand VII, p. 292.

Independently of these reluctant decrees, the Courts in this State have held that the decrees of the Spanish Cortes, except so far as they were incompatible with the new order of things, were in full force in Mexico.

"It is true," says one of the members of the Board of Land Commissioners in this case, "that this decree of the Cortes was, in common with all the other acts of that body, annulled on the restoration of King Ferdinand; but it is also true, that this, like the others, was revived by the revolution of 1820, and was in force at the time the independence of Mexico was achieved." He then asserts that the principle above cited has been "universally admitted, and has been so decided repeatedly by this Commission."—Trans. 111.

After the most careful review, the conclusion to which we have come is, that the Court has jurisdiction.

The next question which, like that of jurisdiction, is a preliminary one, will now be disposed of.

It arises out of the objection that the proceedings in California, in obtaining title to and possession of the mine, being before the Alcalde, the whole were therefore void, not having been made before the mining deputation. It is gathered from the record of the proceedings of the local authority in this case, that there was no mining deputation in the department, and that was the only time since the settlement of Upper California, that a mine had been worked in conformity with the laws; and there being no *Juez de Letras* in the second district, the Alcalde of first nomination, etc. The fact of there being no mining deputation in Upper California was thus announced in a public judicial proceeding in December, 1845, the evidence of which has been on record for years in the archives: the petition in this case was filed in 1852, with the documents of title in which the fact was asserted; and in the years which have intervened not a *scintilla* of evidence has been introduced to contradict the statement thus publicly made in a public document some fifteen years ago. The reasons why Mining Judges and Deputies did not exist in California, will be found in the construction in Mexico of the portion of the Ordinances of 1783, which constructed this somewhat complicated machinery of mining tribunals.

Those reasons are such, even if that portion of those Ordinances was not expressly repealed when Mexico achieved her independence, as would authorize the claimant as a discoverer of a mine which gave him an incipient right, and entitled him under the mining laws in full force to denounce, register, and take possession of the mine before the ordinary judges, there being no Mining Judges.

Even the strictest common law rule adopts the axiom, "*Non cogit ad impossibilia*," and a Court acting on principles of equity as this Court does, would violate them by enforcing a forfeiture solely on that ground. We will now inquire into these reasons.

Gamboa tells us that judicial matters, such as registry, denouncement, the giving possession, and so forth, are the province of the Justices and (by way of appeal) of the royal audiences, as we shall more particularly show in the proper place. —1 Gamboa, p. 149, Sec. 15.

In his Commentary, 2d v. p. 286, Sec. 1, he observes, speaking upon this subject: "This ordinance is not observed in the Indies, nor could it be enforced there, without great damage to the public, and particularly to the miners, etc.—*Ibid.* Sec. 1. He then proceeds to discuss the question, p. 288, Secs 5, 6; and on p. 290, Sec. 10, he states, "Such as *denouncements*, insufficient working, boundaries, questions of the right of possession or property, the proving of entries in the register, the removal of the pillars of support, or the embezzlement of bullion; all which belong to the *Chief Alcaldes or Mayors*, (whom he designates, p. 286, Sec. 2.) 'Ordinary Judges,' and by way of appeal to the Royal Audiences."

Now if special Mining Judges did not exist, and the Ordinance of 1783 on the subject was not observed, nor could be enforced in the Indies (Mexico), how could they have obtained in California without special legislation?

Independently of the fact that no special Mining Judges existed, and the provisions of the Ordinance in relation to them were not enforced in Mexico, there are other reasons for the non-existence of them in California.

To authorize their legal existence here, special legislation was absolutely necessary to organize them in a mode essentially different from that prescribed by the Ordinances of 1783.

Those provisions demanded, previous to any legal organization of "Mining Deputations," a state of things which did not exist in California, a country where the first mine that was ever worked in conformity with laws was the one in controversy.

Title 2d of those Ordinances, Section 2, provides the source from whence the Mining Deputies were to receive their election and authority. It prescribes that "all those, who, for more than one year, shall have worked one or more mines, expending on them, as owners thereof in whole or in part, their capital, their labor, or their personal attention and care, shall be enrolled (matriculados) as miners of that place (lugar), and their names shall be entered in the Book of Enrolled Miners, which shall be kept by the Judge and Notary of that Mining place (*Mineria*).—Mining Laws of Spain and Mexico, 1783; Halleck's translation, p. 201, Art. 2.

By the third article of same law it is prescribed that the miners so enrolled, and certain suppliers being miners, the millers (maquideros) and the owners of *Haciendas*, for grinding and smelting in each place (lugar), shall annually assemble in the beginning of January in each year, in the House of the Judge of Mines, to elect persons who are to fill the office of Deputies of said Mining Place (Mineria).—*Ibid*, pp. 201, 202, Art. 3.

By fourth article of said law it is prescribed that each of the enrolled miners shall be entitled to a vote at such elections, and some qualification is then provided for by this section in relation to the voting by the suppliers, millers, and owners of *Haciendas*. *Ibid*, p. 202, Art. 7.

In the seventh article it is provided that the Judge of each *Real* or *Asiento*, (which words are translated by Rockwell, Mine-town or Establishment; Rockwell, p. 35, Sec. 7), and the Deputies of the preceding year, shall preside over and regulate the election; and in case of disagreement, the casting vote is given to the Judge of Mines.—*Ibid*, p. 202, Art. 7.

By the eighth article it was provided that in each *Real* or *Asiento* of Mines, there shall be a Deputation composed of two Deputies.

According, then, to Gamboa, that portion of the Ordinances which related to Mining Deputations was not of force, and *ex natura rerum* could not practically exist in California, for there was no one *Real* or *Asiento* nor no *Reales* or *Asientos* (enrolled merchants in some or any of those places, who, under the Ordinances, were to elect and organize the "Mining Deputations").

It is urged by the Government, that a complete answer is found to all the above suggestions, inasmuch as the ordinance requires the discoverer to present his written application, if there be no Deputation of Mines in the district in which the mine was discovered, to the *nearest thereunto.*

The most reasonable construction to place on these words, is to refer to the nearest mine-town, mining district or establishment within the limits of the jurisdiction of the department within whose borders the mine was discovered. Under the mining laws the origin of the title by denouncement and registry has always been left to the local authorities; and when these words were used in the Ordinances of 1783, the intention of carrying, under any circumstances, a local jurisdiction into a distant tribunal which might exist in a foreign department, is not to be imputed.

In this case the claimant certainly applied to the proper local authority, the Alcalde. In the case of Mena *vs.* Le Roy (1 Cal. 220) the Supreme Court of this State decided that Alcaldes in Departments of California, New Mexico and Tabasco, had, under the laws of Mexico, the powers of Judges of First Instance, where there was no such Judge of First Instance in the district.

In conclusion, on this point, we refer to what Gamboa says: 'It is to be observed that when a question arises concerning a contract for the purchase or sale of a mine, the right of succession, under a will or otherwise, or any point of like nature, it is competent not only to the Mining Judge and Chief Alcalde, but to the ordinary Justices of the Territory, to entertain the suit, and that it is only upon questions arising under the Ordinances, that the jurisdiction, in the first instance, belongs to the Mining Judge. If there be no such Judge, the question must be tried by the other Justices, as may be noticed in the Ordinances of Peru, above referred to."—Rockwell, p. 362.

The last authority we will refer to on this subject, is that of

Peña y Peña. It seems to be a principle in the jurisprudence of Spain and Mexico, that where cognizance of any particular matter is given to a special tribunal not being in existence, the matter reverts to the ordinary tribunal which had jurisdiction of the same kind of matter, if judicial in its character.

Peña y Peña (2d v. p. 53), says: "A special tribunal is destined to take cognizance only of a certain class of causes, or of particular persons. It is called *special* in contradistinction to the *ordinary*, which is established to take cognizance indiscriminately of all classes, causes and persons, so that a *special* tribunal is an exception to the ordinary tribunals; so that some writers on public law call them exceptional tribunals. From this it is inferred, that an exception being extinguished, the general rule remains in force. So, also, a special tribunal being extinguished, all its jurisdiction returns to the ordinary tribunals as to its source, and remingles with them from the very nature of things, without its being necessary to invest them with the authority of the new tribunal." Peña y Peña, 2, p. 371–2.

Such seems to be the principle of the Spanish Law, although it is not one of the common law.

We have heretofore considered the power of the Alcalde to deliver the juridical possession of the mine in the absence of any Mining Judges in California, in view of the Ordinances of 1783, and the construction placed upon them by Gamboa, and that view has induced us to conclude that he had the power to do so.

But the question may be viewed in another aspect. Since Mexico achieved her independence, we believe that her legislation has expressly transferred the denouncement and registry of mines to the ordinary Judges. By the 7th "Article of the Constitutive Acts of the Mexican Federation," passed 31st January, 1824, by the "Second Constitutive Mexican Congress," the Federation was made to "consist of States, and Territories, the *Californias* belonging to the latter class, and remaining directly subject to the Supreme power."—White's Col. vol. 1, p 375.

By the 18th Article, it was provided that the judicial power should be confided to a Supreme Court of Justice, and to such tribunals as may be established in the several States, and by the 23d Article, that "the judicial power of each State shall be exercised by such tribunal as may be established *by its* Constitution, but that the Legislatures of the different States may provisionally organize their interior government; and until that is done, the laws actually in force shall be observed." White's Col. pp. 378, 379.

In Article 123 of the "Federal Constitution of the United Mexican States," established by said "Constitutional Congress" on the 24th day of October, 1824, it was declared, "the judicial power of the Union shall reside in a Supreme Court of Justice, and in Circuit Courts and District Courts." White's Col. p. 404.

From the time, therefore, of the above "Constitutive Act," and the "Federal Constitution of the Mexican States," the whole judicial power became vested in the "Supreme Court of Justice," in the "Circuit," and "District" Courts, and in such tribunals as the Constitution of each State should "establish." From that time, even if Gamboa should be in error in supposing that the portion of the Ordinances of 1783 in relation to Mining Deputations did not obtain, nor could be enforced in Mexico, it is still certain that since the legislation we have referred to, it could not legally exist in California.

The influence which had been exerted by that legislation is evident from the decree of the Mexican Congress of May, 1826, which in the first article prescribes—

Art. 1. "The Tribunal General of Mining must cease, *according to the General Constitution*, in so far as it relates to the Administration of Justice, with which it was charged."—Decree of 28 May, 1820; Halleck's Mining Laws, 409. And whether such Mining Depuations could exist in the States, depended, from the time of the adoption of their respective constitutions, upon the fact whether they were "established," or "designated" thereunder.

The Federal Constitution of 1824 was overthrown in 1836, and the "Constitutional Laws" adopted in its place, but these did not establish *special* tribunals.

Subsequently, with that kaleidoscopic irregularity which distinguishes the movements of the Mexican Government, Santa Anna, having displaced the said "Constitutional Laws," and assumed dictatorial powers under the plan of Tacubaya, on the 27th day of November, 1841, issued a decree creating a *Junta*, to form and present, as soon as possible, a new project for the *re-establishment* of the *Special Tribunals* of Mining, with the modifications which *the present system of government* requires," etc.—Laras' "*Decretos y Ordenes de Gobierno Provisional.*" Halleck's Mining Laws, p. 425.

"The Establishment of Mining," created by the 2d article of the Mexican Congress (Halleck's Mining Laws, p. 409), continued in existence until the 2d day of December, 1842, when by a decree of that date issued by Nicolas Bravo (*provisionally* substituted for Santa Anna), a new regulation was made, *re-organizing* said Establishment of Mining, under the name of the Board for the Encouragement and Administration of Mining (*Junta de Fomento y Administrativa de Mineria*).

By the 10th Article of Title 1st of said Decree it is provided, that the attributes of this *Junta* shall be those which include an economical and faithful administration of the fund mentioned in this decree, etc., in conformity with the regulation which it shall draw up and transmit to the Supreme Government for its approval. In this regulation, there shall, moreover, be determined: 1st, the manner in which quicksilver shall be obtained, distributed, and sold to those who reduce ores, fixing the cases and mode in which the working of quicksilver mines in the Republic is to be supplied, rewarded, or in other ways stimulated and protected.

2d. Everything relating to the redemption of the debt of the endowment fund, according to what may be directed in the respective title.

3rd. The regulation and direction of the Junta itself; and finally, it shall be an attribute and object of its most efficacious solicitude, to promote the encouragement of the business or branch (*ramo*), its funds, and its College (*Seminario*).

The 16th article of the same title authorizes the said Junta to settle the business pending by the extinguished "Tribunal of Mining," and the "Establishment of Mining." The 24th Article, Title 4, of said Decree provides, "The Governors of Departments, in concert with the Departmental *Juntas*, and with the previous approval of the Supreme Government, will establish in each of them the number of Courts of First Instance which are required within them."

Article 25 of the same title directs that "Each Court shall be composed of three Territorial Deputies, *elected in the manner which is prescribed in the old Ordinance of Mining*, and of these three individuals, the first shall be President of the Court, and the other two, associates."

The foregoing decree is wholly translated in Halleck's Mining Laws, p. 434 (Laras' Dec. y Ord. del Gob. Prov. 1842–3. No. 549, p. 221, 229). On the 24th day of May, 1843, Santa Anna having resumed the functions of Provisional President of the Mexican Republic, issued a decree that "In accordance with my intention to *encourage* whatever may contribute to the *national* aggrandizement and *wealth*, and considering as one of the means *most suitable* for that purpose the granting of rewards and exemptions to the important branch of *Mines of Quicksilver*, so necessary for the reduction of the *precious* metals, the *most important branch of the industry* of the Republic, without which the others can make no progress." After this preamble, the Decree, in its first article, prescribes that the Royal Orders of January 13, 1783, November 12th, 1791, of December 6th, 1796, and of August 8th, 1814, with respect to exemption from excise duties (alcabala), granted to articles of consumption in mining, will be observed with respect to mines of quicksilver in the Republic.

Article 2d. No general municipal impost shall be levied upon quicksilver extracted from the mines of quicksilver of the Republic.

Article 3d. Permits quicksilver to be sold throughout the nation without permits, passes, or other Custom House papers.

Article 4th. Provides there is granted to each one of the first four operators who shall extract in one year from the mines of the Republic 2,000 quintals of quicksilver, a premium of $25,000.

Article 6th. Exempts all operatives in mines of quicksilver from all military service and all personal taxes. (Laras' *Dec. y Ord. del Gob. Prov.* 1842-3. Translated in Halleck's Mining Laws, p. 452, 453.)

On the 5th July, 1843, Santa Anna issued another Decree, having the same object in view—the encouragement of quicksilver mines. (*Ibid*, 1843. Halleck's Mining Laws, p. 454, 455.)

In conclusion, on this point, we consider that it has been shown by the preceding observations, that according to the views of Gamboa, that portion of the Ordinances of 1783 which related to "Mining Deputations," were not enforced in Mexico (New Spain)—that they could not be legally organized in California by reason of there never having been "*Reales*" or "*Asientos*" there, and, as a consequence of none such having existence in Mexico prior to her independence, they could not in California,—that Gamboa lays down the rule, that in the absence of Mining Deputies the ordinary judges may act—that a similar principle is asserted by Peña y Peña—that it is proof that no Mining Deputations existed in California; on the contrary, that the alcaldes acted. Such is the testimony of Mr. Larkin, the United States Consul at Monterey.

In conclusion on this point, "no Courts of First Instance" were ever created in Upper California under the decree of Nicolas Bravo, of December 2d, 1842. Such fact, if it ever existed legally, would be proved and must be to have had the "previous approval of the Supreme Government," and that the election had been made in the manner which is "prescribed in the old Ordinance of Mining," which was the only mode in which such tribunal could be legally constituted—that is, the members must be elected by the miners of each "*Real*" or "*Asiento*," not one of which existed in Upper California.

The Court cannot consider the objection to the jurisdiction of the Alcalde who delivered the juridical possession of the mine an available one.

The next question to be considered is the genuineness of the documentary title presented by the claimant.

There are two classes of this title.

The first consists of documents which are connected with the proceedings which took place in California in relation to the mine.

The other is the evidence of the action of the Supreme Government of Mexico, on which the claimant relies as a ratification and confirmation of the title to the mine, and as curing all defects, if any such exist in it.

It is alleged by the Government that both classes of the documentary title, and each one of them, whether executed in California or Mexico, are forged, and consequently void.

The number of witnesses called to testify in this case, the protracted examinations to which they have been subjected, the mass of immaterial facts which have been elicited, have swollen the transcript to four volumes, amounting to upwards of three thousand printed pages, and has had the effect of presenting to our attention about nine hundred pages of briefs.

In exhibiting such a case the Court is like a man who stands by an immense magazine of wheat. He may take a handful and hold it out to view, but he cannot exhibit each grain in the mass to the eye of any purchaser.

All that is practicable to do is to take a general view of the testimony, save where minuteness to ascertain the weight of testimony to prove the authenticity or genuineness of documents where forgery is alleged.

We shall first consider the evidence given in relation to the documentary title given to the claimant by the local authorities of California, then turn to that connected with that which was obtained by claimant in Mexico.

There are three links in the chain of the claimant's title. The first two are the registry of Castillero in his two communications addressed to the Alcalde, dated respectively November 22d and December 3d, 1845.

Pico, the Alcalde who signed the Act of Possession and gave juridical possession of the mine to Castillero, proves the signatures of the latter to those applications. José Noriega and Antonio Suñol, the attesting witnesses to the Act of Possession, both swear to the genuineness of Castillero's signatures, being acquainted with his handwriting, and having seen him write.

The third link in the chain of claimant's title is the Act of

Possession. The Alcalde, Pico, who signed it, and delivered the juridical possession of the mine, and the two witnesses José Noriega and Antonio Suñol, each and all swear directly each to his own, and the signatures of the two others. The testimony of these witnesses is direct and positive, and if the documents are forged or antedated, each and all are guilty of perjury. Now, the testimony of Antonio Maria Pico has been on the files of this Court in this case nearly three years, and that of the two attesting witnesses, José Noriega and Antonio Suñol, have been five years. No attempt has been made to impeach their testimony by witnesses against their character or their want of reputation for veracity.

So far from such attempt having been made, it appears from the transcript that each and all of these witnesses have been called and examined by the parties in this case at different stages of it, each in its own behalf.

They, then, who have testified as to the genuineness of these documents, stand unimpeached, and must be treated by this Court like all others who stand in the same attitude. So long as the law deems them competent, and the Court finds their testimony not disproved, it must act upon a belief in it. In this case, however, a strict invocation of that rule is not necessary. The forgery or antedating of documents is a fact which is to be proved by those who allege it. Now, the most careful review of the evidence in this case has induced this Court to believe that the testimony of the witnesses is confirmed by other testimony from various quarters—by individuals, by facts and events so nearly simultaneous with, and following so immediately, the proceedings of the local authorities in California, that they cast their shadows upon them.

Those events and facts are proved by the Mexican Archives under the charge of the Surveyor-General of the United States, from the official correspondence of the time, and the judicial records of this State.

Before reference to them we proceed to a witness who was not a party to, nor attested either document. José Fernandez was *Síndico del Juzgado* at one time, at another a second Alcalde. He proves as directly the signatures of Pico, the Alcalde, and those of the two attesting witnesses to that document, as they proved their own. "I know this document," says the witness, "saw it twice, once in 1845, in Court, afterwards when second Alcalde." He further states, "he was Secretary to Alcalde Pico; that this document was handed to him by Gutierrez, who wrote the body and paid him his fee of $3.50."

He is asked, "When handed to him by Gutierrez, what did you do with this document?"

He replied: "It remained there in the Court." He states, "he was in charge of the Archives." We will now refer to some facts.

Alcalde Pico, in the concluding clause of the Act of Possession, after stating the grant by him of three thousand varas to Castillero, declares: "*This* Act of Possession being attached to the Expediente deposited in the Archives under my charge."

Pico's term of office expired one or two days after these proceedings, and he was succeeded by Chavolla as Alcalde. On such occasion, under the law and usages, an inventory of all papers and effects in the Juzgado is taken and signed by the outgoing, and a receipt given for them by the incoming Alcalde. Such inventory produced from the Archives of the City of San José, by the Clerk of that City, Chapman Yates, on the 30th January, 1858, is filed in this case, and it designates, among numerous other papers, one "*Posecion de la Mina de Santa Clara, á D. Andres Castillero.*"

This record of possession must be the one preserved by the Alcalde in the Archives of his Court, and when the old Alcalde system was, in 1850, replaced by the Municipal Authorities of the City of San José, it passed into the Mayor's office, whence, in January, 1851, it was taken to the office of the County Recorder of Santa Clara county and filed, where it has remained to the present time.

Unless these records are forged, and got clandestinely introduced upon the records (of which there is no proof), they confirm the truth of the attesting witnesses to the documentary title. The next circumstance which confirms that truth is this: Under the Mining Laws, the Ordinances of 1783 prescribe, that after the written statement of the discoverer shall be noted in substance in the book, for his security, there be given to him, *as his corresponding title*, a *copia autorizada* of all proceedings in giving possession. The written statements or representations of Castillero in registry were not returned to him, but remained in the Alcalde's office. He, however, received certified copies of them on the 13th January, 1846, signed by Pedro Chavolla, Alcalde, and by José Suñol and Pedro Sainsevain, as attesting witnesses.

The *Copia Autorizada* was delivered to Castillero. This document, which is a copy of the expediente, has been filed in this case.—Transcript, p. 2693.

The history of this document, so far as the evidence goes, is that one Robert Walkinshaw, at the time acting as the agent of Alexander Forbes, part owner and sole lessee of the mine, placed in the hands of a professional legal firm in the City of San Francisco, in January, 1853, this *Copia Autorizada*, to be used in relation to some litigation which had arisen in relation to the possession of the New Almaden mine. It remained in the office of those gentlemen, when it was delivered back to Walkinshaw with other papers, and a receipt taken for them.

These facts are established by Mr. Hall McAllister, one of the said legal firm, and Mr. Reese, in their depositions. —Trans. 2698, 2712.

Walkinshaw, on receiving this document, with the others, from Mr. McAllister, handed them with others to his agent, John Young, with whom he deposited all his papers when he left this country, to which he never returned, as he died in Scotland in April, 1848, and Mr. John Young became the sole executor of his last will and testament. On receiving the papers Mr. Young enveloped and labeled them "Papers relating to disputed *barra* between Walkinshaw," etc.

The time when, and the circumstances under which, this document was discovered, are detailed by the depositions of John Young and Thomas Bell.—Trans. 2684, 2696.

To prove the identity of this *Copia* the Alcalde, Antonio Maria Pico, was examined in July, 1860. He swears that the Act of Possession in this document was signed by him, "Antonio Suñol and José Noriega; these latter signed in my presence as witnesses." He also testifies: "I have no doubt that I delivered this document to Castillero, because I see my receipt thereto for twenty-five dollars, my fee. The body of the document is the handwriting of Gutierrez, and bears my signature, which was placed on the day of date."

In this *Copia* there are two copies, purporting to be such, of the two representations of Castillero to the Alcalde of 1st Nomination. Each of these is signed by Castillero, Pedro Chavolla, Alcalde Pico, and two attesting witnesses, P. Sainsevain and José Suñol. The Alcalde testifies that he is acquainted with the signatures of all of them, and that they are genuine. José Suñol is dead. Pedro Chavoya proves his own signature and those of P. Sainsevain and José Suñol. He was Alcalde at the date, and signed the documents at their respective dates. Pedro Sainsevain proves his own signature, that of Pedro Chavolla and José Suñol.

The document in the *Copia* of the act of possession, the third link in the chain of claimants' title, is the document which Alcalde Pico, as we have seen, has sworn to be genuine, and he has no doubt he delivered it to Castillero, was presented to the witness, José Noriega, one of the attesting witnesses. He testifies the handwriting of the body of the document is that of Gutierrez. "It bears the genuine signature of Antonio Maria Pico, Antonio Suñol and myself; we all signed it at the same time, and in each other's presence, I presume on the date of its date. I have no reason to suppose it was not." He then states, in answer to the inquiry, that the receipt from Antonio Maria Pico to Castillero for twenty-five dollars is in the handwriting of Gutierrez.

At the end of many years, on the procurement of this testimony, the parties in interest, with the single exception of José Suñol, who had died, have been able to invoke every witness who was a party either to the original record or its certified copy.

The earliest act of Castillero, as the evidence informs us, was to communicate to Governor Pio Pico, under date 10th December, 1845: "The mine has been denounced by me, and, between a few, we have formed a company, etc." He also sent a sample of the quicksilver and says: "There is such abundance of quicksilver that eight arrobas of ore give one of metal." In this letter he expresses a wish that the vacant lands near our works be conceded to cut wood, and asks for an order to place him in possession of the Island of Santa Cruz, which had been granted to him by the Supreme Government.

This letter was transmitted by Governor Pico through the Minister of Exterior Relations, with the sample of quicksilver, to the President of Mexico, and he requests his Excellency may be made aware of, and satisfied with, so happy a discovery.

On the margin of this letter is the following order by the Minister: "April 6th, 1846. Received and noted with satisfaction, and with respect to the other matters contained, let him inform attentively as he may think fit."

The letter of the Governor of California, communicating to the Supreme Government the letter of Castillero, giving the discovery of the mine and its denouncement, was dated 13th February, 1846.

The answer made under said order to the Governor is in these words (Trans. p. 1806): "His Excellency, the President *ad interim*, learns with satisfaction, by the letter from Señor Castillero, which your Excellency sent with your official of 13th February last, the important discovery which has been made in that department. His Excellency having seen the sample of that ingredient cited in said letter, and which your Excellency sent me by Don José Maria Covarrubias, I have the honor to say this to you, by supreme order, in reply to the said communication, and with respect to the other matters referred to in Señor Castillero's letter, that Government will please report attentively what it deems convenient."

Now, all this correspondence between the Governor of the Department of California and the Supreme Government of Mexico in relation to the discovery and denouncement by Castillero, unless it was forged and clandestinely introduced into the archives, has been filed among them, since within a few days after the proceedings of the local authority had placed Castillero into juridicial possession of the mine.

These communications between Castillero (voluntarily sought

by him) on one side, and the Territorial Government and Supreme Government on the other, and the approval of both of the discovery and denouncement of the mine, is certainly another circumstance tending to confirm the authenticity of the documents as sworn to by the witnesses.

If Castillero had anything to do with this alleged forgery, crime took an extraordinary course with him, for it impelled him as soon as possible while the event was recent, to give to the public authorities an opportunity of ferreting out his crime, before the dust of the future should have settled upon his foot prints.

Five days after writing to the Governor about the mine, Castillero makes a second commuuication under date of 15th December, 1845. He supposes that his Excellency may not have received his previous letter, and he again informs the Governor of the discovery of the mine. This letter is also in the Mexican archives.—Statement of Hopkins, Trans. 3068; proved by 13th answer of Governor Pio Pico, Transcript, 2533.

Now if this letter has not been forged and clandestinely introduced into the archives, the presumption is that it must have been sent to the Governor at the time it bears date, and such also tends to confirm the testimony of the witnesses.

But the proceedings in giving possession of this mine were not concluded or done in the dark; one of the Mexican functionaries gets hold of them. On the 31st day of December, 1845, Manuel Castro, Prefect of the Second District, makes an official communication to the Secretary of the Departmental Government, that Don Andres Castillero had denounced, and is now working, a quicksilver mine found in the jurisdiction of the town of San José Guadalupe on private property. This letter of the Prefect Castro was answered, the blotter of answer is in the archives; Transcript, 2551. Proved by answer 10th of Pio Pico, Transcript, 2533; Hopkins' statement, 3068.

Having thus satisfied ourselves that the witnesses who have sworn to the genuineness of the documentary title obtained by the petitioner, so far from having suspicion thrown upon their testimony, are confirmed by record evidence from the Mexican archives now in charge of the Surveyor General of the United

States, we will proceed to one or two additional species of testimony.

Capt. John C. Frémont testifies that he paid a short visit to the New Almaden mine; he left it about the 24th January, 1846. This must have been about twenty-four days after the juridical possession of the mine had been given. He visited it in company with Capt. Hinckley, who introduced the witness to the owner, Castillero, who showed him about, and the excavation from which he had taken the ore, showed him two or three heaps of the ore and gave him some specimens, some of which he brought away.

Before visiting the mine, the witness states he had conversed with Capt. Leidesdorff with regard to purchasing the mine. When there, "I spoke slightly with Castillero on the subject, and Mr. Hinckley also said something to him at greater length tending to the same end, but Castillero was not at all disposed to converse about selling. About this time, I think, Castillero was engaged in building a house below in the valley, to be used for the occupation of himself or workmen. I learned from Castillero that he held the mine under a denouncement, and then I, for the first time, became acquainted with the Spanish system of acquiring mines by denouncement."

In an official letter of Thomas O. Larkin, written as Consul of the United States, at Monterey, to the Minister of the United States, at Mexico, under date of the 3d April, 1846, among other things, he states that Don Andres Castillero is going as Commissioner to Mexico from the Military Commander of California, Gen. José Castro. He concludes his communication thus: "Near the town of San José, eighty miles from Monterey, Don Andres Castillero has discovered a quicksilver mine; the ore produces from sixteen to sixty per cent.; I have seen him, from an old gun barrel, in thirty minutes, run out about thirty per cent. in pure quicksilver. This must be a great advantage to California."

On the 4th of May, 1846, Mr. Larkin, as such Consul, addressed to the State Department a long report upon the mineral resources of California. Certain extracts are certified by Lewis Cass, and under the seal of the Department of State, to be true copies of the original dispatches of Thomas O. Larkin, U. S. Consul at Monterey, dated 4th May, 1846, which are to be found in the State Department.—Transcript, 2657.

The Consul communicates to the Department, that "From the town of San José, and near the Mission of Santa Clara, there are mountains of quicksilver ore, discovered by D. Andres Castillero (of Mexico), in 1845, which the undersigned has seen twice produce twenty per cent. pure quicksilver, by simply putting the pounded rock in an old gun-barrel, one end placed in the fire, the other in a pot of water, etc. * * * There appears to be no end to the production from these mountains. Working of the quicksilver is but now commenced, under great disadvantages, from not having any of the materials generally used in extracting that material, etc."—Trans. 2657.

In the same communication Mr. Larkin states what in substance Castillero told Fremont as to the mode of acquiring under Spanish laws a title to a mine by denouncement. Mr. Larkin states that "By the laws and customs of Mexico respecting mining, every person or company, foreign or native, can present themselves to the *nearest authorities*, and denounce any unworked mine; the authorities will, after the proper formalities, put the discoverer in possession of a certain portion, which I believe is according to its extent. The possessor must thereafter occupy and work his mine, or some other may denounce against him," etc.

There are various other official communications, but enough of this species of evidence has been cited. Various others will be found in the Transcript.

If we turn from the record evidence derived from the Archives of Mexico, and that from the official correspondence of a Consular Agent of the United States, to the judicial records of California, we will find historical evidence of the genuineness of the proceedings to which the witnesses have sworn. In March, 1847, in the suit of G. C. Cook vs. James Alexander Forbes and others, the complaint was that Andres Castillero, James A. Forbes and others, were working on the land of plaintiff contrary to law, praying that they be removed. The parties having appeared, the case was continued until the mine was surveyed. Two men made a report, and the case was dismissed. If there were no registry, no act of possession, how could there be a mine in existence to survey?

On the 14th August, 1847, James Alexander Forbes, then British Vice Consul for California, writes to John Burton, Esq., Justice of the Peace of San José, that two persons have commenced digging a pit by the direction of G. Cook, within the limits of the *juridical possession of the said mine.* * * * "Permit me to refer you to the documents which exist in your office, upon which was founded your conviction of the justice of your decision in relation to the claim of Mr. Cook in March last, and to request you will be pleased to adopt such measures for protecting the rights of the owners of the said mine, and of those who are legally interested in the same, as you may deem most conducive to that end."—Trans. 810.

This letter was produced by Chapman Yates, Clerk of the City of San José, from the archives of that place, on the 30th January, 1858 (Trans. 769), and proved by the witness McCutchen (767.)

Subsequently, on the 5th May, 1847, James Alexander Forbes, in a letter to Mr. Alexander Forbes, alludes to the "*juridical possession which was given of the mine by the local authority of this jurisdiction,*" and also of "*the three thousand varas of land given in that possession as a gratification to the discoverers.*"—Trans. p. 842.

Weekes' amended possession, given 21st January, 1848, refers *to the original act of possession*, and declares that "*the right and title of the mine to the mine and land granted as a reward in the original act of possession, shall remain valid.*"

This circumstance, while it confers no title on the claimant, being the act of one who, without the consent of both parties, had no authority, is however the recognition of a *de facto* magistrate of the genuineness of the proceedings to which the witnesses in this case have sworn.

In the action between Walkinshaw vs. Forbes, complaint was filed 18th October, 1849, in which plaintiff claimed "*one-eighth part* of the mine by title derived under the original act of registry."

The mine was denounced on 6th October, 1849, by Mr. Horace Hawes, for *abandonment* and *insufficient registry*, and the mine is described as situated on Berreyesa's rancho, and as known *in its original title of registry as the Mine of Santa Clara.*

In the pleadings of plaintiff it is alleged its last possessors were Andres Castillero, Alexander Forbes, James A. Forbes, Robert Walkinshaw, and the two Robles.

Alcalde May's proclamation, on 23d October, 1849, made in relation to the suit, citing parties to appear, describes it as the quicksilver mine situated in the District of San José, known and designated, *in its original act of registry*, as that of Santa Clara, and now known by the name of New Almaden.—Trans. p. 297.

Now, a great portion of the evidence relied on to sustain the allegation of forgery and antedating of the documentary title of the petitioner, consists of the letters which passed between James Alexander Forbes and other parties in interest, filed by the Government, and which were before the Circuit Court on the argument for injunction, and some letters interchanged between James Alexander Forbes and Alexander Forbes, explanatory of those filed by the Government. Among these numerous letters from various parties in interest there is *one* letter alleged to have been written by Alexander Forbes to James Alexander Forbes, under date of March 28, 1848, which is the only letter where a positive assertion is to be found that forgery or antedating had been actually committed.

Nor is the fact asserted in THIS letter alluded to in any other part of the voluminous correspondence. We refer to it now simply in connection with the documentary titles obtained by Castillero in California.

The letter of 28th March, 1848, speaks a language different from that uttered in any other. The sentences which do so are these: "Were I not already so deeply interested in this negotiation I would never think of investing another dollar in it, but this interest renders it necessary to have the control of all the shares, in order that I may dispose of the whole whenever an opportunity may offer, and save myself from the heavy loss that would ensue, should it unluckily leak out, that in fact, *the documents procured by Castillero in Mexico, as his title to the mine and lands, were all obtained long after the occupation of California by the Americans.* This unfortunate irregularity cannot be easily repaired, and serious objections might be made to our new Act of Possession."

Now, the charge of antedating in this letter, if the Court could believe in its authenticity, apply to only such documents as were obtained in *Mexico* by Castillero, and hence the fact that the charge is not by the letter applied to such as were obtained elsewhere, is one to be considered when determining the truth of witnesses who have sworn directly to their genuineness in California.

The Court, in view of foregoing considerations, must take the authenticity of the documentary title obtained there as satisfactorily proved.

We turn, now, to the documentary title obtained in Mexico, which loses its importance, because, satisfied as the Court is with the proofs in this case of the authenticity of the documentary title obtained from the local authorities of California, and as the title held under them is sufficient to sustain the confirmation of the claim to the title to the mine, no aid is requisite from the action of the Supreme Government to sustain it, save as to the three thousand varas granted by the Alcalde.

No action of an Alcalde can grant to a party such an amount of surface land as "*pertenencias*," or for any other purpose. He, whatever the President can do, has only power to give such extent of "*pertenencias*" as the mining laws allow.

A *discoverer* of one or more mineral hills (*cerros*) absolutely new, may acquire, in the principal vein which they may select, as much as three *pertenencias* continued or interrupted.

As *discoverer* the Alcalde had the power to assign to Castillero three "*pertenencias*" as such.—Halleck's Mining Ordinances, Art. 1; Trans. 54.

Castillero formed a copartnership on the 8th December, 1845, with the two Robles', José Castro, and Padre Real for the working of the mine, and thus working in partnership, the Alcalde had the power to assign to Castillero *four* additional *pertenencias.*—Ordinances of 1783, p. 252; Halleck's Mining Ordinances; Trans. 54.

Thus, seven *pertenencias* is the largest amount an Alcalde is authorized to assign as such, and to *grant* land to any extent was utterly beyond his jurisdiction.

As the appellate Court may take a different view of the evidence of the documentary title obtained in California, and it is made necessary, under the view this Court takes of the action of the Alcalde in relation to his (the Alcalde's) grant of three thousand varas, that we now refer to the documents of title obtained by Castillero in Mexico, we will now do so.

It has been before observed, that the allegation of forgery and antedating as to them, mainly rests upon the letters written between James Alexander Forbes and some of his associates in interest.

Both of the judges on this bench have heard the nature, character and legal effect of those letters as evidence, elaborately argued on two occasions, once on the discussion of the motion for an injunction in the Circuit Court of the United States for this Judicial District, and on the present trial of title in this Court. Everything has been said, and heard by them from both sides, on the question, which learning and research could invoke. With the very elaborate arguments presented, to them on that point, it is deemed needless to travel in detail through the numerous letters offered in evidence on this point. They are all set forth in the Transcript, and open to the appellate tribunal, who, after an examination of them, can correct any error committed by this Court in the conclusion it has come to, upon the character and legal effect of all the letters on which the allegations of forgery and antedating in this case rest.

A careful perusal of these letters has satisfied the Court that, with the exception of the letter of 28th March, A. D. 1848, which James Alexander Forbes swears he received from Alexander Forbes, there is nothing in them which even tends to prove that forgery or antedating had been committed. In other words, there is no proof afforded that the documents presented by claimant are the result of either.

They do show that James Alexander Forbes was *willing* to have titles forged and antedated, for the purpose of curing what he deemed irregularities and defects in the original title. They also prove that he actually suggested to some of his associates the importance of prompt action in the premises, and complained of their delay. This suggestion was made, at one time, by a memorandum left with one of his associates, and at others by letters to one or two of them.

Now, the *willingness* of an owner, or of all the owners, to better their title by forgery and antedating, will not defeat their original title if sufficient to pass the estate.

The question presented to the Court, in cases where forgery is alleged, is whether the titles produced are the fruits and results of such forgery.

In the United States *vs.* West's Heirs (22 How. 315), the Attorney General for the Government contended that a provisional or equitable grant, which may be converted into a legal title, upon the contingency of the approval by the Departmental Assembly, and the performance of other conditions, must be regarded as wholly abandoned, when the conditions are not complied with, and another and a different claim set up under a forged title.

West, the petitioner, died during the pendency of the proceedings before the Board of Land Commissioners; the case was brought up to this Court, which confirmed the claim for a league and a half.

The Supreme Court say on appeal: "We have only to say that the fraudulent attempts to enlarge the grant were made after California had been ceded to the United States; and though the proof of it is undeniable, and was an attempt to dafraud the United States, that cannot take away from the wife and children of West their claim to the grant, which was made to him before California had been transferred by treaty." The Court, therefore, confirmed the claim for one and a half leagues, about the genuineness of which there was no doubt.

Now, in this case, throughout the whole of these letters, there is no evidence to prove either forgery or antedating by any one, even James Alexander Forbes. The only exception in the whole correspondence is the letter of the 28th March, A. D. 1848, which Forbes swears he had received from Alexander Forbes.

We have given a copy of this letter, in which the following language is used: "and save myself from the heavy loss that would ensue, should it unluckily leak out, that in fact, the documents procured by Castillero in Mexico, as his title to the mine and lands, were all obtained long after the occupation of California by the Americans."

From the evidence set out in the Transcript, we believe this letter to be a forgery. The *loss* and *contents* of the original are proved by James Alexander Forbes and Robert Birnie, the former the master spirit and the latter his instrument.

We shall not attempt to go into the details of all the numerous facts which have forced upon the Court the conclusion that the letter of 28th March, 1848, is fabricated, but refer only to a few circumstances.

Forbes, the witness, is impeached by nine witnesses, all of whom resided in the county of Santa Clara, where the witness did. Some of these knew him in 1846, some in 1847, and some in 1850.

This letter of 28th March, 1848, alleges that all the documents procured by Castillero in Mexico, were obtained "long after the occupation of California by the Americans."

A portion of the letters in evidence in this case were purchased by one Henry Laurencel, who, it seems, had some interest in the event of this suit, from James Alexander Forbes, after he had sold out his interest in the mine, and had become hostile to his former friend, Alexander Forbes, for the sum of twenty thousand dollars.

These letters so purchased were deposited at a banker's in this city, subject to the joint order of Forbes and Laurencel, but were brought into this Court by process of this Court, and filed here by the Government.

Subsequently other letters, explanatory of these, and between the same parties, were brought into Court.

The attempt by the production of these letters on the part of the Government, was to prove that forgery had been committed with regard to the title papers, or some of them, in 1850, in accordance with a memorandum made and left by James Alexander Forbes in Tepic in 1849.

This theory was disaffirmed by the letters subsequently filed by the claimants, particularly by the letter of James Alexander Forbes, on 5th May, 1847, to Alexander Forbes, in which he tells the latter of his having seen the copy of the two-league grant, signed by Castillo Lanzas, ordering the Governor of California to put Castillero in possession of the land, and speaks of the juridical possession of the mine of three thousand varas as a gratification. Trans., 842.

Now, when James Alexander Forbes sold for *twenty thousand dollars* the letters filed by the Government, they were sold for the purpose of proving forgery and ante-dating. Where was the letter of 28th March, 1848 (if in existence), at the time of the sale of these letters by Forbes to Laurencel?

It is the only one which gives an express statement of forgery, and is the most important in the whole correspondence.

The account of his reasons when in selling certain documents for a pecuniary consideration, which, for the purpose of enhancing the value of the evidence sold to Laurencel he did not include this original letter among those he sold, are inconsistent.

Forbes, in his answer, in the latter part of it, states in his 197th answer (Trans. 895): "If this letter had been in my possession, or rather, accessible to me, I would have presented it with the others; but it had been mislaid, and I had forgotten where I had put it."

In reply to question 199: "While you were making selections from your correspondence for Mr. Laurencel's inspection and for sale, did it occur to you there was such a letter in your possession as this of the 28th March, 1848?"

The witness had stated that he had *sold* no documents to Mr. Laurencel, but "he paid me for a specific use of those papers."

After reiterating his denial as to a sale, he answers the question as to his recollecting the letter. He says: "I certainly did recollect of the existence of the letter alluded to, and that, having received it a long time before the dates of the correspondence which I allowed Mr. Laurencel to use, I had laid the letter in question aside among some other papers, and I was unable to lay my hands upon it when the aforesaid correspondence was submitted to the inspection of Mr. Laurencel.—Answer 199, Trans. 896.

The attention of the witness was then called to the fact that among the papers sold or delivered for a special use to Laurencel, there were seven letters prior to the letter of 28th March, 1848. His answer is, in the latter part of it: "My answer (200) alluded to those letters which were considered of the greater importance in that correspondence, which were comprised in the dates I have mentioned, 1849 and 1850." And in his answer 201, he says: "In my said answer 199 I had in view the dates of those letters of that correspondence which were considered of the most importance." Now, what paper could he consider more important than the one of 28th March, 1848?

He searched, he says, for that letter when he sold the use of the papers to Laurencel. When asked, "Did you at the time mention to Mr. Laurencel that you had received such a letter?" he replied, "I did not."—Trans. 897, Ans. 205.

In his answer 443 (Trans. 943), he assigns as his reason, and says: "I will *now take occasion to state what was my real* object in retaining that letter in my possession. I knew that the counsel himself had gone to the city of Washington, attended by William E. Barron, for the purpose of getting a bill passed for the taking of testimony in Mexico; and if they should not be successful in obtaining the passage of such a bill, the witnesses would be brought here for the purpose of supporting the title to New Almaden, and after such testimony should have been given in this Court I was determined to exhibit that letter to the authorities of the United States."

He was asked "whether he ever wrote to Mr. Alexander Forbes, requesting any explanation of the expressions in that letter, or to Castillero, requesting some information as to the title?"

The witness answers, he never did, and he considered the attempt useless, inasmuch as he believed the last act of possession was obtained for the purpose of remedying anterior defects in the title (Answer 371, Trans. 919.) In answer 374 he says. "I made no inquiry with regard to the matter set forth in that letter, for the reason already stated, namely: that I considered it next to impossible for me to obtain such information, and more especially as I had been already informed by Mr. Alexander Forbes that in case everything else failed we should fall back upon that possession."

The last reason he assigns for making no inquiry about the letter of 28th March, 1848, was, it stated an important fact which he considered would never be made the subject of correspondence, neither on the part of Mr. Castillero or Mr. Forbes, even had there been a less precarious mode of conveying such correspondence from Mexico to California. Can this witness be deemed to have given a clear account of this letter which he secreted, and made no inquiry for years until June, 1858, a period of ten years from its date, when it is produced by a copy; Forbes swearing the original was stolen from his carpet-bag on the 30th June last?

When asked by counsel for claimant, "Do I understand you to say that on the 30th of June last, and after the conclusion of your examination before Judge Hoffman, and after you had suffered Birnie to make a copy of that letter, the said letter was stolen from your carpet bag at your lodgings, at the Railroad House in this city?" This was on the evening of 30th June, 1858, that Birnie made the copy.

Forbes answered: "That letter was in my possession up to nine o'clock in the evening. I had occasion to leave my lodging for a short time; for security I took that letter out of my coat pocket and put it into my carpet-bag, locked my room door and went up as far as Montgomery Street. I was absent half an hour or more; on my return I found my room door locked as I had left it; and on retiring to bed I had occasion to go to my carpet-bag, when I found that letter was missing."

This is an improbable story, coming from a man who is proved by the correspondence as not only willing to have forged title, but urging his suggestions upon his associates to fabricate it.

There is improbability in the story, as he tells it, as to the manner in which he lost it.

The witness deemed it an important document, and he reserved it *as* his real object, as he tells us, to deliver it to the authorities of the United States, in the event that witnesses from Mexico should be brought here to testify. Why should the character of the letter, as true, depend upon that fact?

It is improbable that a document deemed so important should be left in a carpet-bag in a large public house, and left in a carpet-bag, as the witness swears, for "security."

We shall no further pursue the facts spread out in the Transcript to establish the forgery of the 28th March, 1848, so far as James Alexander Forbes is concerned. His coadjutor, Robert Birnie, takes the stand. Sixteen witnesses, old neighbors of Birnie for four or five years, who knew him well and his reputation, say it is so bad they would not believe him on his oath. Whoever desires to arrive at a knowledge of Birnie's reputation has only to read the testimony of one of these witnesses.

The general fact sworn to by him that he was not to be believed on his oath, is confirmed by all the rest. That witness is Nathaniel Jones, and his testimony to be found in the Transcript, p. 2769.

Mr. Jones is a farmer; held the office of Sheriff in 1850 and 1851; was Public Administrator a short time after; was then elected Supervisor of the County; is the Corresponding Secretary of the Contra Costa Agricultural Society, and Vice-President of the Bay District Agricultural Society, representing Contra Costa county. This witness swears he has lived in the neighborhood of Birnie for some four or five years. He would not believe him on his oath.

On his cross-examination this witness testifies: "It is not from any single act of his, but it is from his want of occupation, and the universal belief among the people that he swears falsely and procures false evidence in land cases; he is utterly worthless, and a loafer."

Now such was the coadjutor of Forbes, who copied the letter of the 28th March, 1848, on the night of 30th June, 1858; who swears he did so, and he swears the letter he copied was one directed to James Alexander Forbes, dated ("I believe") 28th March, 1848, from Monterey.

He swears to every thing about the letter that Forbes did

about his copying, and also he identifies the original letter which he copied as that of Alexander Forbes.

He can say nothing about the stolen letter, as no one was present save Forbes and the carpet bag. But Forbes could not return the favor Birnie had done him.

On his cross-examination, Birnie deposed that the first notice he had to attend as a witness, "was from a subpœna to appear instanter. I came over in the boat yesterday with Mr. Randall, the Deputy Marshal, Mr. Forbes and Mr. Laurencel were in company with the Deputy Marshal. I went to Mr. Forbes and asked him the object of this subpœna so suddenly. He told me that it was to testify about the copy of the letter I got from him," etc. Answer 36; Trans. 861, 862.

Forbes was called some two days after the examination as a witness, and he knew that Mr. Randall, the Deputy Marshal, if he did not tell the truth in despite of his friend Birnie would bring out the truth, as he Randall was cognizant of the falsehood of Birnie; he, Forbes, therefore actually told the truth, and falsified all that his friend had sworn to. "I went," says Forbes, "on our arrival at Oakland or soon thereafter, at the request, or rather by *his consent*, thinking it would be less disagreeable to Birnie to speak to him before the subpœna was served, as the *subpœna* called for his appearance *instanter*." Ans. 35; Trans. 876.

"I told him," says the witness Forbes, "simply, that the United States Marshal was there and had a *subpœna* for him; that he would not be required to go to San Francisco until the following morning, provided he would be ready to go at 5 o'clock in the morning. I then left to see the Marshal, and informed him that Birnie would be on board in the morning." So that what Birnie had deposed to was expressly falsified by Forbes, from the fact of the Deputy Marshal's presence.

Escaping from this forged letter, we will look briefly to the testimony adduced by claimant to sustain the genuineness of the documentary title obtained in Mexico.

Great delay in the administration of justice in this case has taken place, from the fact of the tenacity of the Mexican Government in adhering to the law or regulation inhibiting the use of the great seal to evidence transactions done in Mexico in foreign countries.

The inability of claimant to clothe the evidence he had to procure in a form to be received in the courts of this country under existing laws, has had the effect of making this case drag along at a very slow pace. They have had the means, however, to put an end to further delay, and by the exertion of them to bring their witnesses from their homes in Mexico to give their testimony, and testify personally to the genuineness of the documentary title which the claimant obtained from the Mexican Government. The impression generally has been entertained that titles may be easily forged, and that the Government of Mexico was approachable by clandestine means.

The correspondence of which so much has been said, which was filed by the Government in this case, was naturally calculated to fortify that impression. That correspondence did establish to some extent the settled purpose of one of the parties to have papers antedated and forged, and that one or more of the other parties did not promptly repudiate his suggestions, which the Court consider may have been the result of many motives into which it is unnecessary to inquire, under the view entertained and enunciated by this Court, of the nature, extent, and legal effect of that correspondence as evidence to prove forgery or antedating. The record shows that an application was made by Eustace Barron and Castillero to Castillo Lanzas, while Minister of Relations, to authenticate copies of various documents in the public offices of Mexico relating to the Almaden Mine, with the great seal of the Republic. The application was refused, upon the ground that the uses to which the great seal may be put are defined by law and do not embrace the authentication of copies of public documents. One of the witnesses examined in this case was Lanzas himself, and he gives the laws which regulated the use of the great seal (Trans. 2237.) See the letter of Castillo Lanzas giving the reasons why the President denied the use of the great seal save for the special purposes designated by law (Trans. 2384). Now, it strikes the Court, if the precision and tenacity of the Mexican Government about the great seal are so great that it would not yield to the intercessions of Andres Castillero, he certainly could not have possessed such influence as has been imputed to him to obtain a forgery of documents in all the public offices through which his title passed. This Court has decided that the only evidence produced to prove a fabrication of these documents is the forged letter of 28th March, 1848, and the proof which has been produced to prove the authenticity of these documents is as strong as it can be made. We must come to the conclusion that the archives of the Mexican Government must have been forged, and the eleven witnesses sworn in this country under its law, and examined in the presence of her magistrates, have perjured themselves. The deposition of John Forsyth, United States Minister in Mexico, is in the Transcript, 1111. The following motion in relation to it was passed by this Court:

"THE UNITED STATES vs. ANDRES CASTILLERO. } No. 420.

On motion of Messrs. Peachy and Yale, of counsel for claimants, and by consent of parties, it is ordered by the Court that the deposition of the Hon. John Forsyth, remaining under the seal of R. B. Owen, United States Commissioner, be and the same is hereby published." Trans. 1110.

In his deposition, Mr. Forsyth deposes, he went first to the office of the *Junta de Mineria* in company with the British Consul, John P. Brodie of California, agent of the claimant, and others, and where was produced from the archives of said office an expediente, which, being carefully compared with a copy in the hands of the said Pardo and Brodie, proved to be absolutely alike and correct in every respect, and to that copy he made a certificate dated the 4th day of August, 1858.

"On the 30th July, 1858, I was present in the *College of the Mines* with the same parties, when an expediente was produced from the archives, which was compared, etc., which copy was certified by me on the 4th day of August, 1858. On 29th of July, 1858, I was present at the office of *Foreign Rela-*

tions, when and where the Chief Clerk of the Section of Europe produced from the archives of said office an expediente which was compared, etc., and to that I certified on the 4th day of Aug. 1858. On the same day I was present at the office of the *Ministry* and *Police*, when the Chief Clerk of said office produced from the archives of said office an expediente in the presence of the same parties, which, on being compared," etc.

"On the 30th July, 1858, I was present at the office of the Escribano, or Notary, Don Juan Navarro, where and when said Navarro produced from the archives of his office a book formed of sheets of stamped paper, stitched together and consisting (exclusive of the fly-leaves in the beginning, and the index at the end) of one hundred and twenty leaves, and titled '*Año de* 1846, *Protocolo de Instrumentos del Escribano Don Nazario Fuentes.*' In the presence of the same persons, certain original instruments, contained in said book, were examined, and being compared unto a copy found to be correct and similar, with the exception of two slight omissions, which were certified by the Notary on the same day, and to that copy I certified on the 4th day of August, 1858. These original documents were found in the several offices where they appropriately belonged, and were produced by the officers having the custody of them; and I saw nothing whatever to cause me to doubt that their being genuine originals. I, as Minister, certified each of the copies hereinbefore mentioned, and the facts set forth in those certificates are true, and the certificates are in accordance with the laws of Mexico."

Mr. Forsyth further deposes that the Government will not allow the great seal of Mexico to be attached to copies of such documents, nor will they allow the originals to be withdrawn. "And the manner in which the copies herein mentioned have been authenticated is the only way in which such copies can be authenticated."

Now, all these expedientes, in the archives of the *Junta of the College of the Mines, in the Ministry of Foreign Relations*, in the office of the Ministry of Government and Police—four of the public offices and departments—must have been forged in each one of them, and introduced clandestinely into the archives of each. And the same must have been done in relation to the original documents produced from the archives as sworn and certified to by Mr. Forsyth from the archives of the Escribano. Now this is what is most improbable to do without detection.

The claimants have offered the testimony of the three members of the *Junta de Fomento* in the year 1846, namely: Don José Vicente Segura, Don Maria Flores and Don José Maria Bassoco. The testimony of the first two was taken in Mexico, and that of the testimony of some eleven other witnesses was taken in San Francisco, and proves the genuineness of the documentary title obtained by Castillero; we shall limit ourselves to their testimony.

The following witnesses were sworn in San Francisco, and examined: José Maria Bassoco, before the District Judge; Manuel Couto, who was a Clerk in the *Junta de Fomento* in 1846; remained Clerk until the *Junta* was displaced by the *Administracion del Fondo* in 1853, and since then has remained Clerk in that Administration. He copied Castillero's proposals to the *Junta*. The testimony of Don José Maria Lafragua, Minister of Foreign Relations under President Salas, in the latter part of 1846, proves the *expedientes*.—Trans. 15.

Professor Balcarcel was a Member of the Faculty of the College of Mines in 1846; was present at the meeting of the Faculty when the result of the assay of quicksilver was ordered to be printed.—Trans. 1865.

Another witness, Antonio del Castillo, Member of the National College, kept the minutes of the proceedings of the result of the assay; deposited the specimens sent by Castillero in 1846, in the Cabinet of Minerals.—Trans. 1935.

It is useless to pursue this detailed inquiry into the testimony. It is set out in the Transcript, and direct and positive to the genuineness of the documents obtained by Castillero, in Mexico, in 1846.

The witnesses are eleven in number who have been examined in this country, and they stand free from any impeachment.

In addition to the testimony of Mr. Forsyth, and the eleven witnesses from Mexico, who testify to the genuineness of the documents obtained by Castillero in this city, there is added a document of some significance.

The witness Couto was asked on the cross-examination—"You say, that all these things were done by the *Junta?*" Answer of witness—"I do." "You know they were done by them?" "I do." "Was there a record kept of them—of the discussions of the *Junta* on the subject of such applications?" Answer of witness was—"The result of each day's deliberations was written down in a book, and signed by the gentlemen of the *Junta*. This record was called the book of the *Actas*." "Where is that book now?" Answer—"In my possession, in the office in Mexico. These books are not allowed to leave the office." "Were not the final resolutions of the *Junta* on the subject of any application extended more formally than in the *acts* or journal of their proceedings, to which you have referred?" Answer—"This book of the *acts* is a formal record. The first and last sheets of the book are stamped. The book is stamped in the stamp-paper office, and the pages are all numbered in that office before it is stamped, so that it is the same as if each page was stamped. The same was paid to the stamp office as if each leaf were stamped. The *Administrator de Papel Sellado* certifies on the first page of the book the number of pages in it. This book undergoes all this preparation before it is used by the *Junta*."

"Is this the only book in which the final resolutions of the *Junta* were recorded?"

Ans.—"Yes."

"Why did you not bring a copy of the acts of the *Junta* upon his petition with you, that being as you have said, a mode by which the *Junta* executed contracts and assumed obligations?"

Ans.—"Because my business here was only to prove my own signatures, and to prove that the dates upon the documents in the expediente of which I have spoken were the true dates of those papers, and to prove that they have been in my custody."

The result of this examination was to induce the taking measures to get the *actas* from Mexico, and they will be found in the Transcript, 1682. Here is a whole year's business of the mining body, their proceedings, day after day. Session of 6th May, 1846, there is an entry, stating that Don Andres Castillero appeared and made a verbal report regarding the discovery, denouncement, and actual condition of the quicksilver mine situated in the mining district of Upper California. The Junta resolved that Señor Castillero should present his indications in writing.

Session of 14th May, 1846. "From the Ministry of Justice, of date 9th, acknowledging receipt of the official letter, in which was communicated to it the discovery of the mine of quicksilver in Californias."—Trans. 1684, 1688.

Session 18th May, 1846. "In the third he informs, that of the one thousand two hundred and seventy quicksilver flasks existing in the negotiation, there are only a very few having flaws, and that they ought to be worth three dollars each per piece. Let this difference be represented to Señor Castillero, and to the Government, when his propositions are approved." Trans. p. 1690."

Session of 25th May, 1846. "An official letter from His Excellency the Minister of Justice, dated 20th, approving the propositions of Don Andres Castillero, which the Junta had transmitted to the Supreme Government, and informing that he had sent to the Ministry of Government the petition for two square leagues of land (sitio de ganado mayor) as a colonist, upon his mining property."—Trans. 1694.

"The Junta resolved that the proper judicial agreement be drawn up immediately, and that application be made for the five thousand dollars, on Mazatlan or Guadalajara, to which Castillero agreed; and finally, that by the mail of Wednesday the proper orders be sent to Tasco, that the administrator deliver to the order of Señor Don Tomas Ramon del Moral all the quicksilver flasks in good condition in the store-houses there, at the rate of two dollars."—Trans. p. 1694.

Session 29th May, 1846.—"It was also resolved, in conformity with the report of the Controller's Office, that twenty-five dollars be paid to the notary Calapiz, for the proceedings in the instrument of agreement, which had been made with

Don Andres Castillero, to assist his quicksilver enterprise in the mine of Santa Clara in Upper California, embraced in the official order for the suspension of all payments in this Branch."—Trans. p. 1697.

These "*actas,*" where we see from day to day entries of facts made at the proper dates to meet the acts of Castillero to procure his title, harmonize with the balance of the testimony to negative the idea of forgery or antedating, or that all the documents procured by Castillero in Mexico were all obtained long after the occupation of California by the Americans.

The questions the Court will now discuss are:

1. Whether the claimant has proved satisfactorily a ratification by the Supreme Government, of the act of Alcalde Pico, granting to him three thousand varas of land at the time he delivered to him the mine.

2. Whether the claimant openly and fairly submitted that transaction to the executive supervision, in the propositions he made.

These are important inquiries, in the solution of which, so far as this Court is concerned, the destiny of these three thousand varas depends; for if it be not ascertained beyond reasonable doubt that the grant of the Alcalde was as clearly and fairly presented in the propositions as the other things proposed, or that there is not a direct and distinct ratification of the Alcalde's act, the transaction is a nullity and has no legal existence.

In Castillero's communication to the *Junta de Fomento*, under date of 12th May, 1846, submitting his views as to a contract respecting a mine he had discovered and denounced, he submits various details; among them, he states that he had denounced and taken possession "not only of the said mine of Santa Clara, but also of the extent of three thousand varas in all directions from that point."

The statement of all he had to say as to his past acts and future views having been made, he closes with these words: "My propositions, then, are the following—"

Then follow nine propositions.

The *seventh* only, will now demand our attention.

The original communication of Castillero, with its various propositions, were sent by the *Junta* with their recommendation in favor of it, through the Minister of Justice to the President, who, therefore, had them before him.

The result of his action in the premises is the written order on the margin of the communication he had received from the *Junta*, recommending Castillero's propositions to his approval. It is in these words: "Granted *in the terms* which are proposed; and with respect to the land, let the corresponding order issue to the Minister of Relations for the *proper measures of his office*, with the understanding that the Supreme Government accedes to the proposition."

It is contended that the above order approves *in all its parts* the agreement. Such is the construction placed upon the order by the Minister of Justice in his letter communicating to the President of the *Junta de Fomento* the result of the action of his Excellency the President, under date of 20th May, 1846.

Counsel for claimant places as comprehensive a construction on the order as did Mr. Becerra, the Minister of Justice. He states: "*All* that is proposed is granted, *the two square leagues among other things*."—Mr. Peachy's Brief, 5.

Such are the interpretations placed upon the order of the President by those two gentlemen. The construction to be placed upon this little document now rests upon this Court; if it be erroneous, it is gratifying to feel that there is an appellate tribunal to correct such error.

As to the interpretation of the Minister of Justice, if his only object is to give his inferences and his interpretation of what was done by the President, his course may, in the estimation of many, be deemed correct; but it must strike some, that in the performance of his official duty as a subordinate officer, where large interests were concerned, it would have been as well to issue the order of his superior so that it might speak for itself, without placing his own construction on it. In doing so, he has introduced language not to be found in the order. The words "*in all its parts*," have been seized upon by all who, interested, have written or spoken about the title, and its ratification by the Supreme Government of Mexico, drawing their inference from those words, and the coloring thrown upon the order by them, when in fact no such words exist in the order.

That part of the interpretation of the learned counsel for claimant, which considers the language operates to convey the two square leagues, we will consider hereafter. The inquiry at present is exclusively as to the three thousand varas of surface land.

Castillero was not an illiterate, ignorant man; the evidence shows he was well informed and educated. He was familiar with the mining laws. Colonel Fremont testifies that in the early part of 1846, he obtained from Castillero all the necessary information which he (Fremont) was able to use for his individual benefit subsequently, in obtaining a mining title at Mariposa. It is reasonable to believe, that one so generally intelligent and cognizant of the mining laws, knew precisely what an ignorant Alcalde was doing at his suggestion, as it is reasonable to believe it was, as he was present, and it was done for his benefit. Castillero must have known that in the annals of his country there was no precedent for an Alcalde to grant three thousand varas to a miner, either as "pertenencias," or by way of grant.

It is important, in justice to the President and to all parties in interest, to look with care when such important interests were entrusted to him, to see, when he bound his country and himself, to what extent he did so. It cannot be urged that the interpretation of a subordinate of a Government of the legal document of his superior, is to be conclusive on this Court in the construction of it. We have cited the construction placed upon the order by one of the counsel for the claimant. The first part of it is in these words: "All that is proposed is granted." We do not differ in this from the learned counsel. We merely add the negative to his affirmation and add: "All that is *not* proposed is *not* granted."

We now turn to the proposition of Castillero. He had in his communication to the *Junta de Fomento*, in the statement he had made in it, preliminary to his propositions, informed that body he had "denounced and taken possession not only of said mine of Santa Clara, but also of an extent of three thousand varas in all directions from that point."

Thus far he as an honest man acted; but he looked to his interests, as we shall see.

By the minutes of the *Junta*, it appears that he appeared before them personally, and gave a verbal account of his discovery, and having been requested to make a written statement, he handed, on 12th May, 1846, his written propositions.

These propositions were either prepared or drafted by Castillero, or under his eye. Conversant as he was with the mining laws, he must have known what a gross violation of the ordinances the Alcalde had committed in granting him three thousand varas, which he stated in his communication to the *Junta de Fomento* he had denounced. Now, there is no ground known to the mining laws on which a denouncement could be made of three thousand varas as the appurtenance to a mine.

The statement of Castillero is not only without proof, but is negatived. In the act of possession the reason averred by the Alcalde for delivering the juridical possession of the mine was, that the time between the denouncement and this date had expired. This is the only time the word "denouncement" is used in the act of possession. After stating the delivery of the mine, he says, "I have granted three thousand varas."

Now the assertion of Castillero that he had denounced these three thousand varas, is falsified by the very record on which the title rests which is presented to this Court for confirmation. No man, shrewd and educated like Castillero, makes a false statement, deliberately in writing, without some motive. We can only gather that motive from his subsequent conduct. Now, the principal object of Castillero in invoking the action of the Supreme Government was to procure their ratification of the title to his mine, so as to secure it from any attack that might be made by reason of any serious irregularity that may have been committed in the proceedings; or from want of power in the Alcalde, there being no Mining Deputation in California. Another object was to procure pecuniary means by way of a loan, and the prompt use of retorts, cylinders, and other apparatus to work his mine. These were his principal objects, and with his knowledge of the gross violation of the ordinances which had been committed, he feared that an insertion of the grant by the Alcalde in his proposals, and asking a ratification of his possession under it, would jeopardize or defeat his whole application. He determines, therefore, to insert in the preliminary part of his written statement the Alcalde's grant, and so word his proposition as to possession, that if an unrestricted ratification was given, he might claim under it the land as the subject of the grant as well as that of the mining title.

His seventh proposition, the only one of the nine which asks for the approval of possession, is in these words: "The *Junta* shall represent to the Supreme Government the necessity of approving the possession which has been given me of the mine by the local authorities of California, in the same terms as I now hold it." Not one word is used that conveys an idea of the grant for three thousand varas, or the possession of it. Why was not an approval of the possession of those asked? Was it an inadvertence? It does not look like one. Castillero, if he intended this proposition to cover the land he held as well as the mine, could not be deemed acting inadvertently. He actually converts two separate and different acts of the Alcalde into one and the same transaction; it confounds two different transfers of different species of property, conveyed and regulated by different systems of law. It asks, if so intended, and if not intended why claimed, three thousand varas of surface, together with a mine, "*in the same terms as I now hold it.*"

The statement that he had *denounced* this land must have been introdnced from a desire to aid the effort made in the seventh proposition in converting the two different transactions into one. Now, if Castillero intended to effect his object of obtaining the approval of the Government in this mode, it was a fraud; if it was done through inadvertence, and he only in fact intended to propose for an approval of the possession of the mine, the grant gives him no right to the three thousand varas. The possession of them was never confirmed. To prove the separate character of the two acts, and that Castillero only proposed the approval of the possession of the mine, we submit the following reasons:

As we consider the views expressed by one of the counsel for claimant, as to the character of the delivery of the mine and the transfer of the three thousand varas, as correct, we shall cite them: It seems that in the year 1848, Mr. Alexander Forbes applied to Alcalde Weekes to grant an increase of "*pertenencias*" to the owners of the Almaden Mine, as provided by the ordinances. Upon this, the counsel for the Government objected: "Why, if the original Act of Possession, dated 30th December, 1845, had been in existence in January, 1848, by which the Alcalde granted a mining possession of the three thousand varas in every direction from the mouth of the mine, should Alexander Forbes have prayed the Alcalde to grant an increase of '*pertenencias*' in 1848, as provided by the Ordinances? For three thousand varas, measured in every direction from the mouth of the mine, includes many more "*pertenencias*" than the *seven* which Alcaldes are authorized to grant when a newly discovered mine is worked in partnership." To this objection the counsel replies: "A careful examination of the Act of Possession given by Pico in 1845, discloses the fact that the Alcalde gave possession of the mine, *and also* granted to Castillero three thousand varas in every direction from the mouth of the mine. He first declares that he gives juridical possession of the mine known by the name of Santa Clara, and then proceeding to enumerate several reasons for what follows, he *grants* to Castillero three thousand varas, etc., so that the giving juridical possession of the mine was *one act*, and the granting of three thousand varas was *another act*, both of which were recorded in the same instrument, which is usually called the Act of Possession. What, then, did he mean by the mine? What was its extent? Evidently he intended by the mine such a number of '*pertenencias*' as the discoverer of a mine in a new hill was entitled to by the ordinances, which would be *three* if he worked the mine alone, and *seven* if he worked it in partnership with others."

The distinction between the *judicial* possession of the mine and the grant of three thousand varas, is very explicitly, says the counsel, stated by Castillero in his proposals to the *Junta de Fomento* on the 12th May, 1846. There he says: "Wherefore I have denounced and taken possession, not only of the said mine, but also of three thousand varas," etc.

Now, what we consider is, that Castillero himself being personally conversant with the difference between the possession of the mine and the possession of the land, seeks to sustain his title to the latter upon a grant of the Supreme Government as a ratification, and to which he is not entitled, as he did not propose it for ratification in his seventh proposition, or in any other.

Among the nine propositions submitted by Castillero to the *Junta de Fomento* is the eighth, which is in these words: "It shall also represent [to the Supreme Government] the advantage of there being granted to me, *as a colonist*, two square leagues upon the land of my mining possession, with the object of being able to use the wood for my burnings."

We have seen that the written communication by Castillero, with his nine propositions annexed, were transmitted by that body to the President, with their commendation of the petition of Señor Castillero. That the petition was before the President, with its various propositions, together with the recommendation of the *Junta.* On the margin of this communication from that body the grant for two leagues of land is to be found, as alleged.

Before giving the language which is relied upon as a grant of two leagues of land, the Court will advert to the announcement made of its interpretation by Mr. Becerra of the action by the President, under date 20th May, 1846, which was the first announcement that was made to D. Vicente Segura, President of the *Junta.*

Mr. Becerra wrote that "His Excellency has been pleased to approve, *in all its parts, the agreement* made with that individual." Where is the agreement to which Mr. Becerra alludes?

Castillero had made no agreement with the *Junta de Fomento.* All he did do, he tells us at the end of his proposals: "The subscriber submits this *request* to the deliberation of the *Junta,* which, if accepted, may be made into a formal contract and made legal in the most proper manner."

The communication of the *Junta* was not in form or substance a petition. Petitions are not made from one department to another of the same Government. In the exercise of their powers under the mining laws, the *Junta* exercised their right to recommend the petition of Castillero to the Supreme Government.

There was no agreement before the President; but there was a recommendation of the *Junta*. What they sent for his approval, the only petition made by any one, was the *petition of Castillero* in relation to which the *Junta* say in their letter: "In this view, the *Junta*, in sending up to your Excellency *the petition of Señor Castillero*, has no hesitation of recommending it very efficaciously," etc. Now to what petition did the President accede, and what did he grant? He must have referred to Castillero's petition, the only one before him, and the word *grant* must be of things in the terms proposed by Castillero in his propositions.

On one of the papers before him (the recommendation of the *Junta*), is an *acuerdo* or written order signed with the rubric of the Minister of Justice. It is in these words: "May 20, 1846. Granted in the terms proposed; and with respect to the land, let the corresponding order issue to the Minister of Relations for the proceedings so far as his office is concerned, with the understanding that the Supreme Government accedes to the petition."

It is strenuously urged, that among the nine things proposed, all being proposed are granted, and that the two square leagues among the rest are conveyed.

Now, all that Castillero asked the Junta to recommend to the Supreme Government in his eighth proposition was a colonization grant. His proposal was, in terms, to grant him two leagues *as a colonist.*

This was all the President did; it was all that Castillero proposed. To consider that the word "granted" applied to *all* the propositions of Castillero, and conveys and transfers all property in each, is to do violence to the language, and to annihilate an axiom of almost universal prevalence—"*Expressio unius, est exclusio alterius.*"

The word "granted" applies to all that is proposed, which is not excluded. But after using that word, it excepts the land, and thus qualifies that word in relation to the land. After the word "granted," in the terms proposed, the language as to the land is, "and *in respect to the land*, let the corresponding order issue to the Minister of Relations for the proper measures of his office, with the understanding that the Supreme Government accedes to the petition."

Becerra, the Minister of Justice, in his communication to Vicente Segura, President of the *Junta de Fomento*, states the *agreement* made with Castillero had been approved in all its parts; also states (and it is the only reference he makes to the land) "that on this day the corresponding communication is made to the Minister of Exterior Relations and Government, to issue the proper orders respecting which is contained in the 8th proposition for the grant of lands in that Department."

The Minister of Justice sent a copy of this letter to Castillo Lanzas, the Minister of Relations, and says: "I have the honor to transcribe it to your Excellency, to the end that, with respect to the petition of Señor Castillero, to which His Excellency the President *ad interim* has thought proper to accede, there be granted to him as a colonist two square leagues upon the *land of his mining possession.* Your Excellency will be pleased to issue the orders corresponding."

Lanzas, the Minister of Relations, on the 23d May, 1846, at Mexico, addresses a communication to the Governor of the Department of California.

He premises what he has to say to the Departmental Governor, with a copy of the letter which Mr. Becerra, the Minister of Justice, had written to the *Junta* on 20th May, 1846, and sent one to him, and a copy of the letter which he, Lanzas, had received from the Minister of Relations, and then Lanzas concludes to the Governor in the following language: "Wherefore, I transcribe it to your Excellency, in order that under what is *prescribed by the laws and dispositions upon colonization*, you may put Señor Castillero in possession of the two square leagues which are mentioned."

Now the interpretation by Lanzas of the *grant* was correct. Castillero proposed for a grant as a colonist, in his eighth proposition. It was such, and was the only one proposed that the Supreme Government had granted. The grant was viewed by the Minister in its true light, a colonization grant; he did not direct the Governor to place Castillero in possession, but stated that he had transcribed to His Excellency in order that under what is prescribed *by the laws and dispositions upon colonization, you may put Castillero in possession of the two square leagues* which are mentioned.

In the recommendation made by the *Junta* in favor of his application for land, they declined to express any opinion.

It was, in fact, a subject with which they had nothing to do. The origin of title to mines, and of the granting of lands, and the regulations of them, belonging to two different systems of law; and those laws to the administration of different tribunals. The colonization decree of 1824, and the regulations of 1828, made under that law, had given the granting power to the Governors of departments, subject to the approval of the department, and in special cases to the Supreme Government.

The Minister of Exterior Relations and Government was the functionary through whom the President communicated with the Governors of the departments, and hence the order in this case to the functionary having charge of the granting power where the land was situated, and where the provisions made by the Colonization Decree of 1824, and the regulations of 1828, made for the protection of the rights of third parties, which by the local authorities could be more practically enforced than in Mexico.

Here ends the Expediente of the two square leagues of land. A grant was never issued, nor was a solitary movement made by the Governor towards one, nor a survey of the land made or possession of it given by any Mexican functionary.

The letter addressed to the Governor of California by Castillo Lanzas, was dated in Mexico, May 23d, 1846, some ten days *after* the 13th May 1846, the day the Congress of the United States declared by resolution, and the President announced by proclamation to the people of the United States, that war existed between the United States and Mexico by virtue of the commencement of actual hostilities on the part of the Republic of Mexico.

The Supreme Court of the United States say: From the capture of Monterey on the 7th July, 1846, till the surrender of Los Angeles, and the organization of a Territorial Government by Commodore Stockton, under the United States, there were about six weeks. * * * In the Act of 1851, and the decisions of this Court, that day is referred to as the epoch at which the power of the Governor of California under the authority of Mexico to alienate the public domain terminated.—U. S. vs. Pico, 23 How. 321.

The case of the United States vs. Castillero (23 How. 464), has been cited as a controlling authority in this. If it be in fact so, it will be the duty of this Court so to consider; but in its opinion, it cannot be so regarded, without violating clear principles of judicial construction.

In the case of Cohens vs. Virginia (6 Wheaton, 264), the canons of judicial construction are thus laid down by Chief Justice Marshall: "It is a maxim not to be disregarded, that general expressions in an opinion are to be taken in connection with the case in which those expressions are used. If they be beyond the case, they may be respected, but ought not to control the judgment in a subsequent suit, *when the very point is not presented.* The reason of this maxim is obvious. The question actually before the Court is investigated with care and consideration in its full extent; other principles which may seem to illustrate it, are considered in relation *to the case decided;* but their possible bearing or relation on all other cases is seldom completely investigated."

Now, apply these rules of judicial construction to the case cited.

For years the granting of lands situate in the Department of California, and up to the time of the acquisition of this country by the United States, were regulated by the Colonization Law of 1824, and the Regulations of 1828, of Mexico. Provisions which she deemed called for by her colonization policy, and conditions and terms having for their object the protection of the rights of third parties, were embodied in them.

Among them, one of vast importance confided the granting of lands to the local authorities—the Governors, or, as they once were termed, "Political Chiefs," subject to the approval of the Departmental Assemblies, and, in cases specially provided for, to the action of the Supreme Government.

This was the uniform, almost universal law. This Court is asked to decide in this case, where no grant from the Governor was ever placed upon the record, where no grant by that functionary was ever made, where no "*informe*" was taken, where no solitary provision of the law made for the protection of third parties was observed, to consider the *decision* of the Supreme Court in the above case, which enunciates the principle "that the Colonization Law of Mexico of 1824 and the Regulations of 1828," do not apply to "*islands*" situate on the coast, a *decision authoritative* in this case. If it is to be so treated, this Court will violate all the principles of judicial construction, and decide that *whether or not islands on the coast are subject to the Colonization Law and Regulations* is the "*very point presented*" *in this case, which is, whether those laws and regulations apply to colonization grants of lands in the interior of the country.*

In the case of the United States vs. Osio (23 How. 273), the Supreme Court decided, that an "island" situated in the Bay of San Francisco, not claimed under the Colonization Law of 1824 or Regulations of 1828, but under certain special orders issued to the Governor by the Mexican Government, and the Governor issued a grant under them, the *power* of the Supreme Government to grant the *island* was deemed undoubted; but the claim to it was rejected on other grounds.

At the same term (23 How. 464), between the same parties now before the Court, after making a statement of the petitioner's title, the Court in that case discuss the question of the character of the property claimed. They say (p. 465) "*islands* situated on the coast, it seems, were never granted by the Governors of California or by any of her authorities, under the Colonization Law of 1824 or the Regulations of 1828. From all that has been exhibited in cases of this description, the better opinion is, that the power to grant the lands of the 'islands' was neither claimed nor exercised by the authorities of the Department prior to the twentieth day of July, 1838, as was satisfactorily shown in one or more cases heretofore considered and decided by this Court.

"On that day (July 20, 1838), the Minister of the Interior, by the order of the Mexican President, addressed a communication to Governor Alvarado, authorizing him, *in concurrence with the Departmental Assembly*, to grant and distribute the lands of the desert islands adjacent to that Department, to the citizens of the nation who might solicit the same. * * * * *

"Grants made by the Governor under the power conferred by this dispatch, without the concurrence of the Departmental Assembly, were simply void, for the reason that the power, being a special one, could only be exercised in the manner prescribed. It was so held by this Court in United States vs. Osio, decided at the present term, and we are satisfied that the decision was correct."—*Ibid*, 466.

But the grant in this case (say the Court), was not made under the general authority conferred by that dispatch.

On the same day, the 20th July, 1838, a dispatch of a special character was addressed by the same Cabinet Minister to the Governor.

"By the terms of the communication the Governor is informed, that the President, regarding the services rendered by the claimant, etc., has directed the Minister to recommend strongly to the Governor *and the Departmental Assembly*, that one of the islands, such as the claimant might select, near where he ought to reside with the troops under his command, be assigned to him, before they proceed to grant and distribute such lands under the general authority conferred by the previous dispatch."

It is upon the following remark of Mr. Justice Clifford, upon the facts stated by him above, that reliance is placed to show that the case is decisive upon the question before this Court on a colonization grant. "Beyond question," says Mr. Justice Clifford, "the legal effect of that second communication was to withdraw such one of the islands as should be selected by the claimant from the previous order, and to direct that it be assigned to this claimant." And the following remark is also relied on, which will be found on page 469: "Emanating as the dispatch did from the Supreme power of the nation, it operated of itself to adjudicate the title to the claimant, leaving no discretion to be exercised by the authorities of the Department."

On what was based the assertion of the doctrine the Court had above enunciated? Why, upon the decision made that the granting of "islands" on the coast had never been claimed, nor the power over them exercised by the authorities of the Department; and that the power of granting "islands" was in the Supreme Government, and to which the colonization laws and regulations did not apply.

There is nothing in this case to justify the idea, that if the Court had decided that those laws and regulations DID apply to islands, they would have decided as they have done; and by so doing, thus impute to them a decision which would sanction the repeal by the President of Mexico of all the colonization laws by an act embodied into an order for the violation of them.

Such was not the action of the Court.

Having decided that the Colonization Laws and Regulations did not apply to "islands," they administered to them those laws which in their opinion did.

Now, Spain was an absolute Government, but at the same time a government of laws, and the absolutism of the monarch consisted in the concentration in his person of all the powers of Government, executive and legislative. He could, therefore, change the law at pleasure, but while the law continued in force, he was as much bound by it as was his subject; and any act done by him in contravention of *existing* laws should be disregarded by the Judiciary.

By one of the decrees made as early as the year 1369, it was provided: "*Royal Letters or Warrants* that may be granted contrary to law or contrary to our judicial system, invalid and not to be executed."

This is the caption of the decree which was issued by Henry II, at Tors. The decree itself provides, "That sometimes it happens, that by personal importunities, or in some other manner, we grant Royal Letters or Warrants contrary to law or contrary to our judicial system, by the present decree we ordain that such Royal Letters or Warrants shall be of no value, and shall not be executed, although they may contain a clause *that* they may be complied with, notwithstanding any privilege, order, or ordinance to the contrary."—*Novisima Recopilacion*, Book III, Law 11.

In the year 1448. John II, at Valladolid, issued a decree with the following caption:

"Royal orders to dispossess any person of *his property*, without being first heard and sentenced, invalid and not to be executed." This is the caption.

The Decree itself prescribes: "If it shall happen that we have given, or shall give any Royal Letter *that any* be dispossessed of his property, or public office, and should it be given to others, it is our will and command that such letters may be respected, but not to be executed; and as it is never to be understood that we take property from any person without being first notified and sentenced, and care must be taken that *the laws of the kingdom must be complied with in all* such cases, and the same to be strictly observed in every particular as they are written, etc. * * * * * But if any one in a public office commits any notorious malpractice, the same being certified up to us, the letters we thereupon may give, we command they shall be complied with."—*Ibid,* Book III, Title IV, Law VI.

To repeal a system of laws and regulations made under them, which have prevailed for a series of years "*in* writing," and have become a rule of property in the acquisition of it, from the Government, without any legitimate exercise of the right of previous repeal, is not a principle to be recognized by this Court as existing in a government of laws, unless the appellate tribunal shall so decide. This we do not consider that Court has yet done. In the case cited as controlling this; they have decided that the Colonization Law of 1824, and the Regulations of 1828, "did not apply to islands," and this is all they have decided. We return now to what is called the grant. "If this grant is anything that passes title, it is a colonization grant." Such a one the petitioner applied for in terms, "as a colonist." Such an one, the President, therefore, in the "*acuerdo*" granted, when he directed, with respect to the land, "let the corresponding order issue to the Minister of Exterior Relations for the corresponding measures of his office, with the understanding that the Supreme Government accedes to the petition."

Mr. Becerra, the Minister of Justice, tells the President of the *Junta de Fomento* what kind of grant it was, if we did not learn it from the eighth proposition of Castillero, and the fact displayed on the marginal title by the direction of the President, that the grant to become one must take the course that all colonization grants of lands must take to be valid; that is, must be made under the then *existing* laws.

Mr. Becerra, in his communication of 20th May, 1846, to Castillo Lanzas, transcribes for the latter a copy of foregoing letter he had written on same date to the President of the "*Junta*" to the end "with respect to the petition of Señor Castillero, to which His Excellency the President *ad interim* has thought proper to accede, that there be granted to him, *as a colonist*, two square leagues *upon the land of his mining possession.* Your Excellency will be pleased to issue the orders accordingly."

Lanzas addresses in Mexico, in obedience to foregoing orders, a communication to the Governor of the Department of California, and concludes his letter in these words: "wherefore I transcribe this letter [the copy of which he had received from Mr. Becerra, the Minister of Justice] to your Excellency, in order *that under what is prescribed by the laws and dispositions* upon colonization, you *may* put Señor Castillero in possession of the two square leagues which are mentioned."

Now, this order for two leagues, from the President down to the last functionary, had been treated as a colonization grant, and none other can it be considered; as such it must have applied to it the laws and regulations which govern such, or it is invalid to pass title.

This Castillo Lanzas dispatch was issued, as we have seen, on the 23d May, 1846, in Mexico, so short a time before the day when, by the decision of the highest judicial tribunal in our country, the power of the Government of Mexico to alienate any of its public domain had ceased to exist (23 How. 321). There could not have been a sufficient time within which to place Castillero in possession "in conformity with the laws and dispositions upon colonization."

The condition of California is portrayed by James Alexander Forbes at the time he held an interest in the claim in this case, and was on friendly terms with his associates, in his letter of the 5th May, 1847: "This Departmental Government [he writes to Alexander Forbes] is completely '*acefalo;*' that is to say, it has no Mexican head, or Governor; in consequence of which, the possession of the two *sitios*, ordered by the dispatch of Señor Castillo Lanzas, has not been obtained, nor cannot be obtained, nor even mentioned without imminent risk of opposition on the part of the American Government in this Department."—Trans. 843.

And yet the Government made as much speed as it well could do. On the 20th May, 1846, the alleged grant was made; on the same day it was transmitted by the Minister of Justice, Mr. Becerra, to the Minister of Exterior Relations, and the dispatch drafted in less than three days from the date of the alleged grant. All was vain; the Government found she had undertaken to alienate a large body of land, the *possession* of which she could not deliver under either the laws of colonization or any other.

That the Lanzas dispatch is no grant—that the Supreme Court has confirmed no one claim under what purported on its face to be a colonization grant—and no grant was given, where not a single provision of the law or regulations on colonization had ever been complied with—this Court has been unable to find; and if it existed, the learning and untiring research of the numerous counsel in this case would have detected such. In the case of The United States vs. Osio, and of The United States vs. Castillero, both claimants had grants from the Governor. The title of the claimant in the latter case is thus described by the Judge who delivered the opinion: "All of the documentary evidences of title produced in the case are duly certified copies of originals found in the Mexican Archives, as appears by the certificate of the Surveyor-General, which makes a part of the record. * * They consist of a special dispatch from the Minister of the Interior of Mexico addressed to Governor Alvarado, the petition of the claimant for the same, and the *original grant to the petitioner*, signed by the Governor, and countersigned by the Secretary of State of the Department." The grant was made by the Governor in 1839.

The case of the United States vs. Osio was rejected, mainly on the ground that the granting power was not exercised in strict conformity to the general dispatch from the General Government under which he acted.

The Castillero case was decided in favor of the claimant, on the ground that the Governor acted in strict conformity with the special dispatch on which he based his grant.

They gave precedence in both cases to the action of the Supreme Government in the granting of "*islands*" over the Departmental Governors, as the colonization law and regulations did not apply to them.

We cannot consider The United States vs. Castillero cited as controlling this case. There are some curious features here, apart from those already alluded to. Considering this document in any light, it must be regarded as inchoate and imperfect. It left the President's hands in a direction given to all colonization grants.

The Secretary of Foreign Relations drafted the order to the Governor of California, and closed his communication with these words: "Wherefore I transcribe to your Excellency, in order that in conformity with what is prescribed by the laws and dispositions upon colonization, you may put Señor Castillero in possession of the two square leagues mentioned."

This seems to be plain enough. If the land is not found, and cannot be delivered—if the condition of the country is such that the grantor has lost the power of consummating the contract and delivering the land—it is difficult to perceive how either equitable or legal principles interpose to demand relief in equity in a case where such claim is not legal.

By the insertion by Mr. Becerra, the Minister of Justice, in his letters of the action of the President, we find the land located by Mr. Becerra on the *mining possession of Castillero*. As the mine is described in the act of possession as *situated on the rancho of the retired sergeant José Reyes Berreyesa*, it follows that the location of two square leagues upon the lands of Castillero's mining possession must injuriously affect the rights of a part at least of another man's land for the use of that other. While we admit the power of the Supreme Government to take the land of one and give it to another, under the laws of Spain and Mexico, for mining purposes, and on the terms they prescribe, we cannot acquiesce in its power to take from its owner his lands for any purpose not authorized by law.

This Court have decided in this case, that the three thousand varas granted by the Alcalde to Castillero, was an unwarrantable exercise of authority on the part of the Alcalde. We regarded, however, the merits of the claimant as both discoverer of a mine of quicksilver, and as working it in copartnership; and the Court so considering, concluded to award him the highest number of "*pertenencias*" allowed to any discoverer and partner in working a mine.

The "*pertenencias*," as to mines, are regulated by a separate body of laws, and the Court has awarded the claimant the highest number that those laws concede.

It considers, therefore, that it has gone as far as any equity demands in such a case.

DECREE.

At a stated term of the District Court of the United States for the Northern District of California, held at the Court-room in the City of San Francisco, on Friday, the 18th day of January, A. D. 1861.

Present: Hon. M. HALL MCALLISTER, Circuit Judge.
Hon. OGDEN HOFFMAN, District Judge.

THE UNITED STATES *vs.* ANDRES CASTILLERO.	}	Dist. Court, No. 420. Land Com. No. 366. "New Almaden."

DECREE.

This cause came on to be heard on appeal from the final decision of the Board of Commissioners to ascertain and settle Private Land Claims in the State of California, under the Act of Congress approved March 3d, 1851, upon the transcript of the proceedings and decision of the said Board, and the papers and evidence on which the said decision was founded, and upon the papers and evidence filed in this Court; and it appearing to the Court that the said transcript was duly filed according to law, and counsel for the respective parties having been heard, and due deliberation had in the premises, it is by the Court hereby ordered, adjudged and decreed, that the said decision be, and the same is, reversed and set aside: And it is likewise ordered, adjudged and decreed, that the claim and title of the petitioner, Andres Castillero, to the mine known by the name of New Almaden, in Santa Clara county, Northern District of the State of California, is a good and valid claim and title, and that the said Andres Castillero and his assigns are the owners thereof, and of all the ores and minerals of whatsoever description therein, in fee simple: And it is further adjudged and decreed, that the said mine is a piece of land embracing a superficial area, measured on a horizontal plane, equivalent to seven pertenencias; each pertenencia being a solid of a rectangular base two hundred Castilian varas long, of the width established by the Ordenanzas de Mineria of 1783, and in depth extending from and including the surface, down to the centre of the earth; said pertenencias to be located in such manner as the said Andres Castillero or his assigns may select, subject to the following conditions: first, that the said pertenencias shall be contiguous, that is to say, in one body; and secondly, that within them shall be included the original mouth of the said mine known as "New Almaden."

And it is further ordered, adjudged and decreed, as to all other rights, property and interests set out and claimed in the petition filed in this case, that the same are not valid, and are therefore rejected.

M. HALL McALLISTER,
Judge Circuit Court U. S.

OGDEN HOFFMAN,
District Judge.

January 18, 1861.

Filed January 18, 1861,
W. H. CHEVERS, Clerk.

ORDER GRANTING APPEAL
IN BEHALF OF CLAIMANT.

At a stated term of the District Court of the United States of America, for the Northern District of California, held at the Court-room in the City of San Francisco, on Saturday, the 19th day of January, in the year of our Lord one thousand eight hundred and sixty-one.

Present, the Honorable OGDEN HOFFMAN, District Judge.

THE UNITED STATES
vs.
ANDRES CASTILLERO.
} No. 420.

And now, at this day, on application of A. C. Peachy, in behalf of said claimant, made in open Court, it is ordered, that an appeal from the final decision of the Court rendered in said cause at the present term be, and the same is hereby granted; and that a certified transcript of the pleadings, evidence, depositions and proceedings in the said cause, be sent to the Supreme Court of the United States without delay.

Filed January 19, 1861,
W. H. CHEVERS, Clerk.

ORDER GRANTING APPEAL
IN BEHALF OF THE UNITED STATES.

—

At a stated term of the District Court of the United States of America, for the Northern District of California, held at the Court-room in the City of San Francisco, on Monday, the third day of December, in the year of our Lord one thousand eight hundred and sixty.

Present, the Honorable OGDEN HOFFMAN, District Judge.

THE UNITED STATES *vs.* ANDRES CASTILLERO.	No. 420. Order of Appeal for the United States.

In this case, on the application of Calhoun Benham, United States District Attorney, made in open Court, it is ordered by the Court, that an appeal in behalf of the United States from the final decision of this Court rendered in said cause at the present term be, and the same is hereby granted, and that a certified transcript of the pleadings, evidence, depositions and proceedings in the said cause, be sent to the Supreme Court of the United States without delay.

January 25, 1861.

Filed January 25, 1861,

W. H. CHEVERS, Clerk.

In the United States District Court,

NORTHERN DISTRICT OF CALIFORNIA.

THE UNITED STATES
vs. — No. 420.
ANDRES CASTILLERO

"NEW ALMADEN."

OPINION

OF HIS HONOR

OGDEN HOFFMAN,

District Judge,

DELIVERED, JANUARY 17, 1861.

SAN FRANCISCO:
COMMERCIAL STEAM BOOK AND JOB PRINTING ESTABLISHMENT.
1861.

OPINION

OF

HIS HONOR OGDEN HOFFMAN,

DISTRICT JUDGE,

DELIVERED, JANUARY 17, 1861.

THE UNITED STATES vs. ANDRES CASTILLERO, } "NEW ALMADEN."

The claimant in this case asks the confirmation of his title to,

1st, A mine of quicksilver situated in Santa Clara county, and called the mine of "New Almaden," including three thousand varas of ground measured in all directions from the mouth of the mine.

2d, Two square leagues of land, situated upon the land of the above-mentioned mining possession.

It is objected that this Court has no jurisdiction to decree a confirmation of the former claim. The objections relied on by the United States are two:—

1. That by the Act of 3d March, 1851, the jurisdiction of this Court is confined to cases where the interest claimed is a fee simple interest, or such as in equity should be converted into one, and that such interest was not held by the owner or grantee of a mine by Mexican law.

2. That the subject matter of this claim is not "*land*," within the meaning of the act.

The first objection is not only wholly unsupported by the words of the Act, but it is inconsistent with the sense of justice and the sacredness of treaty stipulations which we must presume to have actuated Congress.

1. The Act declares that every person claiming lands by virtue of *any* right or title derived from the Spanish or Mexican Governments, shall present his claim, etc.

There is thus no specification as to the *nature* of the claim, except that it must be to "*land*," nor of the *estate* or *interest* claimed. It may be leasehold or freehold, conditional or indefeasible, for years or in fee.

2. That such estates constitute property, in its fullest sense, will not be denied.

If, then, claims to such estates in lands cannot be presented for confirmation under the Act of 1851, or if presented are to be rejected, this species of property in land is not merely unprotected, but by the terms of the Act it is confiscated; for the 13th Section provides that all lands, the claims to which have been finally rejected, or which shall not have been presented within two years, shall be deemed held and considered as part of the public domain of the United States.

It cannot, therefore, be imputed to Congress that it meant to declare all lands public property the claims to which were not presented, and at the same time restrict the right of presenting claims to a limited number of cases, thus ignoring and confiscating all rights of property in land but those of a particular description, when all were equally sacred under the laws of nations and the stipulations of the Treaty.

3. But even if the jurisdiction of this Court were limited, as supposed, the estate of the owner of a mine under the Mexican and Spanish Ordinances was of at least as high a degree as a determinable fee at common law, and the concession of a mine conveyed to him full property in the very substance of the mine.

As, however, the nature of the mine owner's estate in a mine cannot be considered without examining and ascertaining what was the nature of a mine itself, and whether it constituted "land," as that term is used in the common law, and in the Act of Congress, the point will be treated of in considering the second objection to the jurisdiction of the Court.

2. Is a mine *land* within the meaning of Act of Congress?

"By the civil law," says Gamboa, "all veins and mineral deposits of gold and silver ore, or of precious stones, belonged, if in public ground, to the Sovereign, and were part of his patrimony, but if on private ground, to the owner of the land, subject to the condition that if worked by the owner he was bound to render a tenth part of the produce to the Prince as a right attaching to his Crown. It subsequently became an established custom in most kingdoms, that all veins of the precious metals, and the produce of such veins, should vest in the Crown and be held to be a part of the patrimony of the King or Sovereign Prince. That this is the case with respect to the Empire of Germany, the Electorates, France, Portugal, Arragon and Catalonia, appears from the laws of those countries, and from the authority of various authors."—1 Heath. Gamboa, 15.

The reasons for attributing to the Sovereign this right of property in a part of the soil of the land of his subjects, it is not necessary to recapitulate; but that the distinction between the ownership of the surface, and that of the mines or minerals beneath, it was recognized at a very early period, appears from the law of the Partida, which declares that the property of the mines shall not pass in a grant of the land by the King, although not excepted out of the grant—and even if included in it, the grant shall be valid as to mines only during the life of the King who made it; and a similar rule prevailed in England with respect to mines of the royal metals, which alone were held to belong to the Sovereign by prerogative.

The distinction thus drawn between the right of property in the surface, and that in the minerals beneath, is founded on the essential difference in the qualities of the various substances of which the earth is composed. It has accordingly been recognized in the jurisprudence of all nations.

In Spain, as we have seen, mines do not pass in grants of land (fundos) by the Sovereign, unless particularly mentioned. (1 Heathf. Gamb. 132.) To the same effect is Solorzano, (Pol. Ind. lib. 6, ch. 1. No. 17,) who says: "So that, although private persons may allege and prove that they possess such lands (tierras) and their appurtenances by special gift and concession of the Prince, no matter how general may be the words in which the grant is made, this will not of itself be of any avail or advantage to them for acquiring or gaining thereby the mines which may be discovered in the lands, unless it is so specially provided and expressed in said grant;" and Colmeira, (Der. Adm. Esp.) observes: "Jurisconsults and Publicists agree in the opinion that it is proper to distinguish in the soil (en el suelo) the right of property in the superficies (de la superficie) from that in the depth (del fondo). Truly, the man who acquires a piece of land, gives not the least labor nor advances the smallest capital in consideration of the riches (las riquezas) which it may conceal. He examines its fertility, its situation, its extent, and all the circumstances which determine its value as a building-lot or agricultural land, but he does not take into account the mines which, perchance, it may conceal in its bowels. There is not, then, the least relation between the proprietor (of the land) and the subterranean matter from which any right can be deduced."

So, Lares, a Mexican author, in his Derecho Administrativo, p. 93, remarks: "The law, then, has not recognized property in the mine to be in the owner of the soil, but has made the property in the mine to consist in the grant which the Nation makes to him who registers or denounces it conformably to the ordinance."

By the French law of 1810, the same distinction is recognized, and all mines of gold, silver, platinum, lead, mercury, etc., are declared not to belong to the owner of the soil, but to be governed by the Mining Laws (Teulet's Sup. to Codes, Verb. Mines). "From the moment" (says the author last cited) "when a mine shall be conceded even to the proprietor of the surface, this property shall be distinguished from that of the surface, and thenceforth considered a new property," and, Le Guay, in his Thesis on Mining Legislation, says he "agrees with Mirabeau in the proposition that any legislation which does not recognize two species of property, one in the surface of the earth, and the other in its depth, would be absurd"—p. 125.

In his work on the General Jurisprudence of Mines, M.

Blavier observes: "It is a truth which has been admitted for a long time, that the preservation and prosperity of mines depend essentially on the adoption of a system of laws calculated to reconcile the interests of the public with that of those who work them. * * * It is these conditions which the governments of ancient and modern states have sought to fulfill in admitting, nearly all of them, that the Sovereign alone has the right to dispose of the public property in mines, or to confer on others the useful enjoyment thereof."

This regalian right, or right reserved by the whole state to dispose of subterranean property as public property, independent of the private ownership of the land which conceals it, seems to have been recognized in Germany from the earliest periods; (Cancrin. Dr. Pub. des Mines en Allemagne, p. 2); and M. Blavier remarks: "It is, therefore, not surprising that this regalian right should have been confirmed by nearly all the governments of the northern countries."

But it is unnecessary to multiply authorities in support of a principle so universally recognized.

From the foregoing citations it is clear—

1. That the grant or ownership of the superficies or land which contains the mine, confers no right of property to mines in the bowels of the earth, or even on its surface; and

2. That this property in the mines is, by common consent of almost all modern States, deemed to be vested in the State represented by the Sovereign, and is a regalian right. Indeed, it would seem that these two propositions are sufficiently demonstrated by the fact that mining ordinances exist in almost every State possessed of considerable mineral wealth; for those ordinances are but the dispositions made by law for the exercise of this regalian right, or right of property in all mines, whether in public or private land.

Having thus established that the right of property in mines is wholly distinct from the ownership of the soil, and that the former remains in the Sovereign, notwithstanding the possession of the latter by a private person, we proceed to inquire—

1. Did this right to the mine constitute a right to "land," as that term is used in the Acts of Congress?

2. If so, did the miner by registry and denouncement acquire an estate in "land," and what was its quantity and quality?

These two inquiries, though in form separate, are nevertheless intimately connected; for the nature of the subject matter of the right of property reserved by the Sovereign, can best be ascertained by inquiring what was the subject matter of the estate conveyed to the subject by registry and denouncement, as also the quality and quantity of such estate. "For how," as remarked by Lares, "can the nation grant that which it has not; or how can it give to one person what belongs to another?"

It will be shown that minerals owned by a title wholly independent of the property in the lands in which they are situated, still form a part of the land itself, and constitute land in the strictly legal acceptation of that term at common law. Independently of authority, it would seem clear that, it being admitted that "various sections of the soil divided horizontally" (1 Eng. Law and Eq. Rep. 249) may belong as separate properties to different persons, the sections which are beneath the surface must constitue "land," as much as those which are upon it. The vein of ore or the bed of coal is "land," as fully and completely as the superficial earth, and he who owns the former owns *land* as much as he who owns the latter. On this point the English authorities are agreed.

Nor is the application of those authorities to the question under consideration at all affected by the circumstance that, in England, the regalian right to mines is held to extend only to mines of the precious metals; for the owner of lands in which mines of the base metals are situated, having alienated to one person the mine, but reserved to himself or alienated to another the lands in which the mine is found, the question whether the owner of the mine is an owner of "land," would be presented precisely as if the severance of the two rights of property had originally existed, and the several properties derived by separate grants from the Crown.

"When the mines form part of the general inheritance, they will, of course, be transferred along with the lands without being expressly mentioned in the conveyance; but when they form a distinct possession or inheritance, a title to them must be established, without reference to the general title of the lands in which they are situated."—Bainbridge on Mines, p. 4; Rockwell, p. 536.

In the latter situation, " they still, of course, retain the qualities of real estate, and will be transferred by conveyances applicable to the particular disposition of them intended to be made."—*Ibid.*

They are capable of livery and being made the subjects of ejectment.—Cro. Jac. 150.

"By the name of *minera*, or *fodina plumbi*, the land itself shall pass in a grant if livery be made, and also be recovered in an assize."—Co. Litt. 6 a.

It is held that mines do not lie in grant, but pass like other hereditaments by livery of seizin.—Chetham vs. Williamson, 4 East, 476; 2 Barn. and Cress. 197.

When a reservation in a deed of feoffment is made in favor of the grantor of mines, no livery is necessary, for the grantor will never have been out of possession; but when the exception is in favor of a stranger to the legal estate, livery cannot be dispensed with.—Co. Litt. 47 a.; 1 Ad. & Ell. 748.

In numerous English cases, the distinction is clearly drawn between a mere license to dig and carry away ores, in land of which the possession is retained by the grantor, and a demise of the mine itself, by which an estate in the land is created; nor is it necessary, to constitute such an estate, that the grantee acquire any right or interest in the surface—for minerals are capable of forming a distinct inheritance in the lands of which they are a part, and consequently, an actual estate may be both created in, and restricted to, any specified kind of minerals.—19 Vesey, 158; 4 East, 469; 2 Barn. and Ald, 724; Bainbridge on Mines, 3, 55, 124, 141, 170.

In Stoughton vs. Leigh, 1 Taunt. 403, it was held, that a grant of a stratum of coal in the land of another is a grant in fee simple of a real hereditament, and that the widow of the grantee was dowable of all the mines of her husband, as well those which were in his own landed estates, as the mines and strata of lead, or lead ore and coal, in the lands of other persons.

In Wilkinson vs. Proud *et als.*, 11 Mees. and Wels. p. 33, it was held that a right to a given substratum of coal lying under a certain close is a right to land, and cannot be claimed by prescription. But a bare right to dig and carry away coal in another man's land may.

An action of ejectment, an assize, a common recovery trespass, an action for use and occupation, and almost all the legal remedies applicable to "land," can be maintained at common law with respect to a *mine*. But they do not lie for an incorporeal thing.—Ad. on Eject. p. 18–20; Co. Litt. 6 a; Bainbridge on Mines, p. 141; 2 Barn. and Ald. 737; 4 Barn. and Ald. 401; 2 Strange, 1142.

The case of the Queen vs. the Earl of Northumberland, Plowden, 310, is relied on by the counsel of the United States to show that in England the right of the King to royal metals in the lands of a subject is a merely incorporeal right or easement, like rights of common, way, estovers, and the like.

But it is not so said in the case. It is merely said, that as the subject may acquire by usage, title or interest in the freehold or inheritance of the King, as common, way, estovers, etc., so the King may possess by prerogative a title or interest in the freehold of the subject. It is not said that the title or interest of the King in royal mines is identical with the title or interest of the subject, who has a right of common, way, estovers, etc., in the King's lands.

If the private owner of the inheritance can, by his grant, sever the right of property in the soil from that in the mine, and vest the ownership of the former in one person, and of the latter in another; and if the mine so owned is "land," it is not easy to perceive why, when the severance is effected by law, and the King is the owner of the mine, he should not also be deemed the owner of "land"; and this, whether the royal right is confined, as in England, to a particular description of mines, or extends, as in Spain, to all mines.

That this was not meant to be decided in the case in Plowden, is clear; for no question was made as to whether the right of the King was a corporeal or incorporeal hereditament, and the judgment of the Court was, "that all ores of gold and silver in the lands of subjects, whether the mines thereof be opened or unopened, with power to dig in the lands of the subjects for the same, and to carry them away, with all other incidents thereto, *belong of right to the Queen.*"

From the foregoing authorities, it is abundantly clear, that the law of England is not guilty of what Mirabeau calls the absurdity of refusing to recognise two species of property, one in the surface of the earth, and the other in its depth.

That in the case of royal mines, the property in the mines remains in the Crown, distinct from the general ownership of the land in which they are situate, and in the case of base mines, the general owner may sever by his own grant the property in the mines from the property in the lands. But in both cases, the mine or strata of minerals form part of the land itself, and are *land* in as strictly legal a sense as the non-metalliferous portions of the soil.

Having thus seen that a property in veins of metals is a property in "*land*," notwithstanding it is entirely independent of the property in the soil which contains them, and that it is so regarded, whether the severance between the two species of property be effected by the grant of the Sovereign to whom royal metals belong, or by the owner of the inheritance, who, by the law of England, was the owner of the base metals within his lands, we will next inquire what was the nature of the estate which was acquired by registry and denouncement under the Spanish and Mexican laws.

This inquiry will necessarily involve a consideration of the nature of a mine under those laws; for it is admitted, as urged by the counsel of the United States, that an estate in a mine may closely resemble an estate in land, while the subject matter of the two estates may remain essentially different.

It will, therefore, be shown not only that the estate of a mine-owner is nearly identical with a fee simple conditional estate in land, at the common law, but that the subject matter of that estate, viz., "a mine," as defined in the Ordinances, is "*land*" in the strictest sense, as proved by its intrinsic nature, and by the application to it in Spanish and Mexican laws and treatises of terms which describe and express what is understood in our jurisprudence by the word "*land*."

In the earlier periods of the Spanish Monarchy, it was established as a principle of Castilian law, that all mines found in the Royal Seignory belonged to the Crown. All persons were prohibited from working them except under royal license.—Part. 3, tit. 28, law 11; Nov. Recop. lib. 9, tit. 18, law 1.

In 1387, permission was given to every person to work mines found within his own inheritance, or in that of other persons with their permission.—Nov. Recop. lib. 9, tit. 18, law 2.

And as early as 1526 and 1551, a general royal license was given to work mines discovered in the Indies, on giving an account thereof to the Governor, and payment of a certain proportion of the products to the Crown.—Recop. de las Ind. lib. 4, tit. 19, 61, 14.

On the 10th January, 1559, Philip II, by a law promulgated on that day, extended to all mines of gold, silver and quicksilver found within the kingdom, whether on royal land or on those of lordships, or of the clergy, or in public, municipal, vacant or private lands, the right of ownership which had previously been vested in the Crown, as to mines found within the Royal Seignory; and all mines of the three metals named were declared "to be resumed and incorporated in the Crown and royal patrimony of the King," excepting only such mines of gold or silver, as had, under previous grants to individuals, been begun to be worked, and were then actually worked.—Nov. Recop. lib. 9, tit. 18, l. 3, Halleck's Mining Laws, pp. 6, 15; Rockwell, 113, 116.

The reasons for this act of high sovereignty are set forth in the preamble of the law. They are chiefly founded on the policy of stimulating the discovery and working of mines, by giving to all subjects of the Crown the right of discovering and acquiring the ownership of them in limited parcels, in order that thereby this source of national wealth might be developed, instead of being locked up, as theretofore, in the hands of a few, who did not or would not develop it.—Rockwell's Gamb. Comm. p. 127; Nov. Recop. lib. 9, tit. 18, l. 3, Halleck's Trans. p. 6–11.

It thus became an established and fundamental principle of

the Spanish law, that all mines of gold, silver, quicksilver and other metals, whether found on public or private lands, belonged to the Sovereign.—Rockw. Gamb. 126, 127; Solorzano Polit. Ind. lib. 6, cap. 1, No. 17; Ordenanzas de Mineria, p. 68; Ordinances of 1783, lib. 5, art. 1; Ordinances of 1584, lib. 6, tit. 13, l. 9.

To carry into effect the policy thus inaugurated, ordinances were established in 1584 (called by Gamboa the Old Ordinances), which continued in force until August 22d, 1783, when the New Ordinances, as they were called by Gamboa, were enacted.

By these new ordinances, all former edicts and ordinances were revoked, so far as they were opposed to the provisions of the new law, except the provisions of the law of 1783, so far as they treated of the incorporation into the royal patrimony of the gold, silver and quicksilver mines of the kingdom.

The second article of the New Ordinances is as follows:

"And in order to benefit and favor our subjects and natives, and all other persons whatsoever, even though foreigners to these our kingdoms, who may work or discover any mines of silver already discovered, or to be discovered, we will and command that they shall have them, and that they shall be their property in possession and ownership, and that they may do with them as with anything their own, observing, as well in regard to what they have to pay by way of duty to us, as in all else, what is prescribed and ordered in this edict." Halleck's Trans. p. 70.

By the laws of the Indies, the Emperor Charles and King Philip II. had made a similar grant to all their subjects, whether Spaniards or Indians (except certain officers), authorizing them to work the mines freely and without impediment, and "making them common to all persons, wheresoever situate;" provided, that the Indians should not be injuriously treated, and that no other parties be prejudiced. 1 Heathf. Gamboa, p. 20, Book IV. Lib. XIX, law 1; Recop. de las Indies, Halleck's Trans. pp. 130, 140.

From the very ample terms of these grants a doubt arose, says Gamboa, whether the mines were still to be regarded as the peculiar right of the Crown, or whether they were to be considered as the absolute property of the subject.

On the one side, Don Mateo de Lagunez and Cardinal de Luca were of opinion, that as all mines in the Indies had been declared "*common*," and all persons were at liberty to try for them, wherever situate, it was to be inferred that the mines were no longer vested in the Crown, and that the effect of the Ordinances was precisely the same as if the mines had been made private property (*particulares*), as they may be by concession and privilege.

This opinion Gamboa combats, and after discussing the question at length, concludes that the mines remained a right of the Crown and annexed to the royal patrimony, and that the laws made the vassals "*participants*" of this right; not giving them the private and absolute property to use at will, but in subjection to the Ordinances; and that although the laws concede to them the ownership and property (dominio y propiedad), it is by "participation," and not by absolute translation (por participacion y no por translacion absoluta); the supreme right of property (alto dominio) remaining in His Majesty.

"The correct opinion then seems to be," says Gamboa, "that the mines remain attached to the Crown (que S. M. mantiene en su corona las minas), and that the King, not being able to work them himself, has admitted his subjects to a share of them (dio parte á los vasallos), under various restrictions and subject to various liabilities."

The opinion advocated by Gamboa appears to have been adopted in the Ordinances of 1783, in which many of the suggestions of Gamboa were embodied.—(Gamboa, cap. 11.)

The first two articles of Title V of that Ordinance are as follows:

"Art. 1. Mines are the property of my Royal Crown, as well by their nature and origin as by their reunion, declared in Law IV Lib. XIII Book VI, of the Nueva Recopilacion.

"Art. 2. Without separating them from my royal patrimony, I grant them to my subjects in property (*en propiedad*) and possession in such manner that they may sell them, exchange them, rent them, donate them, pass them by will, either in the way of inheritance or legacy, or in any other manner alienate the right which in the mines belongs to them, on the same terms on which they themselves possess it, and to persons capable of acquiring it."—Halleck's Trans. p. 222.

It was under these last Ordinances that the rights of Castillero were acquired; but as they adopted the views of Gamboa as to the Ordinances of 1584, his Commentaries on the latter, and definition of the mine-owner's estate acquired under them, are of controlling authority in the interpretation of the Ordinances of 1783.

In Section 24, p. 19, Gamb. Comm. Heathf. Trans. pp. 27–8, he observes: "Having established then the regalian right of His Majesty in the mines, and that this right is entirely consistent with the right of disposition and property (con el dominio y propiedad) of the subjects, it is indisputable, that as a consequence of their passing to the latter with power to dispose of them as their own (como cosa suya) all the incidents of "*property*" (propiedad y dominio) must attach in favor of the proprietor, and that they may therefore be exchanged, sold, leased or alienated by contract, donation and inheritance, may be given as a portion in marriage, or may be charged with a rent, and that interest may be demanded for the purchase-money while remaining unpaid," etc., etc.

"There passes, then, to the subject the direct dominion or property, as also the right to the use (dominio directo ó propiedad, y tambien el util), by virtue of the favor and grant of the Sovereign—which we hesitate not to call a gift on conditions (una modal donacion), as will appear upon considering the rules by which that species of grant is defined in law: that is to say, that it be a free and complete act, which being perfected, a charge attaches on the donee from that time forth (and the being worded as a condition makes no difference), and that upon the failure of the modification limited by the donor in his own favor or in that of a third person, or of the kingdom or republic, the gift determines, as will be seen by reference to various text and doctors."—Ch. 11, Sec. 25.

Throughout the Commentaries of Gamboa, the estate of the mine owner in the mine is called the "dominio y propiedad;" and in Sec. 15, ch. xxi. he says: "Another privilege treated of by the authorities above alluded to, is that of appropriating nine parts of the metal, paying only the tenth to the King as an acknowledgment of his having shared with his subjects the direct and superficial ownership (de aver participado á sus vassallos el *dominio util y directo*), of so valuable a kind of property (*de tan preciosos fundos*).

We have already seen that the "alto dominio" of eminent domain is reserved by the King, by virtue of which and of the conditions on which the grant is made, a new grant may be made if the first be forfeited by breach of the conditions, viz., the payment of the fifth, and the observance of the Ordinances, which Gamboa calls the modo, or gravamen with which the donation is charged, and which make it a "donacion modal."—Chap. 11, Art. 26.

From the foregoing citations the nature of the estate in the mine acquired by registry and denouncement under the Spanish laws, might seem sufficiently clear.

It will, however, not be inappropriate to show what in Spanish law is the meaning of the various terms which are used in the Ordinances, and by Gamboa, as descriptive of his estate. Those terms are, "propiedad," "propiedad y posesion," "propiedad y usufructo," "propiedad y utilidad," "plena propiedad," "dominio," "pleno dominio," "dominio absoluto y perpetuo," "dominio directo y util."

"Propiedad," is the right to enjoy and dispose freely of things which are ours, so far as the laws forbid not. Commonly, the dominion which is not accompanied by the usufruct is called "propiedad," or "nuda propiedad," and the dominion which is accompanied by the usufruct is called "plena propiedad."—Escriche's Dict. verb. "Propiedad."

"Dominio," is the right or power to dispose freely of a thing, if the law, the will of the testator, or some agreement does not prevent.

It is divided into the full, and the less than full—"dominio pleno y menos pleno."

"Dominio pleno y absoluto," is the power which one has over anything to alienate independently of another—to receive its fruits—to exclude all others from its use.

"Dominio directo," is the right a person has to control the disposition of a thing, the use (utilidad) of which he has ceded.

"Dominio util," is the right to receive all the fruits of a thing, subject to some contribution or tribute, which is paid to him who reserves in it the "dominium directum."—Escriche's Dict. verb "Dominio;" Alvares' Institutes, vol. ii, pp. 23–33.

It appears, therefore, that the words which the laws and commentators apply to the miner's estate, viz., "propiedad y usufructo," "plena propiedad," "pleno dominio," "dominio directo y util," etc., are descriptive in the Spanish law of the highest estate or right of property in land which the subject can acquire.

It would seem to be higher than any right of property in land enjoyed in England; for, says Blackstone, "A subject has only usufruct and not the absolute property of the soil;" or, as Sir Edward Coke expresses, "he hath the '*dominium utile*,' but not the '*dominium directum*.'"—Comm. book 2, ch. 7.

The ownership of the subject in the mine remains, however, subordinate to the dominium altum, or eminens, or dominio alto of the Crown.

But this is merely the right tacitly reserved by the Government or sovereign authority of a State, and to which all individual rights of property are subject, to take possession of the property in the manner directed by the Constitution and laws of the State, whenever the public interest may require.—Per Ch. Walworth; Beekman vs. Saratoga R. R. Co., 3 Paige R. p. 45.

The fact that the estate of the mine owner may be forfeited or determined by a failure to comply with those of the Ordinances which impose that penalty, *i. e.* by breach of the conditions subsequent attached to the gift, in nowise affects its nature as an estate in fee.

Every estate held by feudal tenure was subject to forfeiture for breach of the conditions on which it was granted. Nor were these conditions always expressed in the grant; for every act of the vassal which amounted to a breach of his allegiance, or the tie which bound him to his lord, operated a forfeiture of the land.

In a grant of a mine the principal conditions, the breach of which worked a forfeiture, were, 1st, the payment to the

Crown of its proportion of the products; and 2d, the working it according to the laws.

If no breach of these conditions occurred, the estate of the miner would continue forever.

It therefore closely resembled a conditional or determinable fee at common law (1 Prest. on Est. 480). The number of cases where, by the Ordinances of 1584, the mine was declared forfeited (perdida), was no less than fifteen, but many of these causes of forfeiture were totally abolished, and the rigor of the laws was essentially mitigated by the Ordinances of 1783, and subsequent laws.

The decree of the Cortes of Spain, of January 26, 1811, exempted the miners of quicksilver from the payment "of all *duties*, even including *the duty* of the fifth, or the proportion *which* the miner is bound to pay;" and annulled "all dispositions which opposed free trade *in said* mineral, and the security of *absolute and perpetual* ownership (dominio absoluto y perpetuo) of the miner; provided, that in managing and working them, he observe the general rules established on the subject."

In the communication from the Secretary of the Regency, transmitting to the Royal Tribunal General of New Spain this decree, he says:

"Under this date I notify your Excellency, that the prerogative of Seignory, which from remote times the Vice-Royal Treasury has reserved to itself with respect to mines of quicksilver, has been annulled by the General and Extraordinary Cortes, in consequence of the resolution and manifestation of the Council of Regency, enacting at the same time that the said mines shall be worked under the same rules and ordinances as those of gold, silver and other metals, and that their possessors shall preserve their ownership and usufruct (propiedad y usufructo), and in no case shall they be obliged to alienate them to the State—giving them permission moreover to sell their products to any one who will pay the highest price for them. This measure affirms in a manner *inviolable* the ownership (propiedad) and profits (utilidad) of this kind of real estate (de talis juicas), and dispels the reasonable fears which prevented individuals from taking them under their care."

It would not be easy to find words to express more strongly the notion of a permanent and inviolable right of property, than the language of this decree and communication.—Orden. de Min. pp. 79–82; Halleck's Mining Laws, pp. 381–385.

One of the principal causes of forfeiture established by the Ordinances of 1584, was the failure to work the mine for four successive months. But even this provision, so necessary to secure the great object of the Crown in conceding the mine, viz., the development of the mineral resources of the country, was regarded as highly penal, and therefore to be strictly construed. It was thus easily evaded, for by working the mine for a few days every four months the forfeiture was avoided. This defect was remedied in the Ordinances of 1783, which declared that the omission to work for eight months, whether successive or interrupted, in any one year, should forfeit the mine; but the Judge before whom the mine was denounced, was authorised to admit, in addition to the excuses allowed by the former law, any "other just causes which, combined with their former merit, might render the miners worthy of equitable consideration."—Halleck Trans. 244; Ord. of 1783, Art. 15; Heathf. Gamboa, p. 80–86.

Some of the other causes of forfeiture were for the breach of directions in the Ordinances clearly necessary for the security of the mine or of the operatives—such as removing the pillars or supports of the mine, or in other ways neglecting to comply with the rules of the Ordinances, or the directions of the competent authorities, as to their security and preservation, and their better working.—Ord. of 1783, Arts. 7, 10, 11, 12, 13, 14, 15; Halleck's Trans. 242 *et seq.*

In some cases these forfeitures were only imposed after repeated offences (Art. 10), and all the facts were to be judicially ascertained in appropriate proceedings.

From the foregoing it results, that although "the observance of the Ordinances" was in strictness a condition of the concession, yet in fact the forfeiture of the mine was a punishment or penalty imposed for violations of salutary and necessary rules, incurred in but a limited number of cases, and when no equitable considerations existed for mitigating their rigor.

A determinable fee, however, determined by the act or event expressed in its limitation, usually beyond the control of the tenant, and the occurrence of which could not be imputed to him as a fault, still less as a violation of law.

The annexing of conditions subsequent to grants, seems to have been a policy by no means peculiar to Spanish jurisprudence with regard to mines. It was intended to provide a security for the attainment of those objects which formed the consideration for the grant.

These, in colonization grants, were generally occupation and cultivation. In Florida and Louisiana, the settlements were made under a preliminary concession or warrants of survey, and on the fulfillment of the conditions, the final or perfect title was issued.

In Mexico, however, under the law of 1824, and the regulations of 1828, a perfect title issued in the first instance, but charged with conditions as to occupation and settlement; and the grant provided that "if the grantee failed to comply with the conditions, he should lose his right to the land, and it might be denounced by another."

I am unable to perceive any substantial difference between the estate of the colonization grantee and that of the mine owner. Both were liable to denouncement and forfeiture for noncompliance with the conditions expressed in or annexed by law to the grant; and in both cases, equitable excuses could, in general, be received for the omission.

In the California land cases, the existence in the grants of conditions subsequent has never been regarded as presenting any obstacle to their confirmation—where no forfeiture had accrued and been judicially ascertained under the former Government.

Having thus ascertained the nature of the mine owner's estate in a mine, we are next to consider what was the nature of a mine itself, considered as a subject of property.

From the quality of the estate, and the fact that it possessed all the incidents of an estate in *land*—that it was alienable, devisable, and inheritable—that it could be leased, charged with a rent, mortgaged, and given as a portion in marriage, etc., it might reasonably be concluded that the Spanish law, like the English, regarded the mine as "*land.*" But all doubt is removed by recurring to the terms which were applied to it. It is a "fundo," or land (1 Gamb. ch. v, sec. 5 : id. ch. xxi, sec. 15). It is "bienes raices," or real estate (id. ch. xvii, sec. 22). It is "bienes immobles," or immovables (id. ch. xxiii, sec. 6). All the rules relating to petitory and possessory actions for lands are applied to it (ch. xxiii, secs. 1, 6, 19); and in these cases the proceeding may be for the ore, or merely to substantiate the right of property in the mine (sobre los metales ó puramente sobre la propiedad).—ch. xxiii, secs. 20, 21.

The fact that a judicial delivery of possession was authorized to be made, indicates that the thing to which the title or right of property had attached was a corporeal hereditament. It is laid down by Alvarez (Inst. v. 11, p. 23 *et seq.*) as an axiom of the civil law, that "only things corporeal can be delivered, since only these can be transferred by corporeal act from one to another." And this corresponds with the rule of the common law, that corporeal hereditaments pass by livery of seizin, while incorporeal hereditaments lie only in grant.

As the Ordinances gave to the miner the right of property in and possession of the mine, the judicial delivery of the mine, after the title *to* the thing had been acquired by denouncement and registry, vested the mine owner with the right *in* the thing, and gave him the *full* dominion—the title to and the right in the thing, the right to dispose of it, and the right to use—*dominio directo y util.* It is unnecessary to pursue this discussion further.

No one, I think, who examines the Ordinances and the Commentaries of Gamboa with care, can avoid the conclusion that the mine owner, by denouncement and registry, and the delivery of the possession to him, obtained, under the Mexican law, a right to and property in the mine, and that the mine so acquired consisted of the very substance of the minerals of which it was composed—was a corporeal hereditament like a mine or stratum of ores in the English law, and was land in the strictest and fullest sense of the term.

The case of Frémont *vs.* the United States, 17 How. 565, is relied on by the counsel of the United States as sustaining his objection to the jurisdiction. In that case the Court say: "In relation to that part of the argument which disputes the right on the ground that the grant embraced mines of gold and silver, it is sufficient to say, that under the mining laws of Spain, the discovery of a mine of gold or silver did not destroy the title of the individual to the land granted. *The only question before the Court is the validity of the title.* And whether there be any mines on this land, and if there be any, what are the rights of sovereignty in them, are questions which must be discussed in another form of proceeding, and are not subjected to the jurisdiction of the Commissioners or the Court, by the Act of 1851."

It will be seen, that the only question before the Court was whether the fact that mines of gold and silver had been discovered on land which had been previously granted under the colonization laws, destroyed the title of the individual to the land granted. It was of course held that it did not. Whether there were any outstanding rights of sovereignty in such mines, and, if so, what was their nature, and in what sovereign they were vested, the Court declared itself without jurisdiction to determine.

But it was not decided, nor could it have been intended to be decided, that a private grantee of a mine under Mexican law could not present his claim for confirmation under the Act of 1851. That question was not before the Court. And even if the language is to be construed as extending to a right to a mine acquired under Mexican law by a private individual, and is not to be restricted, as its terms imply, to rights of sovereignty outstanding in the United States or the State of California, it must be regarded as *obitar dictum*—not necessary to the determination of the case, and relating to a question which did not and could not have been presented by it.

A similar decision of the point presented for determination had already been made by the Supreme Court in Delassus vs. The United States (IX Pet. 117). The claim in that case was, like that of Fremont, for the confirmation of a concession of a tract of land. It differed from the case of Fremont in the circumstance that the motive of the petitioner in soliciting the grant was declared to be to make explorations for lead mines.

But the Court held that the concession was for land; and as all claims for lands were submitted to its jurisdiction, without any reservation in the Act of Congress of claims for lands containing lead mines, a confirmation could not be withheld.

In Fremont's case the same objection—viz., that the land granted contained mines—was taken, and the Court held that that circumstance did not impair the validity of the grant, or divest its jurisdiction over it as of a claim to land. Such, it seems clear to me, was the whole scope and effect of the decision.

The nature of a *claim* to a mine, as distinguished from the right of an agricultural grantee to the surface, and the jurisdiction over such a claim confided to the Commissioners by the Act of 1851, were not considered, or intended to be decided.

My conclusions, then, on the question of jurisdiction, are—

1. That the jurisdiction of this Court is not restricted to those cases where the estate claimed in lands in California is a fee—or such as in equity should be converted into a fee.

2. That even if it were, the estate of a mine-owner in a mine is, under Mexican and Spanish law, a fee with conditions subsequent; closely resembling that of a grantee under the colonization laws.

3. That by the common law of England the ownership of the surface may be vested in one person, and that of a mine, or substratum of mineral, in another; and that the latter is "land," in its strictest sense.

4. That by the Spanish and Mexican laws a mine considered as a subject of property, and distinct from the ownership of the surface, is "land"—or rather, that terms are applied to it in those laws which are the precise equivalents of the common law term "*land*."

And that therefore this Court has jurisdiction to ascertain and settle a private land claim to a mine, the title to which is derived from the Spanish or Mexican Governments, as of a claim to "*land*."

Having thus ascertained that a mine is "*land*," and that the estate of the mine-owner closely resembled a conditional fee at common law, we will next briefly inquire by what proceedings and in what manner that estate was acquired.

Such an inquiry naturally precedes an examination into the genuineness of the alleged title to the mine, and the determination of its legal effect if found to be genuine.

In Art. 4, Tit. VI, of the Ordinances of 1783, entitled "Of the modes of acquiring mines—of new discoveries, registries of veins, and denouncement of mines abandoned or forfeited," is as follows:

"Those mentioned in the preceding articles must appear with a written statement before the Deputation of Mining of that Territory, or the one nearest if there should be none there, stating in it their names and those of their partners, if they have any; the place of their birth, their residence, profession and employment; and the most particular and distinguishing features of the place (*sitio*), hill (*cerro*), or vein, of which they ask the adjudication. All of which circumstances, and the hour at which the discoverer presents himself, shall be noted in a book of registry, which the Deputation and Notary (*Escribano*) of Mines, if there be one, shall keep, and this being done, his written statement shall be returned to the discoverer for his due security, and notices shall be fixed to the doors of the church, Government houses, and other public places of the town, for due information. And I order that within ninety days he shall have made in the vein or veins of his registry, a pit (*poso*) of a vara and a half wide, or in diameter, and ten varas down, or in depth; and that as soon as this is done, one of the Deputies shall personally go, accompanied by the Notary, if there be one—and if there be none, by two assisting witnesses and a professional mining expert of that Territory—to inspect the course and direction of the vein, its width, its inclination to the horizon, which is called *echado* or *recuesto*, its hardness or softness, the greater or less firmness of its sides, and the species or principal indications of the mineral—taking an exact account of all this, in order that it may be added to the corresponding part of its registry, with the evidence (*fé*) of possession, which shall immediately be given in my Royal name, measuring to him his *pertenencias*, and causing him to fix stakes (*estacas*) in his boundaries, as will hereafter be mentioned; which being done, there will be delivered to him an attested copy of the proceedings as a corresponding title.

"Art. 5. If during the ninety days any one shall appear pretending to have a right to said discovery, he shall have a brief hearing in Court, and it shall be adjudicated to the one who best proves his claim; but if he appear after that time, he shall not be heard."

Art. 7 provides, that when a question shall arise as to who is the first discoverer, he shall be held such who shall have first found metal in the vein—and in case of doubt, he who shall first have registered it.

It is well said by one of the counsel for the claimants, that the object of the various Mining Ordinances of Spain was to stimulate and promote to the utmost extent the discovery of mines and the development of their riches.

The means adopted to stimulate *discovery* were to give to the discoverer the mine he might discover—the State reserving for itself a small part of its products. The means by which the development of the riches of the mine was secured, consisted in making the continuance of the right of property dependent on working it to the extent and in the manner prescribed by law.

Discovery, therefore, was recognized by those laws as the real foundation of the right, and the true consideration for the grant of property in a mine.

Gamboa considered the discoverers of mines as entitled even to greater encouragement than the inventors of useful arts (Heathf. Gamb. 1, p. 259), and in all the Mining Ordinances of Europe the right of the discoverer is recognized by the promise of a grant of the mine as a reward for the discovery.

In the *Sala Mejicana* (vol. 2, p. 58), discovery is declared to be one of the modes of acquiring mines, and it is recognized as a species of occupation, and as constituting a source of title, like the finding of a buried treasure, precious stones, and the like.

In the jurisprudence of Spain and the continental nations, discovery, with regard to mines, as a source of title and consideration for the grant, corresponds with occupation and cultivation under our own pre-emption laws with regard to vacant public lands; and in all the claims to lands in Florida and

Louisiana, submitted to the Supreme Court, the fact of settlement, under a contract for or with a just right to expect a title, has been regarded as a valuable consideration secured by the State, and creating an equitable obligation to confer the title.

But this inchoate right created by discovery must be perfected in the manner prescribed by law.

It is therefore required that the mine be registered; and without registry, says Gamboa, no mine could be lawfully worked, and it remains liable to be registered by any other person—the form of the ordinance not having been complied with. (Chap. V, sec. 3.) The reasonableness of this regulation, he remarks, is evident. It is not necessary to recapitulate the various arguments by which he vindicates its policy and necessity.

As the revenue was interested in a portion of the product of the mine; as the public policy required that an account should be taken of so important a part of the national wealth; as the mine-owner, from the moment of registry, became subject to laws designed to secure the development of the mine he had discovered; for these and other reasons, it was indispensable to provide a mode in which the discovery should be formally made known to competent authority, and the right of the discoverer judicially declared and defined. Registry is, therefore, said by Gamboa to be the fundamental title, or the base of the title (el titulo fundamental de las minas), and the attributive cause of the subject's right of property in it; for the Crown has conceded the mines, and made them common, subject to this burden or condition (gravamen.)—Ch. v, sec. 2.

And in other places, he speaks of registry as "el titulo fundamental de el dominio de las minas."—Ch. xi, sec. 2; ch. vii, sec. 3.

The registry, however, did not constitute the *title* to the mine in the sense of being itself a concession or grant.

The registry, or formal declaration of the discovery in the manner prescribed by law, was the fulfillment of the condition imposed, in the general grant by the Sovereign, on the discoverer of mines, and on its performance the law itself annexed the title as the legal consequence of discovery and registry.

It still remained subject to further conditions—some to be performed before a formal delivery of possession could be given, and some to be perpetually observed under penalty of forfeiture.

The registry was therefore required to be made before a judicial tribunal, and not before an administrative officer, who, like the Governors of California, might exercise a discretion as to whether or not the concession should be made. The declaration of the ownership of the discoverer was termed an "adjudication," and even after registry, and at any time during the period allowed the person registering for digging the pit, any person pretending to have a right to the discovery, was entitled to a hearing "*in Court*," and the mine was *adjudicated* to him who best proved his claim.

In sec. 14 of chap. V, Gamboa states that "registries of mines in the Indies are not to be made before royal officers, but before the justices of the Department alone; and the oath of the discoverer, that he will bring in to be stamped all the gold and silver, etc., is to ensure the due levying of duties, and is not to be made upon the registry of a mine, *the title to which comes under the cognizance of the justice.*"

In section 15, he says: "But *judicial matters*, such as registry, denouncement, the giving possession, etc., are the province of the justices, and, by way of appeal, of the royal audiences."

In sec. 24, he defines registry to be "any judicial order, or proceedings (autos ó diligencias) which authenticate and afford evidence of some judicial act."

The whole proceeding thus seems to have been the judicial recognition and declaration of a previously existing right, asserted and established in the manner required by law, rather than the creation and conferring of a title which had no previous existence.

Thus, Gamboa, in speaking of the necessity of registering says: "And therefore, the discoverer, *if he would preserve his right*, should give notice of his discovery, and make himself known."—Ch. V, 4.

So, in sec. 17 of the same chapter—"If, after the expiration of the term of twenty days, some other person should come forward and register, the discoverer loses *his right*, this being the penalty he is liable to pay for his culpable default, in neglecting to register *his mine*, and thus frustrating the ends of the Ordinances; for a mine which is worked without being registered, is not properly to be called a mine, and does not merit the name, even though it should yield good ore.

"The Ordinances give the name of mines to such only as are registered, because the registry is the fundamental title (el titulo fundamental) to every mine; and because the omitting to make registry evinces a vicious intention to dispose of the ore or silver clandestinely, in fraud of the right of the Crown, and to put impediments in the way of other individuals who might wish to take mines upon the same vein, or at the same spot."—1 Heathf. Gamb. p. 150.

In neither the Ordinances of 1584, nor in those of 1783, is mention made of any title paper to be delivered to the party, containing, in terms, any grant or concession of the mine.

His "statement" is, by the Ordinances of 1783, to be returned to him, after its contents have been duly noted in the register, "for his security;" and when he shall have dug his pit he is to be put in possession, if no adverse claim be interposed. But the only evidence of his title consists in the judicial ascertainment and record of the fact that he declared his discovery in the form required by law, which being done, the law itself gave the title on the conditions fixed by the Ordinances.—When the pit had been dug, and judicial possession given, the 4th Article of the Ordinances of 1783, directed, as we have seen, that an authorized copy of all the proceedings (de las diligencias) should be given him as his corresponding title (como titulo correspondiente). To perfect and secure the right to a mine, registry was, in all cases, indispensable. In making it, all that was necessary was "to manifest the person, the place, and the ore." But if the mine had previously been worked, and was denounced for abandonment or other cause of forfeiture, a preliminary proceeding called a denouncement was required. This was in the nature of an accusation against the former owner, charging him with having left the mine unworked, or having come within some other ground of forfeiture. Upon this question a summary judgment was had, after notice to neighbors, proclamation, etc. If the mine was declared forfeited, the denouncer was, nevertheless, obliged to register it and go through the same proceedings as the discoverer had done.—Heathf. Gamboa, ch. v, secs. 21–22, p. 153.

That the discoverer of a mine was recognized by the law as having acquired a right to it even before registry, is further shown by the terms of the twentieth Ordinance of 1584, which enacts that, "No person shall presume to register, or to enter in the register a mine which is not *his own property*, under the penalty of one thousand ducats," etc.

Among the cases mentioned by Gamboa, to which this law applies, is that of a person other than the discoverer of the ore registering a mine before the expiration of the twenty days allowed by the Ordinance to a discoverer for registering his mine.—1 Heathf. Gamboa, p. 156–8.

The discoverer is thus treated as the true owner of the mine, until, by failing to register within the time prescribed, his rights are lost; and the registry of the mine within that time by any person in his own name is considered and punished as a fraudulent attempt to acquire the property of another, like a similar act by the mortgagee of the mine, or by the curator or tutor of a minor, and other cases mentioned by Gamboa.

From these and many other provisions in the Ordinances, and passages in the Commentaries of Gamboa, the nature and effect of registry is unmistakably manifested; nor can our views of it depend upon the meaning we attach to a single phrase of Gamboa (el titulo fundamental), as to the correct translation of which a question was raised at the Bar.

The foundation of the right to a mine was discovery. But this right was lost unless the discoverer made known the fact before the judicial tribunal authorised to receive such declarations. The proceeding for the purpose was entirely *ex-parte*, and consisted merely of a production of the ore, a description of the place where it was discovered, and the person of the discoverer. These facts being duly made known and recorded, the title passed by operation of law, unless within the time limited some one having a better right appeared.

The foundation of the denouncer's right was in principle the same.

Having brought to the notice of the court, and established that a mine was abandoned or had been forfeited, the law gave him the right to register it in his own name, like a new mine, except that not being an original discoverer, the mining space (pertenencias) to be assigned to him was more limited.

The registry cannot be regarded as the base of the title to the mine, in the sense that without registry a right of property could, in no case, be asserted to it; for as we have seen, even after registry by an alleged discoverer, and after he had dug his pit, obtained judicial possession, and even had his pertenencias measured, any one pretending to have a right to the discovery, could within ninety days from the date of the registry assert his claim and procure the mine to be adjudicated to him; but where no objection was interposed, and the person registering was the true discoverer of the mine, the causing it to be registered, or the formal declaration of his discovery, was the fulfillment of the condition established by law upon which his inchoate right as a discoverer became a perfect right of property (although until the pit was dug he was not entitled to a judicial delivery of possession, nor could he alienate the mine). And the register itself, or the record of the procedure, became his fundamental title *paper*, or *evidence of his title*, for it established judicially the fact of the discovery, and the fact that he had declared it as required by law.

I have thus endeavored to arrive at a clear conception of the nature and effect of the registry of a mine, and of the rights of a discoverer, that we may be the better able to judge of the validity of the alleged registry by Castillero, and correctly to estimate the equitable or inchoate rights he may have acquired as the discoverer of a new mine.

Having premised this much as to the rights of a discoverer, and the nature, objects and effect of a registry, we will now particularly consider the provisions of the articles under examination.

It will be observed, that to effect a registry, and to entitle himself to the judicial delivery of possession of the mine, the only acts required of the discoverer were: 1st, his appearance before the deputation, with a written statement of the facts necessary to be set forth; and 2d, that he should within ninety days thereafter, make a pit in the vein of his registry of the required dimensions.

The noting in the book of registry of the contents of the statement, the hour at which the discoverer presented himself, the notices for due information, etc., were acts to be performed by the Judge or Deputation. It would seem that any injurious effect on the discoverer's right by the omission of the first of these formalities, was intended to be provided against by the direction that the written statement should "be returned to him for his due security," and it is presumed, with a certificate annexed of the fact and time of its presentation.

We learn from Gamboa, that in 1727, the Viceroy Marquess of Casa Forte issued an order, dated at Mexico, commanding the royal officers and justices to send an account of the mines within their several districts, whether at work or abandoned, etc.; and in case they should have no book of registry for the mines which might have been registered there, to form one with all possible dispatch, that an account might thus be obtained of all the mines in the kingdom, from which a general book might be made up, etc. "But we are not aware," says Gamboa, "that this order, so agreeable to the spirit of the Ordinances now under consideration, and so important to the interests of the revenue in a public, and of the subject in a private point of view, was soon carried into effect." We find here no intimation that the failure of the officer to enter the registry in a book in any way impaired its validity, and this, notwithstanding that the Ordinance he was considering (art. xix.) expressly requires the Mining Administrators of each district to keep a book in which all registries made in such district were to be entered, and for this purpose the miners were required to send authenticated copies of such registry to that officer.

Gamboa goes on to observe, that "the original proceedings (diligencias) ought *not* to be given into the custody of the owners until the registry, etc., be made in the proper book; for otherwise these important instruments would be exposed to the contingencies alluded to above, and very serious difficulties may arise in subsequent dealings in ascertaining whether the registry or denouncement was made with due solemnity, the time and manner of making it, the greater or less antiquity of the mine," etc.

This suggestion appears to have been adopted in the Ordinances of 1783, which provide, as we have seen, that the contents of the statement and the hour at which it was presented, shall first be noted in the book of registry, and the statement then returned to the discoverer "for his due *security*."

The registry having been effected, the working of the mine could lawfully be commenced at once; and within ninety days the pit was required to be dug. As soon as this was done formal possession could be given.

If, however, the party who had registered the mine failed to do this, his rights were, by the Ordinances of 1584, forfeited, and the mine might be denounced by and adjudged to another. As, however, from the nature of the vein, a pit of the depth of three estados might be wholly unnecessary, or the hardness of the rock, the caving in of the pit, the breaking out of springs of water, etc., might prevent the digging of the pit within the time limited, it was provided by the Ordinance, that the justice, on an application made to him, and an investigation, might dispense with this requirement of the law, or enlarge the time for its fulfillment, as might be necessary.

In the Ordinances of 1783, the penalty of forfeiture is not in terms imposed for the omission to dig the pit within ninety days. It is presumed, however, that a breach of so positive and important a provision of law would, if without excuse, under the later Ordinances have rendered the mine liable to denouncement.

By Art. 4, of the Ordinances of 1783, the pertenencias of the mine were required to be measured, and the stakes fixed in the boundaries at the time possession is given.

By the Ordinances of 1584, the miner was not obliged to do this until cited by some neighboring miner "who asked for stakes." Ten days were then allowed him to select the ground he might prefer, or, using our mode of expression, to *locate* his pertenencias—subject, however, to the condition that his original pit of possession, or fixed stake, should be within the limits of the boundaries he might select. If he failed to make his selection within ten days, his boundaries were established by the Justice. But this designation of his pertenencias was not final; for he might at any time afterwards, on discovering the true course of the vein; etc., apply to have his stakes bettered (*mejorar las estacas*), and his boundaries might be altered in any way not injurious to the neighbor, between whose mine and his own the boundaries had already been established.—1 Heathf. Gamb. pp. 297, 8, 325; 2d Heathf. Gamb. p. 10.

The frauds and litigation to which this practice gave rise, led Gamboa to suggest that every one should, by positive ordinance, be required to set out his boundaries at the time possession was given, under pain of incurring the forfeiture of the mine and of being *ipso facto* deprived of it, even though not denounced by any other party.—2 Heathf. pp. 10, 11.

This suggestion was adopted in the Ordinances of 1783, except that the omission to set out the boundaries at the time possession is given was not declared to forfeit the mine, *ipso facto*.

But, notwithstanding this establishment of boundaries, the miner could improve the location of his stakes (mejorar las estacas) or change his boundaries by the authority of the deputation of the district, provided it could be done without injury to his neighbors, who were to be summoned and heard in the matter.—Ord. of 1783, Art. 11, Halleck's Trans. p. 236.

It thus appears that the giving possession was, under the Ordinances, a formal proceeding, like a livery of seizin, of which the measurement of pertenencias, or the establishment of boundaries, did not of necessity form a part; and that although these acts were required to be done by the Ordinances of 1784, their omission did not forfeit the right of property acquired by the registry; still less does it appear that their performance was a condition precedent to the vesting of the title.

In all cases where land was granted under the Mexican colonization laws, a formal judicial delivery of possession was in strictness required. But it has never been held by the Supreme Court, that this formality was necessary to vest the title or right of property; and in the majority of cases passed upon by this Court it was not given. In its own nature, it was an act which supposed the existence of a title acquired, but had no effect in conferring one; that having already been done by the concession or grant. I see no reason to make any distinction between the judicial delivery of possession of an agricultural grant and that of a mine, or for considering that the omission of that formality would be fatal to a right of property already acquired in the one case more than in the other.

We shall see, however, when we examine the proofs in this case, that the Alcalde, accompanied by witnesses of assistance, gave to Castillero judicial possession of the mine he had discovered, as well as of three thousand varas in all directions from it, which he undertook to grant him; that no pertenencias were measured or stakes fixed, but that Castillero had already commenced working the mine and had dug a pit, the precise dimensions of which do not appear; that he continued in possession, and working his mine, with the full knowledge, not only of the authorities of the Department to whom he made known his discovery, but of the American Consul, and the inhabitants generally; and that this possession has been retained by his assigns and representatives to this day.

I proceed to consider the evidence as to the registry of the mine by Andres Castillero, and having ascertained what was in fact done by him, to determine its validity and effect.

The documents relied on by the claimants as constituting the registry of the mine of New Almaden are:

1st. A written statement by Andres Castillero, addressed to the Alcalde of the First Nomination of the Pueblo of San José, dated Nov. 22, 1845, setting forth his name, office and residence, and the fact that he had discovered a vein of silver with a *ley* of gold on the land of the retired sergeant José Reyes Berreyesa, which he desired to work in company. He therefore requests the Alcalde to fix up the proper notices in order that his right might be made sure when the time for giving the judicial possession should arrive.

2d. A second statement addressed by Castillero to the same officer, and dated December 3d, 1845, setting forth that on opening the mine previously denounced by him he had taken out, besides silver with a ley of gold, liquid quicksilver. He therefore asks the Alcalde to unite this representation to the previous denouncement and to place it on file.

3d. The act of juridical possession. In this document the Alcalde recites:

"There being no Mining Deputation in the Department of California, and this being the only time since the settlement of Upper California that a mine has been worked in conformity with the laws, and there being no Juez de Letras, (professional judge) in the Second District, I, the Alcalde of First Nomination, citizen Antonio Maria Pico, accompanied by two assisting witnesses, have resolved to act in virtue of my office, for want of a Notary Public, there being none, for the purpose of giving juridical possession of the mine known by the name of Santa Clara, in this jurisdiction, situated on the land of the retired Sergeant José Reyes Berreyesa, the time having expired which is designated in the Ordinance of Mining for citizen Don Andres Castillero to show his right, and also for others to allege a better right between the time of denouncement and this date; and the mine being found with abundance of metals discovered, the shaft made according to the rules of art, and the working of the mine producing a large quantity of liquid quicksilver, as shown by the specimens which the Court has; and as the laws now in force so strongly recommend the protection of an article so necessary for the amalgamation of gold and silver in the Republic, I have granted three thousand varas of land in all directions, subject to what the general Ordinance of Mines may direct, it being worked in company, to which I certify, the witnesses signing with me; this Act of Possession being attached to the rest of the expediente, deposited in the Archives under my charge; this not going on stamped paper because there is none, as prescribed by law.

Juzgado of San José Guadalupe, Dec. 1845.

ANTONIO MARIA PICO.

Assisting witnesses—ANTONIO SUNOL,
JOSE NORIEGA."

These documents are produced from the Recorder's office of Santa Clara county. They were found in the Mayor's office at that place in 1850, by Capt. H. W. Halleck, and by the Mayor transferred to the Recorder's office, where they now remain. It appears that a large portion of the archives or papers belonging to the former Alcalde's office in San José, were deposited in the Mayor's office of that place, a knowledge of which circumstance induced Capt. Halleck to institute a search in the latter place, which resulted in the discovery of the document.

The Expediente thus produced contains several papers besides those enumerated above. These papers will hereafter be referred to.

In investigating the genuineness of these documents it will be convenient to consider—

1st. The proof of the signatures attached to them;

2d. The evidence tending to establish that the documents were executed at the time they bear date, and were filed or archived in the Alcalde's office—and,

3d. The proofs which show that, in fact, the mine was discovered and denounced, and the judicial possession given, as stated in the Act of Possession.

The Act of Possession is signed by Antonio Maria Pico, as Alcalde, and by José Noriega and Antonio Suñol, assisting witnesses. All these persons have been sworn, and testify to the genuineness of their signatures, and that they were affixed on the day the instrument bears date. The genuineness of these signatures, and that of Castillero, is also proved by other witnesses.

I do not understand that the fact that the instruments were signed by the parties whose names they bear, is seriously questioned, as all of them, except Castillero were produced, and testified not only to their own signatures, but to the facts which the documents recite. The theory of the Government which supposes these statements to be false, must admit their readiness to affix their names to antedated documents.

The real questions, therefore, are *when* were these documents prepared and signed, and *when* were they placed in the Alcalde's office?

2. On this point we have, as before stated, the evidence of

Pico, the Alcalde, and the two subscribing witnesses, Suñol and Noriega.

We have, also, the testimony of José Fernandez, who was Sindico del Juzgado and Escribano of the Court in 1845. This witness not only swears to the handwriting of the documents and the genuineness of the signatures, but states that he saw the Expediente in 1845, when it was brought to him by Gutierrez, in whose handwriting the body of the decree is. He also swears that he saw it again in 1849, among the archives of the office, he being then second Alcalde.

James W. Weekes, a witness called by the Government, testifies that he saw the Expediente in 1846–7, when Burton was Alcalde, and in 1848, when he was himself Alcalde. He is unable, however, positively to identify the Expediente now produced with the "little book" which he saw in the office when Burton was Alcalde. This witness, in 1848, certified at the request of Mr. James Alex. Forbes, to a copy of the Expediente. The copy was prepared by James Alexander Forbes, from a document handed to him by Alexander Forbes of Tepic (who was then in California), not from the original in the Alcalde's office. The certificate of Weekes, that it was a faithful and literal copy of the latter document, was obtained, but no comparison was made of the copy made by Mr. Forbes with the original; the witness supposing, as he states, that it was correct. The copy thus certified to by Weekes differs from that found in the Alcalde's office in several particulars, which will hereafter be noticed.

The claimants have also produced the original inventory of papers and effects in the Alcalde's office, which, as was customary, was on the 1st of January, 1846, signed by the outgoing Alcalde, Pico, and receipted by the incoming Alcalde, Chavolla.

This document is produced by the Clerk of the City of San José. It was found amongst other papers which had accumulated under the government of the Alcaldes of the Pueblo, and which now form part of the Archives of the City. The signatures and rubrics of Pico and Chavolla to the inventory are proved; and the document itself is in the handwriting of José Fernandez, the Escribano or Secretary of the Juzgado.

In this inventory, amongst nearly a hundred entries of papers and records, and of the smallest objects belonging to the office—such as candlesticks, old knives, tables and benches—is found a note of a document entitled "Posecion de la mina de Santa Clara á Don Andrés Castillero."

No attempt has been made to impeach the genuineness of the signatures to this document, nor is it said that there is anything in the handwriting of the important entry in question, or in its position on this list, which could suggest the idea of a possible interpolation. Unless this entry be forged, it would seem conclusive evidence that a document of the kind described was on file in the Alcalde's office on the 1st January, 1846.

Another inventory, of a similar kind, made on the 10th day of November, 1846, about nine months subsequently to the former, has been produced by the United States from the Archives of San José.

In this inventory, no entry of the document in question is found. If the lists were in other respects similar, the omission of this one item might possess much significance. But the two lists seem to be different in many particulars; and though some of the entries are alike in both, several which appear in the first are wanting in the second. That all the papers mentioned in the first inventory must have existed on the files of the office, and should have been noted in the second inventory, is evident. When, therefore, we find not only the "Posecíon" of the Mina de Santa Clara, but several other documents, omitted in the second inventory, we necessarily conclude that the latter was prepared in the loose and inaccurate manner in which the public business of such offices was usually conducted in those primitive times.

A striking illustration of the incompleteness of the second inventory is presented by the evidence offered by the United States. Three documents are produced from the Archives of San José, purporting to be orders by the Alcalde for the publication of the denouncements of mines—one on the lands of Justo Larios, another on those of José de J. Vallejo, and a third on the rancho de Ojo de la Coche. These orders are dated in April, March and June, 1846. I find no one of them noted in the inventory made in the succeeding November. And yet, both the documents and the inventory are produced as genuine by the United States.

I can see nothing, therefore, in the omission of the entry of the possession of the mine of Santa Clara in the second inventory which, in the absence of any suggestion that the handwriting or the color of the ink of the entry in the first inventory differs from those of the rest of the document (as would be the case if it had been interpolated after any considerable interval), or that the position of the entry on the list would have rendered such an interpolation possible, should weaken the force of the evidence afforded by the inventory that a document, purporting to be the possession of the mine of Santa Clara, was on file in the Archives of the Alcalde's office of San José on the 1st January, 1846.

3d. As to the proofs which show the facts of denouncement, judicial possession and working of the mine about the time indicated by the documents.

We have already seen that the subscribing witnesses, the Alcalde and José Fernandez, testify to the fact that the possession was given as described by them.

It is shown, however, by evidence which is uncontroverted, that in December, 1845, and early in 1846, Castillero and his partners were notoriously known to be working a mine of quicksilver, of which they claimed to be the owners by denouncement. That these facts were made known to the Governor of the Department, and by him communicated to the Supreme Government. That they were known to the United States Consul, and by him communicated to his own Government, and also to a Cabinet Minister of the Government of the Sandwich Islands, with whom he corresponded, and by whom his letter was published in a Hawaiian newspaper of the date of July 25th, 1846, a copy of which is produced. That the mine was in December, 1845, worked by Indians under the superintendence of Chard, an American employed by Castillero, who is produced as a witness, and whose employment continued until about the middle of 1846. That in December, 1845, it was visited and examined by Col. J. C. Frémont, to whom Castillero, who claimed to own it, explained the Mexican mode of acquiring titles to mines by denouncement and registry, but declined some overtures for its purchase made to him by Frémont.

Other allusions to and recognition of the possession and grant of three thousand varas, in letters, judicial proceedings, etc., at a late period, but previous to the supposed date of their fabrication, will subsequently be considered.

Among the documents alleged to have come from the City of Mexico, traced copies of which are exhibited, was a communication from Pio Pico, Governor of the Californias, to the Minister of Relations, dated February 13, 1846. In this letter Governor Pico states that he incloses a letter from Don Andres Castillero, apprising him of the important discovery of a quicksilver mine, and transmitting a sample of the quicksilver. He therefore begs the Minister of Relations to bring it to the superior knowledge of His Excellency the President, etc.

On the margin of this letter is the usual note, stating its reception on the 6th April, 1846, and that it is noted with satisfaction, etc. The letter of Castillero alluded to in the foregoing, dated 10th December, 1845, is also produced from the Mexican Archives. On searching the Archives in this City, for records of this correspondence, there was found the borrador, or office copy, of the letter from Pio Pico, and a letter from Castillero—not the one inclosed by the Governor in his communication to the Minister of Relations, for of that, he states in that communication, he sends the original; but a subsequent letter, dated December 15th, 1845. In this letter he states: "I have the satisfaction of informing you, if you have not received my other letter through the Prefecture, that I have discovered," etc., repeating substantially the contents of the former letter, which had, in fact, been received and inclosed to the Minister of Relations. It is not disputed that these documents are in the Archives. The borrador, or draft, of Pico's communication to the Minister, is in the handwriting of Olvera, who was Secretary of the Assembly at the time.

There was also found, at the same time, by Mr. Hopkins, the Keeper of the Archives, amongst those records, a letter from

Manuel Castro, the Prefect of the Second District, to the Secretary of the Departmental Government, dated December 31, 1845. In this letter he states that "Castillero has denounced and is now working a quicksilver mine;" and after felicitating the Secretary, and through him the Governor, on so beneficent a discovery, he adds that he incloses a petition by Castillero for two square leagues of land adjacent to his mine. A borrador, or draft of the reply by the Secretary to this letter is also found in the archives, but it appears to have been cancelled by black lines drawn transversely.

The handwriting and the signatures of these documents are proved by Pio Pico, who also states his recollection of having dispatched J. M. Covarrubias with his letter of the 13th February, 1846, to the Minister of Relations, and the bottle of quicksilver sent to him by Castillero.

The testimony of Pico on this point is corroborated by that of José M. Covarrubias, who swears that he left San Pedro in the schooner Juanita, Capt. Snook, on the 14th February, 1846, taking with him the Governor's dispatch, Castillero's letter, and a bottle of quicksilver, all of which he delivered on his arrival at Mexico to the Minister of Relations, Mr. Castillo y Lanzas. Files of the "Diaro Oficial," the Government newspaper, published in Mexico, and of the "Monitor Republicano," and the "Republicano," also published in Mexico, are produced, and under the heading of marine news there are found notices of the arrival of the "Juanita," Capt. Snook, at Mazatlan, on the 2d March, 1846, twelve days from San Diego; of her departure on the 12th of the same month from Mazatlan for San Blas, having on board as passengers José Maria Covarrubias and others.

It cannot, therefore, be doubted that the letter of Castillero of the 10th December, 1845, a traced copy of which is produced from the Mexican Archives, was, in fact, sent to Governor Pico, and by him transmitted in February, 1846, to Castillo Lanzas, Minister of Relations, together with the dispatch, the borrador of which is found in the Archives in this City, and a bottle of quicksilver.

There are also produced by the claimants two letters from Castillero to M. G. Vallejo, of Sonoma.

In the first of these, dated December 2d, 1845, Castillero says: "While waiting for the time of my departure, I have employed myself as a miner, having extracted from the same vein quicksilver, silver, and gold, in surpassing quantities."

In the second letter, dated December 21, 1845, he, amongst other things, informs Vallejo that he has found such abundance of quicksilver that he has extracted twenty pounds of it from twenty arrobas of ore, etc.

These letters are produced and proved by Gen. Vallejo. I do not understand that the genuineness of Castillero's signatures to them is disputed. As Castillero left California in the spring of 1846, and has never returned, they must have been written about the time they are dated, unless we suppose they have since, and after an interval of many years, been written in Mexico, antedated, and sent on to Vallejo to be produced by him—a supposition which the contents of the letters and the allusions in them to personal matters and contemporaneous events of slight importance render wholly inadmissible. There is also produced, by the claimants, a copy of the *Polynesian*, of the date of July 25th, 1846, which contains a letter from G. P. Judd, the Minister of Finance of the Hawaiian Kingdom, to the editor of the newspaper, inclosing a letter received by the Minister from Thos. O. Larkin, U. S. Consul at Monterey, dated June 24, 1846. In this letter, which is also published in the *Polynesian*, Mr. Larkin informs Mr. Judd of the discovery of a quicksilver mine, seventy miles north of Monterey, and states, that in 1845, "a Mexican being in the vicinity examined the rock and immediately denounced the place before the nearest Alcalde, and then made known what it contained. The owner, with a priest, in a small and imperfect manner has commenced extracting the metal." After describing the process adopted by them, he adds: "They obtain about fifteen per cent. of the metal."

The receipt of this letter in the Sandwich Islands is sworn to by the editor and publisher of the newspaper. He is wholly unimpeached. The fact that it was published in the newspaper on the day alleged, is sworn to by a gentleman of San Francisco, who read it, and whose attention was particularly drawn to it. A copy of the paper is produced and filed. From among the papers of the late Mr. Larkin is produced, by his son, the reply of Mr. Judd to his father's communication. The reply is dated July 20, 1846. It acknowledges the receipt of Larkin's letter of the 24th ultimo (June) with a specimen of the ore, and it states that he had sent it to the editor of the *Polynesian* for insertion. The handwriting and signature of Mr. Judd are proved by persons intimate with him.

It has already been mentioned that files of several Mexican newspapers, published in 1846, have been produced by the claimants. In the "Diario del Gobierno de la Republica Mexicana" of the 27th December, 1846, we find credited to the "Espia de la Frontera," a newspaper not produced, a notice of the account given in the "Polynesian" of the 24th July, of the discovery of a quicksilver mine seventy miles north of Monterey, and the same notice purporting to be taken from the same paper, the "Spy of the Frontier," is found in the "Republicano" of the 9th December, and in the "Monitor Republicano" of the 6th.

It is unnecessary, however, to dwell on these incidental corroborations; for the fact of the reception of Larkin's letter by Mr. Judd, and its publication in the *Polynesian* cannot be doubted.

The claimants have also produced from the files of the State Department at Washington, extracts from official despatches of Mr. Thos. O. Larkin to the then Secretary of State. These extracts are certified by Mr. Cass, November 28, 1859.

The first dispatch of Mr. Larkin is dated May 4, 1846.

The extract produced states that—

"Near the Mission of Santa Clara there are mountains with veins of quicksilver ore, discovered by D. Andres Castillero, of Mexico, in 1845, which the undersigned has twice seen produce twenty per cent. of fine quicksilver, etc. * * * * * * * * By the laws and customs of Mexico respecting mining, every person or company, foreign or native, can present themselves to the nearest authorities and denounce any unworked mine. The authorities will then, after the proper formalities, put the discoverer in possession, etc. * * Up to the present time there are few or no persons in California with sufficient energy or capital to carry on mining, although a Mexican officer of the Army, a Padre, and a native of New York, are on a very small scale extracting quicksilver from the San José Mine."

There is also produced from the consular book of Mr. Larkin a dispatch addressed by him to the American Minister at Mexico, dated April 3, 1846.

After mentioning the intended departure of Don Andres Castillero from this port (Monterey) in a few days, for Acapulco, on board the Hawaiian barque Don Quixote, as Commissioner to Mexico from Gen. José Castro, and that he would arrive in Mexico by the 25th or 30th of this month (April), Mr. Larkin says: "At the town of San José, eighty miles from Monterey, Don Andres Castillero has discovered a quicksilver mine. The ore produces from fifteen to sixty per cent. I have seen him, from an old gun-barrel, in thirty minutes run out about thirty per cent. in pure quicksilver. This must be a great advantage to California." In a letter to Capt. Montgomery, of the U. S. Ship Portsmouth, dated May 2, 1846, Mr. Larkin communicates to that officer substantially the same information.

These extracts from Mr. Larkin's correspondence are important, not only as showing that the mine had been discovered, and was being worked in the spring of 1846, but that the mode of acquiring a mine, as understood by Larkin, was precisely that alleged to have been adopted in this case. And further, that "the officer," the "Padre," and "the native of New York," spoken of as working the mine, were undoubtedly Castillero, Padre Real, and William G. Chard, as will hereafter appear.

The whole official dispatch of Mr. Larkin to the Secretary of State, is produced by the son of the former from the letter book of his father. The portions extracted and certified to by Mr. Cass are all that it is important to notice.

That the mine was worked by Castillero in January, 1846, is shown by the deposition of Col. Frémont.

That gentleman states, that in January, 1846, he visited the mine in company with Capt. Hinckley; that the latter introduced him to Castillero, the *owner* of the mine, who showed him the excavations, the heaps of ore, etc., and explained the process of extracting the metal. Impressed with the value of the mine, he spoke slightly to him about purchasing it; but Castillero was not disposed to converse on the subject. Castillero informed him that he had acquired his mine by denouncement, and explained the nature of the proceeding. Acting on this information, Col. F. subsequently denounced the mines upon his own property of Mariposa.

He also adds, that Capt. Leidesdorff, with whom he had spoken as to the purchase of the mine, supposed it might be effected for $30,000, "an immense sum of money in California in those days."

The working of the mine, so far back as December, 1845, is also proved by Mr. Wm. G. Chard. This witness testifies that he was employed by Castillero and the priest Don José Maria Real; that he went there to open the mine in November or December, 1845; that the metal was extracted by heating the ore in gun-barrels; that while working in this way, the possession was given, in December, 1845 or January, 1846. The witness enumerates among those present on that occasion, the Alcalde Pico, Suñol, Noriega, Fernandez and the old man Berreyesa. He does not recollect to have seen Castillero on the ground when possession was given—a circumstance, as observed by counsel, not surprising—for Castillero, Chard states, was constantly coming and going, and on one visit stayed there eight days; but the statement indicates the good faith of the witness in declining to swear to what he did not recollect.

Mr. Chard describes the operations at the mine. They were conducted by himself, another white man, a blacksmith whom they called Old Billy, and some Indians. He built a furnace and smelted the ore in some large whaler's try-pots, capable of holding three or four tons of ore. He remained in this employment until August or September, 1846.

Chard states himself to be a native of Columbia county, New York. He is evidently the "native of New York" to whom Mr. Larkin refers in his dispatch. His testimony is uncontradicted, and his character unimpeached.

There is much other testimony which corroborates the foregoing on various points, but which it is unnecessary to notice. It relates chiefly to the first visit of Castillero to the mine; his first experiments with the ore; his trip to Sutter's Fort and visit to Vallejo, at the baptism of whose child he was godfather, and who thus became his compadre, by which title he addressed him in his letters already cited; his return to Santa Clara, and the formation of the partnership between himself, Castro, Father Real and the two Robles', in November, 1845. As this writing of partnership is conceded to be genuine, and as it relates to the working of the mine of silver, gold and quicksilver on the land of José Reyes Berreyesa, the fact that the mine was discovered at that time must be taken to be admitted.

We thus find that early in December, 1845, the discovery and denouncement of the mine was made known to the Governor of California, and the information, with a sample of the quicksilver produced, by him transmitted to Mexico. That in May, 1846, Mr. Larkin officially communicated the fact of the discovery and the working of the mine, with an explanation of the mode of acquiring title to it under Mexican laws, to our own Government.

That in June of the same year, he informed Mr. Judd of the discovery of the mine in 1845, and the fact that it had *been immediately* denounced.

That in January 1846, Col. Frémont visited the works, and conversed with the "*owner;*" that its reputed value was then about $30,000.

That it had been worked from the November or December preceding by a person employed by Castillero, and continued to be worked by the same person until August or September, 1846.

It has appeared to me incredible that Castillero, a Mexican acquainted with mining laws, should, on discovering so valuable a mine, have omitted to denounce it. That he knew the necessity of the proceeding, we learn from Frémont, as did also Larkin, *a foreigner*, as is shown by his dispatch.

To suppose that Castillero, with a knowledge of the great value of the mine, of the necessity and efficacy of a denouncement, neglected, notwithstanding his statements to the Governor, to take the simple proceedings he is alleged to have done, and that Larkin was entirely mistaken as to the fact of his having done so, is to suppose what I cannot but consider a moral impossibility.

I am aware that the fact that the mine was denounced by Castillero, and claimed and worked by him as owner, does not necessarily show that a juridical possession of it was given, or that the record of that possession is genuine. It is shown, however, by evidence in part introduced by the United States, that the juridical possession, as alleged to have been given, was recognized and alluded to in the correspondence of the parties, and in official acts of Alcaldes, before the date at which, on the hypothesis of the United States, the forgery was committed.

So early as January 30th, 1846, James Alexander Forbes, in a letter to Eustace Barron, of Tepic, apprises the latter that "D. Andres Castillero, a sort of Commissioner from the Mexican Government, is now working a quicksilver mine near the Mission of Santa Clara, which has yielded forty per cent. upon the assay of mineral employed;" and on the 5th May, 1847, the same person, who had in the interval been placed in charge of the mine, in a letter to Alexander Forbes, who had become a part owner, urges him "to obtain from the Government of Mexico the unqualified ratification of the *judicial possession which was given of the mine by the local authority of this jurisdiction, including, if possible, the three thousand varas of land given in that possession as a gratification to the discoverer*." The fraudulent nature of this suggestion is obvious, but it nevertheless implies that a juridical possession and a gratification of three thousand varas had already been given, a ratification of which was thought necessary.

In the preceding March, the same person, together with Castillero, Castro, Real and the Robles', had been sued by the owner of an adjoining rancho for working on it contrary to law. It would seem from the imperfect record of that suit, produced from the archives of the Alcalde's office, that a survey of the mine was ordered and the plaintiffs mulcted in costs; a result which could hardly have occurred if the persons working the mine on the lands of another had been destitute of record evidence of their rights. The fact that a survey of the mine was ordered would seem to be a recognition of the mine-owner's right to his mine, and that the boundaries of his possession were capable of being ascertained.

On the 14th August, 1847, allusion to this suit, and a still more explicit reference to the juridical possession, is made in an official letter of James Alexander Forbes, then H. B. M. Vice-Consul for California, to John Burton, Alcalde.

In this letter Mr. Forbes informs the Alcalde that "two persons have commenced digging a pit, by the direction of Mr. Cook (the plaintiff in the former suit), within *the limits of the juridical possession of the mine.*" He adds, "Permit me to *refer you to the documents which exist in your office*, upon which was founded your conviction of the justice of your decision in March last in relation to the claim of Mr. Cook, and to request that you will be pleased to adopt such measures for the protection of the owners of the mine, and of those who are legally interested in the same, as you may deem most conducive to that end."

The genuineness of this letter is not disputed. It will be observed that, though written in August, 1847, it refers the Alcalde to documents existing in his office upon which a decision rendered in the March preceding was based.

On the 19th January, 1848, Alexander Forbes, who had come to California, made a petition to the Alcalde, Weekes, to "visit and inspect the mine, as required by the Ordinances, and to determine the direction and inclination of the vein, for the purpose of reforming and correcting (since there is occasion for it) the *boundaries of the former Act of Possession*, and to correct such other mistakes as may appear in it."

In conformity with this petition the Alcalde proceeded to inspect the works, and having ascertained the true course of the vein, and admitted the right of the owner to an improvement of stakes (mejora de estacas), he established his boundaries, assigning to him *four pertenencias*, the location of which he designates; but without prejudice to the right and title of the mine (siendo constante el derecho y titulo de la mina) to the gratification or gift (gracia) of land conceded in the original Act of Possession.

Whether the four pertenencias, which were thus designated, were all which under the Ordinances a discoverer, though working in company, was entitled to, we will hereafter consider. The only purpose for which the proceeding is now referred to, is to show that at its date (January, 1848) and about the supposed time of the alleged forgery, "an original Act of Possession," containing a "gracia" of land of a much larger extent, is plainly alluded to as existing; nor is the force of this fact weakened by the circumstance that Weekes, the Alcalde, may have been ignorant and willing to comply with all that Alexander Forbes required; for the fact that the latter inserted such an allusion in the document he may have caused Weekes to execute, is at least evidence that at that early day he claimed that there was in existence an original Act of Possession, including a *gracia* of an extensive tract. It will be observed that the petition of Alexander Forbes to Weekes is dated January 19th, 1848. Its object was to procure the judicial ascertainment of the inclination and depth of the vein, to correct the boundaries of the former Act of Possession, and to decide upon an increase of pertenencias and the square corresponding to them. But if, at that very time, Forbes had already fabricated, or was about to fabricate, an Act of Possession, which was to be antedated and placed in the archives, where no document of the kind had hitherto existed, the application to Weekes, and the designation of pertenencias by him, would be wholly superfluous, if not absurd; for in the forged paper which was to serve as the original Act of Possession by the Mexican authorities, the designation of pertenencias might have been inserted and the boundaries established in any way the forger might desire. All objections or doubts as to the authority of an American Alcalde to act under Mexican Mining Ordinances would thus have been avoided; and the same Alcalde who, according to the theory of the United States, was induced to recognize and affirm the existence of an Act of Possession, either not in his office, or recently forged and placed there, could, with equal facility, have been brought to recognize an Act of Possession which should be free from the errors and uncertainties which he was called on to correct, and which should contain as many pertenencias as he was desired to designate.

Similar allusions to the original registry and act of possession are found in various judicial proceedings during the year 1849. On the 18th October of that year, Robert Walkinshaw, between whom and James Alexander Forbes a contest for the possession of the mine had arisen, filed a complaint against the latter, averring himself to be "the owner of one-eighth of the mine, *by title derived under the original act of registry*."

Previously to the filing of this complaint, Mr. Horace Hawes, a lawyer of much acuteness, and very familiar with Mexican law, had denounced the mine before the Alcalde for abandonment. In this denouncement he describes it "as known in *its original title of registry* as the Mine of Santa Clara."

In the proclamation issued thereupon, on the 23d October, 1849, the Alcalde describes the mine "as known and designated in its *original act of registry* as that of Santa Clara, and now known by the name of New Almaden." On the refusal of the Alcalde to take jurisdiction of the proceeding, Mr. Hawes files a protest, dated 15th November, setting forth that, "besides having failed to work the mine, the alleged owners had never acquired any title thereto, by *reason of the insufficient registry thereof*, which he stands ready to prove in Court by witnesses, records and documents," etc.

As Walkinshaw, though he had originally obtained possession of the mine as the agent of Castillero and his assigns, was in these proceedings endeavoring to acquire the mine for himself, and co-operating with and assisted by his attorney, Mr. Hawes, these references to the original act or title of registry conclusively show that such a document was on all sides admitted to exist, though Mr. Hawes maintained it to be "*insufficient;*" nor is it conceivable that when Hawes and Walkinshaw, by possessory suits, by denouncement for abandonment, by purchasing the lands of adjoining rancheros, etc., were struggling with so great pertinacity to obtain the mine for themselves, they should have utterly failed to disclose the fact, of which Walkinshaw could not have been ignorant, that no registry had ever been made, and that the document purporting to be the act of possession was a recent forgery, then lately interpolated among the archives.

In the correspondence between James Alexander Forbes and Alexander Forbes, and the other owners of the mine, the necessity of procuring fraudulent and antedated title papers from Mexico, is repeatedly and urgently represented. But the fraud recommended is the fabrication of an absolute grant of two *sitios* of land, and a ratification of "the acts done by the Alcalde in the possession given by him of the quicksilver mine in his jurisdiction."

Such are the very terms of the memorandum of "documents to be procured by Castillero," alleged by James Alexander Forbes to have been left by him in Tepic in May, 1849. And in October of the same year, chagrined, it would seem, that his suggestions had not yet been acted on, he complains that he is obliged to depend on "the precarious and illegal possession of the mine granted by the Alcalde of this district to Castillero, who was in reality the judge of the quantity of land given by the Alcalde."

Whether the doubts here expressed as to the legality of the possession be well or ill-founded, it is clear that in this most confidential communication, where no motive existed for suppressing or distorting the facts of the case, that possession is treated as having actually been given, and the record of it as actually existing and genuine, though the possession itself is considered precarious and illegal.

Having thus reviewed the evidence which establishes the genuineness of the act of possession by the testimony of the witnesses to the document; of those who were present when the possession was given, and who testify to the fact; by the production of documents from the Archives of California, and the correspondence, both private and official, of the United States Consul at Monterey; by the testimony of unimpeached witnesses that the mine was, early in 1846, claimed by and recognized as belonging to Castillero, and worked by him as such; by the proofs afforded, by a correspondence admitted to be genuine, that the act of possession was treated and spoken of by the parties, when writing in the most confidential manner, at a time when they could not have been ignorant of the facts, as genuine, though perhaps invalid, and was so recognized in various judicial proceedings by persons who would certainly have discovered and denounced any forgery which might have been committed; and finally, the intrinsic improbability of the supposition that Castillero would have omitted to denounce a mine, of the great value of which he was fully aware, and the means of acquiring a title to which, under the Mining Ordinances, he was well acquainted with, as was also Mr. Larkin, a resident foreigner—I shall next consider more particularly the nature and contents of the documents produced by the claimants, as well as the principal objections to them urged by the counsel for the United States.

These documents are four in number. The first is the original Expediente, produced from the Archives of the City of San José, and discovered by Capt. Halleck in 1851, among the Archives of the old Alcalde's office, in the office of Belden, the Mayor of San José.

This document contains the two representations of Castillero, and the act of possession with the original signatures of Castillero, the Alcalde and the assisting witnesses.

It also contains a petition of José Castro, dated June 27, 1846. In this petition, Castro states that he represents the person and rights of Capt. D. Andrés Castillero, and other individuals composing the company in the quicksilver mine which Señor Castillero denounced on the 3d day of December, 1845, and of which possession was given on the 30th of the same month and year. He therefore claims that, in conformity with the Mining Laws, there be given three pertenencias in continuation of the first; and that this petition be attached to the Expediente of denouncement, and remain among the Archives. On the margin of this petition is an order signed "Pacheco," directing it to be archived as prayed for.

If this document be genuine, it affords important evidence of the date of the judicial possession.

The handwriting of the petition is sworn by Fernandez to be that of Benito Diaz. The signature of Castro is proved by himself, and he testifies that it was signed at its date, having been prepared from a draft left with him by Castillero. He also states that the handwriting of the marginal order is that of Salvio Pacheco, and the signature that of Dolores Pacheco; and that after sending the petition to the Alcalde he left Santa Clara, but was informed, on his return, by the Alcalde, in presence of Manuel Castro and Juan B. Alvarado, that his petition had been granted.

Salvio Pacheco is also produced, and testifies to his own handwriting and the signature of his brother the Alcalde. He swears that the order was written at its date. As this witness has been produced by the United States to sustain the character of a witness impeached by the claimants, it is presumed that his own character is not liable to the imputations from which the United States rely on him to shield another.

The only evidence tending to show that the petition was not written at its date is that of Benito Diaz. He does not precisely specify the time at which it was written; but it can be gathered from his testimony that it was at the end of 1847 or the beginning of 1848.

But this witness is, unfortunately, too well known to the Court to permit any reliance to be placed upon his unsupported declarations. I have not been able exactly to understand on what theory this petition is supposed to be forged.

If the act of possession be genuine, it is immaterial, so far as the rights of the parties are concerned, whether the petition be antedated or not; but it is not easy to imagine the motive of the parties in fabricating a petition addressed to an Alcalde who had ceased to be in office, and whose antedated marginal order did not even purport to convey any rights. Had the marginal order contained a grant of an increased number of pertenencias, some motive for fabricating it would have existed; but it merely directs the petition to be archived; and the application for an increase of pertenencias is in substance renewed in the petition of Alexander Forbes to Weekes, made in 1848, at the very time when we are asked to suppose this petition of Castro was fabricated. If the act of possession be genuine, and the Castro petition be antedated, the conduct of the parties seems to me inconsistent, absurd and inexplicable. But if the act of possession was itself fabricated in 1848, and did not exist at the date when the Castro petition was fabricated, the acts of the parties are equally incomprehensible. Of what use could it be to have a petition for an increase of pertenencias included among archives which contained no registry or denouncement, or grant of any pertenencias whatever? It may be said that the fabrication of the act was then in contemplation; but if so, why not make that document contain all that was desired as to the number of pertenencias, the designation of boundaries, etc. Why accumulate superfluous forgeries, involving the necessity of new perjuries, and largely increasing the risks of detection; and why resort to the proceedings had before Weekes for the ascertainment of boundaries and an increase of pertenencias, when it was known that the fundamental title to the mine had yet to be forged, and might be framed in any way to suit the interests of the parties?

The counsel for the United States has urged upon the Court the inconsistency between the petition of Castro for three additional pertenencias and the supposition that a concession of three thousand varas in all directions, amounting to nine hundred pertenencias, had already been obtained. But this objection, whatever be its force, seems to admit the genuineness of the Castro petition; or, it attributes to the fabricators the absurdity of contriving at the same time two forged documents repugnant to each other. That an act of possession, either genuine or forged, was in existence when Castro's petition was drawn, is evident, for the dates of the denouncement and of the judicial possession are given. To what end then, file a petition which could have no other purpose than to furnish a plausible argument against the genuineness of the previous concession of three thousand varas?

From all the evidence, and on consideration of all the circumstances connected with this petition, I confess myself unable to discover any sufficient reasons for considering it forged.

The claimants have also produced a document alleged by

them to have been delivered to Castillero shortly after the date of the judicial possession.

It contains certified copies of the two representations of Castillero, purporting to have been made on the 13th January, 1846, and a duplicate original of the act of possession signed by Antonio Maria Pico. Appended to these is a receipt by Pico for twenty-five dollars, dated December 30, 1846.

This document was recently found among the papers of Robert Walkinshaw, deceased. It is shown in the deposition of Hall McAllister, Esq., who was counsel for Walkinshaw in a suit respecting a share of the mine, that the document was placed in his possession by Mr. Walkinshaw early in the year 1853, and that it remained in his office until May, 1858, when he delivered it to Walkinshaw.

The genuineness of the signatures is testified to by all the assisting and subscribing witnesses, except José Suñol, who is dead.

This expediente does not contain the petition of Castro, for the certified copies appear to have been made on the 13th January, while the Castro petition was not filed until the June following. There is one circumstance, however, which, though entirely accidental, affords important proof of its genuineness. In copying the first representation of Castillero, it appears that a line was omitted. This has been supplied by another hand, and the handwriting is that of Castillero. As Castillero left California early in 1846, and has never returned, we must suppose that this interlineation was made before he left, or else that the document was fabricated here at a later period, sent on to Castillero in Mexico, interlined by him, and returned to Walkinshaw's possession before January, 1853, when he delivered it amongst other papers to Mr. McAllister. But that he was in possession of "some important papers of the original registry of the mine," in 1849, appears from James Alex. Forbes' letter of October 28th of that year; and it also appears from the certified copy made out by Weekes on the 20th January, 1848, that the expediente we are now considering must have been the document which James Alex. Forbes copied, and which Weekes erroneously certified to be a literal copy of the original in his office. The year in which it is supposed by the Government that these titles were fabricated, is 1848. How, then, could this expediente have been made, certified to by the subscribing witnesses, sent on to Castillero, interlined by him, and returned to California in time to be copied by Forbes and certified to by Weekes, on the 20th January, 1848? And why, if the Castro petition had then recently been written by Benito Diaz, and antedated, was it not included in this expediente, concerning which so much pains were taken?

The omission of the Castro petition in this expediente seems to me, I confess, an important corroboration of the statements of the witnesses who prove the genuineness of the signatures.

There is also produced the copy of this expediente, by Weekes, already alluded to. Weekes himself swears that he made it, and he is corroborated by James Alex. Forbes. I do not understand this fact to be disputed.

The claimants have also produced a copy of the original expediente, certified by Pedro Chabolla, on the 13th August, 1846, to be a literal copy of the original acts (autos) in the archives of his office.

The whole of this copy is proved by Salvio Pacheco to be in his own handwriting, and to have been made at its date. It contains the Castro petition which had been made in the preceding June, and attached to the original on file, and it even omits, like the original, the date of the act of possession—that date being on both papers December, — 1845, and not December 30th, as in the testimonio or duplicate original given to Castillero. That the document was scrupulously compared is further evident from the fact that in the copy the date of Castillero's first representation is Mission de Santa Clara, Nov. 22, de 845, instead of 1845—and on turning to the original, we find in the printed copy that the first figure of the date is separated by a comma from the three other figures. I have not had an opportunity to examine the original, which remains at San José; but it would seem from the printed Transcript that the date is written in an unusual manner, which has been either exactly reproduced or has led to the omission of the first figure in the copy.

It has been earnestly contended by the counsel for the United States, that the non-existence of the original expediente in the archives at San José, even so late as December 23d, 1850, is proved by the affidavit of Mr. Halleck, at that time, and since, Superintendent of the Mine, and counsel for James Alexander Forbes in a suit brought against him and Walkinshaw by the Berreyesa's. In this affidavit, which was made in answer to an order obtained by the plaintiffs in the suit upon the defendants, requiring them to produce the papers on which they intended to rely as a defense, or copies thereof, Mr. Halleck swears:

"That the defendants have exercised all due diligence to procure and produce said papers in Court, by writing immediately on the receipt of the above mentioned order to the *parties in Mexico who hold them.* But to this date the defendants have not received them. * * * * And the defendants specify among others the following papers and documents as absolutely necessary to them before they can proceed with the trial of this cause, viz.:

"1st. *The original denouncement of the mine of New Almaden, and the juridical possession given of the same in the year* 1845.

"2d. The confirmation of said denouncement and possession by the Supreme Government in 1846, and prior to the late declaration of war by the United States against Mexico.

"3d. The original grant of land, including said mining possession, made by the Supreme Government of Mexico prior to the declaration of war as aforesaid, to the owners of said mine."

It is obvious that this affidavit states that the original denouncement and judicial possession of the mine was then in Mexico, and not in the Alcalde's office. That Mr. Halleck, then lately appointed Superintendendent of the Mine, might have been ignorant of the fact that those papers were among the archives of the Alcalde's office, is conceivable; and he may also have accepted the assurances of his client, Jas. Alex. Forbes, that he had exercised all due diligence to procure them, as sufficient to authorize his affidavit of that fact; but it cannot be supposed that Jas. Alex. Forbes could have labored under a similar misapprehension. We have already seen that in August, 1847, he had, in an official letter to Alcalde Burton, referred him to the documents existing in his office, upon which was founded his conviction of the justice of his decision in relation to the claim of Mr. Cook, in the preceding March.

In January, 1848, he had himself copied and procured Weekes to certify to the expediente containing certified copies of Castillero's representations and a duplicate original of the act of possession. This expediente has since been produced from among Walkinshaw's papers, and its possession by him, or his counsel, is traced back as far as 1853. The circumstance that it is interlined in the handwriting of Castillero, proves it to have been at some time in his possession. As Castillero left California early in 1846, it is in a high degree improbable that the document could have been fabricated here, sent on to him in Mexico, and returned before January, 1848, when it was copied by Forbes and certified; nor does such a hypothesis comport with the theory of the United States, which supposes the forgeries to have been committed about the time of the Weekes certificate. It is almost equally improbable that this document, after being copied by Weekes, should have been again sent to Mexico, and returned to Walkinshaw in time to be found among his papers in 1853.

There is no reason to presume that the Weekes copy ever left this State. It was produced by the claimants when proceedings were first instituted before the Board of Commissioners, in 1852. It is clear, therefore, that as the order of the Court called for the documents on which the defendants relied, or *copies* thereof, it was easy for Mr. Forbes to have satisfied the order by furnishing the copy required.

It also appears from his own letters that he had already received a notarial copy of the Lanzas dispatch on which they rely. A copy of this could also have been furnished. We are thus compelled to seek for some other motive for withholding those copies which the order required, and which, on any theory of the case, he could readily have furnished. That motive seems to me apparent. From the 5th of May, 1847, up to the 26th February, 1850, James Alexander Forbes had not ceased to urge upon his associates the necessity of obtaining fabricated documents of title. In his letter of February 26th, 1850, he again dwells upon the necessity of carrying his suggestions into effect, and specifies the required documents as follows:

"1. A full and complete ratification of all the acts of the Alcalde of this jurisdiction, in the possession of the mine.

"2. A full and unconditional grant to Castillero, of two *sitios* of land, covering that mining possession, expressing the boundaries stated by me in the memorandum I left with you at Tepic. Both of *these documents* to be of the proper date and *placed in the proper Governmental custody in Mexico.*"

On the 7th April, 1850, Alexander Forbes, of Tepic, writes to James Alexander Forbes: "Mr. Barron and Castillero have arrived in Mexico, and have every prospect of finding the document you are aware of, and which will, of course, be forwarded as soon as possible."

When, therefore, in December, 1850, James Alexander Forbes represented to Mr. Halleck, that papers had been sent for, and were daily expected from Mexico, it cannot be doubted that he referred to the documents, the fabrication of which he had so urgently recommended. The description of the expected documents in the affidavit, in no respect applies to the Lanzas dispatch; for the ratification, and the grant of two *sitios*, are evidently described as two separate instruments, and they are spoken of as "*of the proper date,*" viz., "prior to the late declaration of war by the United States against Mexico;" that is, prior to the 13th May, 1846; whereas, the Lanzas dispatch is dated on the 23d May.

We have already shown that James Alexander Forbes could readily have complied with the order of the Court, by furnishing "*copies*" of the denouncement and registry, and of the Lanzas dispatch, both of which he must have then had in his possession. The statements, therefore, which he made to his counsel, and on which the affidavit was founded, were evidently made for delay, and to enable him to receive the more full and explicit documents he so much desired. Such being the motive and intent of Mr. Forbes, the allegation that "the original denouncement of the mine was in Mexico," may well be taken as made in furtherance of the same object, and to give increased force to his showing, for the postponement which he was so anxious to obtain.

That Mr. Halleck should have embodied in an affidavit these representations of Forbes will, perhaps, not be surprising to any one acquainted with the facility, often too great, with which counsel receive and adopt in affidavits statements made by their clients in the progress of a cause. Such has seemed to me the more probable explanation of this affidavit. For, whatever may have been therein sworn to, I can see no reason for concluding, on the strength of that affidavit alone, and in the face of the mass of testimony which has been adduced to the contrary, that the expediente of the mining possession was not then in the Alcalde's office.

It was also strongly urged by the counsel of the United States, on the hearing, that the non-existence of the alleged act of possession and concesssion of three thousand varas was proved by the acts of the parties themselves, and their dealings with each other.

The circumstances chiefly relied on were—

1. The fact that in the Castro petition, drafted by Castillero, three pertenencias are asked for, in continuation of the one already obtained.

2. That in the power of attorney executed by Castro, on the 12th June, 1846, to McNamara; in the contract by McNamara, under the power of attorney, executed in Tepic on the 28th November, 1846; and in the ratification of that instrument by Castillero in Mexico, on the 17th December, 1846, the mine is spoken of as consisting of only three pertenencias, while the grant of three thousand varas is not mentioned.

3. That in the numerous deeds and acts of sale by which barras or shares in the mine were transferred, the writing of partnership executed by the original owners of the mine is the only document referred to, no allusion is made to the possession of three thousand varas, or to any tract of land whatsover; and the Expediente of Registration is, for the first time, mentioned in the deed from Padre Real to Walkinshaw, dated August 9th, 1849.

These objections, so far as they relate to the mention of the pertenencias, will more conveniently be considered in treating of the legal effect of the judicial possession; but in respect to the omission of all allusion in the deeds either to the registration or to "lands," it is to be observed that to many of these deeds James Alexander Forbes was a party. We find by his letter of May 5th, 1847, that at that date the "ratification" of three thousand varas of land given by the Alcalde, and the concession of two sitios of land to Castillero, were known to and spoken of by him as having actually been made, the object of that letter being to urge the necessity of obtaining an unqualified ratification of the mining possession, and a positive, formal, and unconditional grant of the two sitios. Similar references to the act of possession, and the order of Castillo Lanzas, occur throughout his correspondence. And, as we have already seen, the former document and the fact of registration is referred to in various legal proceedings by Forbes and Walkinshaw in the years 1847–8; as also in Alexander Forbes' petition to Weekes, in January, 1848.

The inferences, therefore, which might otherwise be drawn from the silence of the deeds on this point, seem to be repelled by the fact that, in letters and various judicial proceedings, the registration, the grant of three thousand varas, and the concession of two leagues, are frequently spoken of and claimed to have been made.

It will be noticed that in the first representation of Castillero, dated November 22, 1845, the mine is described as a vein of silver with a ley of gold; and, by his second representation, it appears that he subsequently, and at some time previous to December 3d, discovered it to contain quicksilver. The writing of partnership, however, dated November 2d, describes a silver mine with a *ley* of *gold and quicksilver*, showing that twenty days previous to his first denouncement he must have been aware of the existence of quicksilver in the vein. This discrepancy is forcibly urged by the counsel of the United States, as affording conclusive proof that the alleged denouncements are forgeries.

It seems, from the evidence produced by the claimants, that

in the month of October, 1845, Castillero and Castro set out from Monterey, to visit General Vallejo at Sonoma, and General Sutter, at Sutter's Fort. On their way, Castillero, at the suggestion of Castro, examined the spot which, as the latter told him, had for a long period been reported to be a mine, but of what kind was unknown. He assayed the ore, and found a little gold and silver, and a small quantity of quicksilver. The latter he considered of no importance.

The party proceeded to Sonoma, and thence to Sutter's Fort, and set out on their return on the 12th November. On reach- Santa Clara, Castillero made further assays of the mineral. "He then discovered," says Castro, "abundance of quicksilver, denounced the mine as a quicksilver mine and formed a company to work it." But this statement is evidently erroneous, for the writing of partnership is dated November 2d, and, if executed at its date, must have been made when the parties were on their way to Sonoma and Sutter's Fort, and not on their return from the latter. But if Catsillero, at the latter date, had discovered quicksilver in large quantities, how can we account for his first representation, which omits all mention of that metal, nor for his second representation, which announces the discovery of it, as having been made after the date of his first representation, *i. e.* after November 22d?

It is quite probable, however, that Castro may be in error, if he means to state that Castillero discovered the abundance of quicksilver and denounced the mine as a quicksilver mine immediately on his return to Santa Clara. Castro himself went on, as he states, to Monterey. The party, having left Sutter's Fort on the 12th November, must have reached Santa Clara between the 16th and 20th. It may well be, therefore, that Castillero made his first representation immediately on his arrival, and subsequently made the further assays spoken of by Castro, in consequence of which he prepared his second or amended denouncement. On this hypothesis we can account for his omission to mention the existence of quicksilver in his first denouncement, as he did not then know that it existed in sufficient quantities to deserve attention. But whatever explanation of this discrepancy be offered or conjectured, I have been unable to perceive how it furnishes the conclusive evidence of the fraudulent character of the denouncement, which the counsel for the United States supposes it to afford.

If these papers were forged about the year 1848, they must have been forged at a time when the character and great value of the mine were well understood. What motive, then, can be suggested for fabricating two representations, in one of which the existence of quicksilver in the vein was entirely ignored?

Impressed, as the parties must then have been, with the great value of the mine, as a mine of quicksilver, can it be supposed that, merely to give the appearance of truth to the documents they were fabricating, they caused them to express a pretended ignorance of the nature of the vein?

Even if so refined and subtle a cunning could be attributed to them, the same cunning would not have suffered them to overlook the fact that the writing of partnership existed, which fixed upon them the knowledge of the existence of quicksilver in the vein twenty days before the date of the first denouncement.

I confess that, though unable to demonstrate how this discrepancy has occurred, I perceive in it rather what the ingenious counsel for the United States has on another occasion characterized as the "deshabille of truth," than that meretricious ostentation of consistency, which falsehood would not have neglected to display.

But the circumstance that I have found most difficult to account for, and which most strongly suggests suspicions as to the genuineness of the Act of Possession, is the failure of Castillero to exhibit, or even mention, it during his protracted negotiation with Mr. Negrete. The correspondence of the latter with his principal, Alexander Forbes, shows that Castillero was called on to exhibit his title papers. He responded by producing the writing of partnership and, after a little delay, the Lanzas dispatch. He not only does not exhibit the "copia autorizada," which, if genuine, he must have received before his departure from California, but he does not even mention that he has been put in possession and received a concession from the Alcalde of a tract of three thousand varas in extent, nor that any such possession had been ratified by the Supreme Government. So far as appears from the instrument executed at that time, and the testimony of Mr. Negrete, the writing of partnership was the only document of title to the mine relied on, and an interest in the two-league grant during the term of the lease is added as a kind of voluntary cession to the aviadores. In all the deeds which passed between the parties for several years, no allusion whatsoever to the Act of Possession occurs, but the writing of partnership and a mine of three pertenencias are alone spoken of.

That the parties were ignorant of the precise number of pertenencias allowed by the law is not improbable, and that they should have treated the mine as consisting of the number of pertenencias assignable to a discoverer, can be reconciled with the facts as they are claimed to have existed. But the omission of Castillero to exhibit the Act of Possession, which constituted his only title paper for the mine, and the only evidence of his denouncement and registry, and which alone showed that the persons, or any of them mentioned in the writing of partnership, had any rights whatever in the subject matter of their contract, is a circumstance which I have found it impossible to account for.

Even on the hypothesis that he had neglected to bring with him, through accident or otherwise, a copy of the act of possession, it would still seem almost inevitable that he should have given to Mr. Negrete some information of the existence of such a paper, and at least mentioned the Alcalde's concession of three thousand varas. No explanation of this circumstance is offered by the claimants. I have been much impressed with its significance. It might well seem to justify the inference, not that the mine was not discovered and worked as alleged, or that it was not in some manner denounced, or that the Alcalde did not give a possession, as sworn to by the witnesses—for of these facts, there can, I think, be no doubt—but that the record of the Act of Possession has been since fabricated and antedated.

But when we consider the vast number of perjuries and complicated forgeries which such a supposition involves, and the grave and almost insuperable objectionswhich present them selves to any theory of forgery, no matter what date be assigned for its commission, and especially if we accept the date fixed by the counsel for the United States (viz., subsequent to February 2, 1848), it seems impossible under the proofs to adopt the hypothesis of the United States.

Our daily experience apprises us that events are constantly occurring which would *a priori* be pronounced in the highest degree improbable. That which is true does not always present the appearance of truth, and it is not usually safe to discredit positive testimony to a fact on an estimate of what would be likely to have happened.

But in this case, if, reasoning on the extreme improbability that Castillero would have failed to produce to Mr. Negrete the Act of Possession, if it existed, we adopt the conclusion that it did not then exist, we encounter improbabilities greater than those we are seeking to avoid.

For admitting, as we must admit, that he discovered the mine; that its great value, estimated by Col. Fremont at $30,000, was known to him; that he denounced it in some form, as is stated by Mr. Larkin to Mr. Judd, and to his own Government by himself, and by Castro to the Governor of California, as was notorious throughout the country—it is, as before observed, almost incredible that he should not have made the denouncement in writing, and substantially as is now claimed. If the papers were fabricated after February, 1848, how can we account for the copy certified to by Weekes in January of that year, and which must then have contained the interlineation in Castillero's handwriting, who was absent in Mexico?

How can we account for the useless forgery of the Castro petition, which the counsel for the United States allege to have been made by Diaz in 1847, and which speaks of the denouncement and act of possession, which had not yet been fabricated? Why, if the parties were then preparing their spurious documents, did they, at the same time, ask for additional pertenencias, when the documents they were forging could be made to express all that they desired?

How shall we explain the absence, throughout James Alexander Forbes' voluminous correspondence, of any reproach, or even regret, that the forgery had been so clumsily effected as to leave the Act of Possession "precarious and illegal;" how account for the allusions to the documents of registry in judicial proceedings, official letters of Forbes to the Alcalde, and the entire absence of any accusation or hint of forgery, when Walkinshaw, who must have been in the secret, was struggling so fiercely with Forbes for the acquisition of the mine?

These, and many other considerations which might be offered, are sufficient, without now alluding to the large number of witnesses who swear to the genuineness of the documents, to apprise us that the theory of forgery is beset with greater difficulties than the supposition that Castillero, for some unexplained reason, omitted to produce or allude to the Act of Possession, which, nevertheless, existed.

Assuming, then, the genuineness of the Act of Possession, I proceed to inquire what this document and the proofs show to have actually been done—and what was its legal effect.

In the first place, it appears that the written statement required by Art. 4, Tit. VI of the Ordinances to be made by the discoverer, was in fact presented by Castillero. The vein was described as situated on the lands of Berreyesa, and the discoverer declared that he wished to work it in company. The names of his partners were not stated as enjoined by the Ordinances.

It does not appear that the corresponding entry, in a book kept by the Alcalde, was made, nor was the "statement returned to the discoverer for his" due security.

Whether notices were affixed, as required by the Ordinances, is perhaps not clearly established, although some witnesses testify to the fact.

It further appears, that within the ninety days limited by law a pit had been dug, the mine opened and working commenced, and that at the expiration of thirty-eight days from the date of the denouncement, judicial possession of the mine was given.

It is not pretended that any number of pertenencias were measured to Castillero when possession was given, nor that he was caused to fix stakes in his boundaries, as directed by the Ordinance.

It has already been shown that, by the spirit and terms of the Ordinances, discovery was recognized as the true foundation of title to a mine. That the registration was but a formal announcement of the fact of discovery, and only required, in the language of Gamboa, three things to be manifested—"the person, the place, and the ore." That upon this declaration the law itself annexed the title, and the mine was said not to be granted, but to be *adjudicated*, by the judicial tribunal which had jurisdiction in such matters. That from the moment of denouncement the discoverer had the legal right to commence working his mine, and was required within ninety days to dig a pit of certain dimensions under penalty of forfeiture of his right, but that as soon as this was done he was entitled to receive judicial possession of his mine; and this, notwithstanding that the period allowed for others to show a better right, had not elapsed.

It was also shown that under the old Ordinances no measurement of pertenencias was required to be made when possession was given; and though this was required by the Ordinances of 1783, the penalty of forfeiture was not by those Ordinances annexed to the omission to do so, though such a provision had been recommended by Gamboa. It was also shown that the requirement of the Ordinances with regard to noting the contents of a statement in a book, etc., was directory to the Alcalde, and that his neglect of duty in that respect ought not to impair the vested right of the discoverer any more than the omission to record a colonization grant should effect the title of a *bonâ fide* grantee of land. The duty of recording registries in a book to be kept for the purpose was also imposed by Ordinances prior to the time of Gamboa; but that author, though he strongly urges its policy and convenience, nowhere intimates that a failure by the mining tribunals to comply with this requirement of the law affected the title of the mine owner—whose rights were evidenced by the copy of the "diligencias" or proceedings which was delivered to him—which corresponded to the "attested copy of the proceedings" which, by the Ordinances of 1783, was required "to be delivered to the party as his corresponding title."

If these views be correct, it follows that all the provisions of the Ordinances indispensably necessary to vest the title in the discoverer of a mine have in this instance been followed. And Castillero, by his denouncement, the digging of a pit within ninety days, and the judicial possession given, acquired by law a right to his mine, with the number of pertenencias allowed to a discoverer working in company.

But it appears, from the loose and informal document executed by the Alcalde, that in addition to the juridical possession which he was empowered to give, that officer "concluded to grant to Castillero three thousand varas in all directions, subject to that which the general Mining Ordinance indicates."

It will be observed, that the Alcalde does not here pretend to adjudicate the mine to the discoverer, nor to put him in possession of any designated number of pertenencias, but to grant him a large tract of land about his mine. It is unnecessary to say that such a concession even of public land by an Alcalde was wholly void, and as against either the Sovereign or a private owner conveyed no rights whatever.

It is insisted by the counsel for the United States, that this distinction between the possession of the mine and the *gracia*, or gift, of three thousand varas, is due entirely to the ingenuity of counsel for claimants, and is not found in the words or sense of the instrument.

We have already seen that the same distinction is clearly taken in James Alexander Forbes' letter of May 5th, 1847. We will hereafter see that it is alluded to in Castillero's communication to the Junta, in which he states "that he has taken possession not only of said mine, but also of an extent of three thousand varas in all directions from that point."

The distinction is not, therefore, a recent suggestion of ingenious counsel.

That it is very clearly expressed in the very inartificial document called the Act of Possession, is not pretended.

Taken literally, that document merely states that, "I, the Alcalde, have resolved to act, by virtue of my office, in order to give juridical possession of a mine known by the name of Santa Clara, and [after sundry recitals] have concluded to grant him three thousand varas in all directions, subject to what the Mining Ordinance indicates."

Nor is it pretended that any possession of the mine, as distinguished from the three thousand varas, was given. That the Alcalde, the assisting witnesses and others, went to the mine for the purpose of giving a possession of some kind, is clear; but he made no measurements, and fixed no stakes. He probably told Castillero that he gave him possession of the mine, and added that he might take three thousand varas in every direction around it.

The explanations of this act given by Pico himself, are confused and contradictory.

In his first deposition taken on behalf of the claimants, Pico says: "Castillero told me he required three thousand varas in all directions, and I told him to take them. He told me he had that right by reason of his being the first discoverer of the metal. According to what Castillero told me, I believed that I, as Alcalde, had authority to do that, there being no other *Juez de Letras.* He was a man learned on all those subjects." He adds: "I do not know whether it [the land of which he gave possession] was round or square, because I made the division in different directions, as I proposed to Castillero; that is to say, that he should take it where it was vacant, or in the mountains, because the rancheros would not have mountains—they wanted plains only."

In a subsequent deposition, taken some three years afterwards—viz., in 1860—Pico admits that he had stated in a deposition taken in another case, that he pointed out the boundaries which Castillero was to take, but gave him no fixed possession; that there was a question between Castillero and Berreyesa—Berreyesa would not consent that possession should be given to Castillero unless he would admit that he (Berreyesa) should have an interest in the mine. In consequence of this, I did not give any fixed possession of the land."

In a subsequent part of the deposition of 1860 (Ans. 17), Pico says: "I intended to grant only what was intended by the Ordinance around the mine, and the rest to be taken on public land." "I never intended to grant another man's land" (Ans. 16). When reminded by the counsel for the United States that the mine was at some distance from the nearest body of public land, recognized as such at that time by himself, and asked how he could have granted three thousand varas, to be measured in all directions from the mouth of the mine, if he intended only to grant public land, the witness replies: "It was because Berreyesa agreed with Castillero at the time, and told me I might grant the land, provided I did not include the land needed for cultivation; and therefore I made the grant." And in Answer 24, he admits that three thousand varas in all directions would include part of Berreyesa's land as well as public land.

Amidst these confused and contradictory statements it is, I think, not difficult to perceive what was really done, or attempted to be done, by the Alcalde. He was aware, and so informed Berreyesa, as he states, that "the Ordinances authorized a certain quantity of land around the mine to be granted, whether on public or private land—that is, that to the discoverer and to the one working in company, a certain number of pertenencias were to be assigned. How many, neither he nor probably Castillero knew. But in addition to these pertenencias, which determined the extent of the mine, he also undertook to grant a tract of three thousand varas to be taken on public, or on Berreyesa's land, if the latter consented. The transaction with regard to this grant seems to justify the observation of James Alexander Forbes, in his letter of October 30th, 1849, that the possession of the mine granted by the Alcalde to Castillero "was precarious and illegal; the latter being in reality the judge of the quantity of land given by the Alcalde."

The concession of the three thousand varas is by its terms provisional, for it is declared to be "subject to what the general Mining Ordinance indicates"; nor does it purport to be a concession of the tract described, as of so many pertenencias of the mine, but rather a grant of land as a gracia or gift.

The difference is important, for in the one case Castillero (if the grant were valid) would, under the Mexican laws, have been the owner, not only of the large tract conceded to him, but of all the mines which might be discovered within it; in the other, he would merely have owned the mineral veins within the pertenencias allowed by law, while all others within the three thousand vara tract would have remained liable to denouncement by any one who might discover and be ready to work them.

Castillero himself seems to have understood that he was entitled to four pertenencias, for in the petition of Castro, which was drafted by himself, there is asked "three pertenencias in continuation of the first." It is not, however, very clear, that three additional pertenencies are here asked for, or perhaps Castro may have misunderstood Castillero's instructions, for it appears that in his ratification of McNamara's contract, made at Mexico, Dec. 17, 1846, Castillero describes the mine as of three pertenencias only. That such was understood to be the dimensions of the mine by all parties in 1846 and 1847, appears not only from their acts of sale, but from the testimony of James Alexander Forbes.

This witness states, that "in 1846 he received from Padre Real possession of the mine, the hacienda, about a mile distant from it, the mining utensils and some ores. No definite extent of land was specified. It was understood that the mine contained three pertenencias at that time, but the hacienda was not understood to be within the three pertenencias." The possession transferred by him to Walkinshaw in 1847, and again on his return from Tepic received back from Walkinshaw, is stated to have been like the original possession received from Padre Real. It comprised the hacienda and the mine, but no definite tract of land.

In a subsequent part of his examination, he says: "There was only one act of possession which I understood to have been given. This embraced three pertenencias, so far as regarded the mine. Three pertenencias, and also lands about the hacienda, I understood to have been given to Castillero in 1845."

"These lands were understood to be of the extent of three

thousand varas," he adds; and in the deed received by him from the Robles', and which conveys "all their rights and shares in each one of the *three pertenencias* of the mine," their interest in the lands and hacienda passed, for "by Mexican custom, a sale of barras in a mine includes an interest in the hacienda."

In the possession obtained by Alexander Forbes from Weekes, Alcalde, in January, 1848, four pertenencias seem to have been considered the number to which the parties were entitled, and a tract two hundred varas long and eight hundred wide, comprising exactly four pertenencias, is designated by the Alcalde.

It is, I think, apparent, that from 1846 down to a late period, the parties interested considered themselves the owners of a mine, with the number of pertenencias allowed by law—whether three or four they seem to have been uncertain—and that this ownership, acquired by registration and denouncement, carried with it the ownership of the hacienda or reducing establishment. But they seem to have attached, so far as we can discover from their acts of sale, proceedings at law, etc., little importance to the rights in the large tract six thousand varas square, which the Alcalde assumed to grant them partly on public, and partly on private land. On no other hypothesis can we account for the omission to mention the word "lands" in any of their conveyances during so long a period.

No inference of fraud can, however, be justly drawn from this circumstance, for the same omission is observable in deeds and proceedings *after* the date of the supposed forgery, as before it.

But whatever may have been the notions of the parties as to their rights, it is, I think, plain, that the possession given to Castillero should legally be treated as having included two entirely distinct objects: one, the *mine* properly so called, comprising the pit or "pozo de posecion," with the limited number of pertenencias allowed by the Ordinances; and the other, a tract of land six thousand varas square, which the Alcalde assumed to grant, but of which no "*fixed possession was given.*"

This distinction between the judicial possession of the mine, which the tribunal having jurisdiction of mining matters had a right to give, and a grant of more than a square league of *land*, partly on private and partly on public, is not only admitted but insisted on by the counsel for the claimant. Its importance will appear hereafter when we come to consider the alleged "*ratification*" of the mining possession by the Supreme Government.

By Art. 14, Tit. VI, of the Ordinances of 1783, it is provided, that "Any one may discover and denounce a vein or mine, not only in common land, but also in the private lands of any individual; provided he pays for the land of which he occupies the surface, and the damage which immediately ensues therefrom, according to the valuation of experts appointed by both parties, and a third in case of disagreement. The same being understood with respect to him who denounces a place (sitio) or waters for establishing works, and working the machines necessary for the reduction of ores, which are called haciendas; provided, they do not include more land nor use more water than may be necessary."

It is not pretended that in this case any denouncement of a sitio or of waters for a hacienda, was made. Had such been the case, and the title to a hacienda duly acquired by the mine-owners, it may well be that, as affirmed by Mr. Forbes, a sale of a barra, or share, in the mine would have conveyed, by Mexican usages, a corresponding interest in the hacienda. But the parties appear to have selected and taken possession of this hacienda, relying on the grant of six thousand varas square, made by the Alcalde, or on some arrangement made with Berreyesa, the reputed owner of the land.

Antonio Maria Pico, in his last deposition, seems to desire it to be understood that Berreyesa gave an oral assent to the grant, provided it did not include the level land; and in a previous deposition before the Land Commissioners, the same witness testified that a written contract was entered into between Berreyesa and the owners of the mine, by which the former was to have a share in the mine, and was to be paid for the wood and limestone used in the establishment. After the possession was given, Berreyesa demanded a compliance with the contract, and wrote to Padre Real to that effect. It does not appear that any arrangement was entered into with him. He was soon after killed by the Americans, and the hacienda and tract of three thousand varas in every direction has remained in the possession of the New Almaden Company to this day.

I am unable to perceive how, under these circumstances, and in the absence of any denouncement of the sitio used for a hacienda, Castillero can be deemed to have acquired, by the attempted concession of the Alcalde, any title whatever to lands beyond the limits of the pertenencias which the law allowed.

With respect to these, the Ordinance does not expressly declare whether the denouncer of a vein or mine on private land is required to pay for the land of which he occupies the surface as a condition precedent to the vesting of his title. It would seem that he is not.

The denouncement or formal declaration of discovery is evidently the first step to be taken; the pit is then to be dug and the discoverer is, by the terms of Art. IV, to be put at once into possession. No provision is here made for a suspension of proceedings until the land can be valued and paid for, nor until the expiration of ninety days can the denouncee be sure that some one having a better right will not present himself; as the private land-owner is only to be paid for the surface land which is occupied, and the ensuing damage, it would seem impossible to ascertain what amount of surface-land is to be occupied until the number of pertenencias is fixed and their boundaries marked out, which can only be done when judicial possession is given.

The Ordinance seems to contemplate a claim or proceeding instituted by the land-owner; for it provides that the land shall be appraised by experts appointed by both parties. If, therefore, the payment for the land, and satisfaction for damages, are conditions precedent to the vesting of any title to the mine, the land-owner might, by refusing to appoint an expert, indefinitely postpone its acquisition—thus defeating the policy of the Mining Laws, as well as the right of the discoverer, which those laws so fully recognized, and so amply protected. For it is not to be forgotten, that under the Spanish as well as all other Mining Ordinances, the discoverer was considered the true owner and creator of the wealth he had discovered; and that the grantee of the superficies had merely the right to an indemnity for actual damage done by the occupation of a small portion of the surface. But to hold, under the circumstances of this case, that no title vested by denouncement and discovery, because the claim of Berreyesa was not first satisfied, would be peculiarly inequitable.

The mine, it is true, was generally understood to be on his land, and denounced as such. But no judicial measurement of his land had been made, nor were his boundaries established. It is to this day unsettled, whether in fact the mine is within the boundaries of Berreyesa, or those of his neighbor, Justo Larios, or without both, on public land. Had Berreyesa been paid, as required by the Ordinances, it may yet prove that the payment was unnecessary, because the mine was on public land; or to the wrong person, because it is on the land of Justo Larios. Where land was gratuitously distributed in tracts from one to eleven square leagues in extent, the indemnity which experts would have awarded for the occupation of a few rectangles of the surface, two hundred varas long by two hundred wide, would have been little more than a nominal sum; and to defeat the meritorious title of a discoverer, because under such circumstances the indemnity was not paid, would seem unjust and absurd.

The next question to be considered on this branch of the case is, whether the Alcalde had, under Mexican laws, jurisdiction to receive denouncements, make registrations, and adjudicate the titles of mines.

By Art. IV, Tit. VI, of the Ordinances of 1783, the discoverer was required to present himself before the Deputation of that Territory (territorio), or the one nearest if there should be none there.

It is unnecessary particularly to examine the nature and organization of the Special Tribunals to which the Ordinance refers. It is sufficient to say that they were composed of deputies chosen by the enrolled miners of each mining territory, who themselves were members of the great mining corporation or body of matriculated miners throughout the Kingdom of New Spain.

These Special Tribunals were in the Federal territory abolished by the Constitution of 1826 and by the law of 1837.

But by the law of December 2d, 1842, Courts of First Instance, composed of three Territorial Deputies elected in the manner prescribed in the Ordinance, were required to be established in each of the Departments by the Governor, in concert with the Departmental Junta, and with the previous approval of the Supreme Government.

To these Courts were given substantially the powers formerly possessed by the Territorial Deputations under the Ordinances.

Under this law mining tribunals were established in various departments—but none were ever organized in the Californias.

It is contended, that inasmuch as these tribunals existed in some parts of Mexico, it was the duty of the discoverer to address himself to the one nearest to his mine; and that that tribunal alone had jurisdiction in the premises.

It is not disputed, as a general principle of Mexican law, that in default of any of the authorized special tribunals, their functions devolve upon the Courts of general jurisdiction. The question then is, was there any special tribunal to which the discoverer of a mine in the Department of the Californias could address himself.

It seems to be considered by the counsel for the United States, that the provision of the Ordinance which directs the discoverer to the Deputation of the nearest territorio, in default of any deputation within his own territorio, of necessity directs him to the nearest Departmental Court organized under the law of 1842, if there be none in his own department. But this provision of the Ordinance is not adopted or alluded to in the law of 1842. The small territorios, in each of which a mining deputation was by the Ordinances of 1783 to be established, in no respect corresponded to the great divisions of the Mexican Republic called "Departments," in a single one of which both the Californias were included. The Deputies under the Ordinances were to be elected in each Real or Asiento of mines by the matriculated miners "*of that place*" (*lugar*), whose names were embraced in a book kept by the Judge and Notary of that mining place (Mineria).

I am not informed what were the ordinary territorial limits of the Reales or Asientos of mines here spoken of; but it is obvious that they could not have been larger than would be consistent with the convenience of the miners who were required to enroll their names, and every year to vote at elections. When, therefore, it happened that in a newly discovered mining district, no deputation had been elected, the discoverer was reasonably directed to the nearest deputation, which would ordinarily be at no great distance.

But to send him to a remote Department of Mexico on such an errand would be absurd, and a practical denial to him of any right whatever to register his mine. Nor could in such case the other provisions of the Ordinances be complied with; for to what purpose affix notices on the doors of the churches in Chihuahua, that an individual had discovered or denounced a mine in Upper California, and how could one of the Deputies personally go, and within ninety days inspect the mine, examine the pit, and give the possession as enjoined by the Ordinances?

That no such proceedings could have been contemplated by the law of 1842, is also clear from the terms of the law itself. The Governor and Junta of each Department were to establish, as we have seen, as many Courts of First Instance as were required within their limits.

It is to be presumed that the jurisdiction of each of these Courts was restricted to the territorial limits assigned to it; but it certainly did not extend beyond the boundaries of the Department. Art. 26 provides, that "Each one of these Courts shall exercise *within its territory* the executive and economical powers given by the old Ordinance," etc. How, then, can it be supposed that, organized under the authority of the Department, and with its jurisdiction restricted to its territory, it could take cognizance of mining matters in another department, separated from it by hundreds of leagues, and with which communications were rare and difficult.

It is, I think, beyond doubt, that at the time of the discovery of this mine, there were not only no special tribunals in Californias which had jurisdiction in mining matters, but there were none anywhere established in Mexico which possessed jurisdiction to make a registration of a mine discovered in California. On the principle of Mexican law already referred, the functions of the special tribunals, under these circumstances, devolved on the Courts of ordinary jurisdiction—or rather the jurisdiction remained in them, as it had done in the Federal Territory from the adoption of the Constitution of 1824 and the law of 1837; it never having been divested in this Department by the establishment of special tribunals under the law of 1842.

If, then, the ordinary Courts had cognizance of mining matters in the Californias, the jurisdiction must have been vested in the Alcaldes, for no ordinary Courts of First Instance existed.

By the decree of May 23, 1839, the judiciary of each of the departments was to be composed of Justices of the Peace, Alcaldes, Judges of First Instance and a Superior Tribunal.

But this organization was not perfected in the Californias, and the Alcaldes in this department appear to have exercised the functions and jurisdiction which would otherwise have belonged to the Courts of First Instance.

In the decree of March 2, 1843, it is stated, that in the Californias there had been no Courts of Second and Third Instance established in the Californias, New Mexico and Tabasco; and by Act 28th, the Governors of those departments are ordered "to take care that justice is punctually and completely administered in first instance by Judges of that grade, if there be such, or by Alcaldes or Judges of the Peace. But even if the authority of an Alcalde to take cognizance of mining matters were doubtful, it ought, I think, to be sustained as that of a *de facto* officer exercising an undisputed jurisdiction.

The registration of a mine was a simple proceeding, of which the principal objects were to apprise the Government of its existence, so that it might secure its portion of the produce—to give an opportunity to other persons to show a better right than that of the alleged discoverer or denouncer, and to subject the latter to the salutary rules of the Ordinance as to its working and preservation. When the discoverer had made known his discovery in the manner prescribed by law, and dug his pit of possession, the law itself gave him the title.

It would seem, therefore, that if the discoverer addressed himself to the only judicial authority of the country, and if that authority, with the knowledge and acquiescence of the Governor and inhabitants of the Department, took cognizance of the matter and adjudicated the mine to him, a presumption in favor of the rightful exercise of power ought to be indulged. Nor should we affirm the Alcalde's acts to be void, except on the clearest proofs that he was wholly without jurisdiction in the premises.

Having thus seen that by the discovery and registration of his mine, and by the possession given by the Alcalde, Castillero acquired a right to the mine, with the number of pertenencias allowed by law, I proceed to inquire what number of pertenencias he thus became entitled to.

By Art. 1, Tit. VI, of the Ordinances, it is provided that "the discoverers of one or more mineral hills (*cerros*), absolutely new, in which there is no mine nor trial pit open, may acquire in the principal vein which they may select, as many as three pertenencias, continuous or interrupted, according to the measurements which are hereinafter prescribed; and if they have discovered more veins, they may have one pertenencia in each vein, said pertenencias being discovered and marked out within the term of ten days."

Art. 2, Tit. XI, provides: "Although by these Ordinances I prohibit any individual miner who works in the ordinary limits from denouncing two contiguous veins on the same vein—notwithstanding this, I grant to those who work in company, although they be not discoverers, and without prejudice to the right which by reason thereof they may have in case they are such, (y sin perjuicio del derecho que por este titulo deban tener en caso de que lo sean), the right to denounce four new pertenencias or mines which have been worked and abandoned, even when they are contiguous and on the same course."

It is claimed that, under these provisions, Castillero was entitled as a discoverer to three pertenencias, and as one working in company, to four additional pertenencias, making seven in all.

The counsel for the United States contends that the allowances of pertenencias mentioned in the foregoing articles, are not cumulative; and that four, or the number given to him who works in company, are all that can be acquired.

The determination of this question entirely depends upon the true meaning of the Spanish text. The counsel for the United States insists that the Spanish phrase quoted above reads, when properly translated: "And without prejudice to the right which by this title [viz, that of discoverers] they may be entitled to have in case that they may be such."

That is, that although it was prohibited to an individual miner to denounce two contiguous mines, yet those who work in company may denounce four pertenencias, either new and unopened, or old and abandoned mines; and this right is not to prejudice their rights as discoverers, in case at any time thereafter they may *become* such.

If, however, we attribute a present and not a future signification to the verb, the meaning of the phrase would be, that all persons working in company shall have four pertenencias, and shall enjoy this right without prejudice to their rights as discoverers, in case *they are such;* that is, all persons working in company shall have four pertenencias, and if they be also discoverers, their rights as such shall remain to them.

I do not pretend to be able to determine, from a knowledge of the Spanish, which is the true translation of the phrase. I have therefore addressed myself for information to several persons skilled in that language. They all concur in adopting the translation suggested by the claimants.

There are some general considerations which serve to strengthen my belief in its correctness.

The object of the law was to determine the extent of mining spaces, or pertenencias, to be allowed to miners. As the merit of the discoverer was greater than that of one who merely denounced a forfeited mine, and as the policy of the law was to encourage and reward discoveries, it gave to the discoverer, though working alone, three pertenencias, if his discovery were of an absolutely new hill, in which no mine had been opened; but if the discovery were of a new vein in a hill known and worked in other parts, he was allowed to acquire two pertenencias. It was also the policy of the law to promote the development of mines by encouraging the formation of companies, the associated capital of which would enable them to prosecute the works on a larger scale and with greater efficiency. Title XI, in the 1st Article, enjoins upon the Viceroy to encourage, promote and protect all such partnerships by all convenient measures.

In furtherance of the same policy, Article 2d gives to all those who work in company the right to acquire either four new and unopened pertenencias, or four mines (of one pertenencia each) which have been worked and abandoned. The law which prohibited the acquisition of two contiguous pertenencias by an ordinary miner, is *pro tanto* repealed, while those which determined the rights of discoverers are allowed to remain.

Such would seem the natural mode of carrying out the evident policy of the law-giver.

For why should the ordinary miner be rewarded with three additional pertenencias, because he works in company, and the discoverer only be allowed one additional pertenencia?

The law recognizes two species of merit—that of discovery, and that of working in company. If a miner possesses both, ought he not to receive the rewards allowed for both?

The phrase "and without prejudice to their rights as discoverers," etc., is evidently intended to guard against an interpretation of the provision prejudicial to the rights of the discoverers. It is the exclusion of a possible conclusion which might otherwise have been drawn.

But could it have been supposed that because persons working in company are to have four pertenencias, no one of them could have the rights of a discoverer, if at any future time, and perhaps at a distance from his mine, he discovered an entirely new hill? Such a construction of the provisions in favor of partnerships would have been wholly unwarranted. This could not, therefore, have been the conclusion intended to be excluded. But it might have been supposed that the law, in giving to partnerships four pertenencias, meant to fix the maximum number of contiguous mines which the same individuals could in any case acquire. If, as discoverers, they were already entitled to three, the formation of the partnership would give them but one more. Both provisions would thus have been satisfied. Three of the pertenencias would be held by a double title,—that of discovery, and that of working in partnership—while the fourth would be given for the latter reason alone.

To guard against this construction, the provision was inserted that the allowance to partners should be without prejudice to their rights as discoverers, in case they were such; and in this view the provision was sensible, and perhaps necessary. It left to each kind of merit its appropriate reward, and gave to the miner who united both in himself all the privileges which the law attached to each.

I am, therefore, of opinion, that under the Mining Ordinances referred, Castillero, as a discoverer and as one working in company, was entitled to seven pertenencias.

Having thus ascertained what acts were done and rights acquired by Castillero in California, we will next consider the title claimed to have been obtained by him from the Supreme Government of Mexico.

The facts as alleged by the claimants, are as follows:

Early in 1846, and while he was yet in California, Castillero, impressed with the importance of the brilliant discovery he had made, communicated the fact in two letters addressed to J. J. De Herrera, former President of Mexico, dated at the Mission of Santa Clara on the 19th and 22d February 1846, respectively, and also in another letter written on the last mentioned day to Don Tomas Ramon del Moral, at Mexico.

These letters, together with some specimens of cinnabar and a small flask of quicksilver, were sent by the hands of Lazaro Piña, who sailed from Monterey for Mazatlan in the brig Hannah, in the early part of March, 1846.

Extracts from the two letters to Herrera were, it appears,

furnished by him to Señor Moral, and a note embodying these extracts, together with a copy of Castillero's letter to himself, were, by Señor Moral, about the middle of April, 1846, communicated to the Junta de Fomenta y Administrativa de Mineria, a body charged with the development and encouragement of mining interests in Mexico, and the administration of certain funds connected with the same object.

There were also transmitted by Moral to the Junta, at the same time, some specimens of cinnabar which had been delivered to him by Piña.

On the 21st April, 1846, the Junta addressed a letter to José Maria Tornel, Director of the College of Mining, transmitting to him copies of Castillero's letters, and the specimens of cinnabar, and requesting that an assay might be made of the latter.

On the 23d April, Tornel, by an order on the margin of the letter of the Junta, directed the specimens to be sent to the Junta Facultativa, or Faculty of the College, for assay.

The result of the assay was communicated to Tornel by Señor Moral, President of the Junta Facultativa, on the 24th April, 1846, and the receipt of this letter was acknowledged on the 29th of the same month by Tornel, who, on the same day transmitted the letter of Moral of the 24th, announcing the result of the assay, to the Junta de Fomento. His official communication on the subject was received by them on the 3d May, and on the 4th, ordered by an "acuerdo" or marginal order to be sent to the Government. On the succeeding day, viz., the 5th, a communication signed by Vicente Segura, President of the Junta, and Isidro R. Gondra, First Clerk, was accordingly addressed to the Minister of Justice, in which was stated the reception of the specimens, and their transmission to the Director of the College for assay. A copy of the communication of the Director of the College, stating the result of the assay, was also embodied in the Junta's letter, and the Minister of Justice was informed that the Junta had already asked Castillero what kind of aid or protection he needed for the encouragement of his brilliant enterprise, etc., etc.

This letter was received by the Minister of Justice on the 9th May, as shown by the marginal note of its contents and reception, and on the same day the Minister formally acknowledged its receipt in a dispatch addressed to the Junta.

On the 12th May, 1846, Castillero, who had sailed from California in the barque Don Quixote, and arrived in Mexico, submitted to the Junta nine propositions in writing, in which he indicated the kind of aid and protection he required. He had previously however appeared before them, given a verbal account of his discovery, and been requested by the Junta to furnish a written statement as to what aid he required.

On the 14th May, 1846, the Junta transmitted the written statement of Castillero to the Minister of Justice, retaining a copy in their own office. In this communication of the Junta, the Government is urged to accept the propositions of Castillero.

Among the propositions thus made by Castillero to the Junta, and by the latter transmitted to the Minister of Justice, were the following:

"7th. The Junta shall represent to the Supreme Government the necessity of approving the possession which has been given me of the mine by the local authorities, in the same terms as those which I now hold it."

"8th. It shall also represent the advantage of there being granted to me as a colonist two square leagues upon the land of my mining possession, with the object of being able to use the wood for my burnings."

The communication of the Junta, inclosing the propositions of Castillero, and urging their acceptance, was received by the Minister of Justice, and on the 20th May, the following "acuerdo" was noted on the margin:

"Granted in the terms which are proposed, and with respect to the land, let the corresponding order issue to the Minister of Relations for the proper measures of his office, with the understanding that the Supreme Government accedes to the petition."

This "acuerdo" is signed with the rubric of Becerra, Minister of Justice.

On the same day (May 20th) Becerra addressed to the President of the Junta an official dispatch, as follows:

"MINISTRY OF JUSTICE
AND PUBLIC INSTRUCTION. }

"Most Excellent Sir:—Having reported to His Excellency the President *ad interim* of the Republic, your Excellency's communication of the 14th inst., with which you were pleased to transmit with a recommendation the petition of Señor Don Andres Castillero, for the encouragement of a quicksilver mine which he has discovered in the Mission of Santa Clara, in Upper California, His Excellency has been pleased to approve in all its parts the agreement made with that individual in order to commence the working of said mine, and on this day the corresponding communication is made to the Minister of Exterior Relations and Government, to issue the proper orders with respect to that which is contained in the 8th proposition for the grant of lands in that department.

"I repeat to your Excellency the assurance of my esteem.

"God and Liberty. Mexico, 20th May, 1846.

"BECERRA.

"To His Excellency, D. VICENTE SEGURA,
President of the Junta de Fomento de Mineria." }

On the same day Becerra addressed to the Minister of Relations an official communication, in which he transcribes the foregoing dispatch to Segura, and adds:

"And I have the honor to transcribe to your Excellency, to the end that with respect to the petition of Señor Castillero to which His Excellency the President *ad interim* has thought proper to accede, that there be granted to him as a colonist two square leagues upon the land of his mining possession, your Excellency will be pleased to issue the orders corresponding."

In obedience to these orders, the Minister of Exterior Relations, Castillo Lanzas, on the 23d May, 1846, directed an official dispatch to Pio Pico, Governor of California, in which, after transcribing the foregoing communication of Becerra, he says:

"Wherefore I transcribe it to your Excellency, in order that in conformity with what is prescribed by the laws and dispositions upon colonization, you may put Señor Castillero in possession of the two square leagues which are mentioned.

"God and Liberty. Mexico, May 23d, 1846.

"CASTILLO LANZAS.

"To His Excellency the Governor of
the Department of the Californias." }

Upon these last dispatches, viz., that from Becerra to the Junta de Fomento of May 20th, and that from Castillo Lanzas to the Governor of California of May 23d, the claimants rely, as constituting a ratification of the grant by the Alcalde of 3000 varas in every direction, and a concession of two square leagues of land. They also claim that the "acuerdo" or marginal order found in the communication of the Junta of May 14th, 1846, amounts, in equity at least, to a concession of all that the Junta recommended.

The proofs of the foregoing allegations consist of a large number of expedientes from various public offices in Mexico—of certified copies of the actas or minutes of the proceedings of the two Juntas, viz., the Junta de Fomento and the Junta Facultativa of the College of Mining—of certified transcripts of entries in official books of the Ministers, and of the parol testimony of Members of the Juntas, Clerks in the offices by whom the documents were written, and of the Minister himself, Castillo Lanzas, who was the author of the dispatch of May 23d. These witnesses swear, not only to the existence of the archives, the handwriting and the genuineness of the various documents, traced copies of which are produced, and to the accuracy of those copies, but also to the facts stated to have occurred, or to such parts of the transactions as each was personally concerned in.

Some of these witnesses, who were brought from Mexico by the claimants at great expense, have held distinguished official positions, are of advanced years, and independent fortunes.

The United States aver that their testimony is false and perjured; and that the documents sworn to by them are forged and antedated. To arrive at a just estimate of the force of the evidence and reasons on which the United States rely to support this accusation, a brief statement of the nature and amount of the proofs offered by the claimant is necessary.

In the short summary of the evidence which I propose to give, I shall follow rather the chronological sequence of the events alleged to have occurred, than the order in which the various documents were produced.

It will be remembered that the Junta de Fomento was first

notified of the discovery of a mine of quicksilver by Castillero, by receiving from Don Tomas Ramon del Moral a communication, containing copies of letters from Castillero, that this communication was sent to the Director of the College that an assay might be made, and by him referred to the President of the Faculty, who in due time reported the result of the assay to the President, who in turn communicated it to the Junta, and the latter to the Minister of Justice.

There is accordingly produced from the Archives of the College of Mining the original communication of the Junta de Fomento, signed by the President, Vicente Segura, and addressed to the Director of the College, together with copies of the communication of Moral, which, with the specimens of cinnabar, were sent to the Director. The accuracy of the traced copy produced, and the existence of the original in the College of Mining, is testified to by José Maria de Bassoco, for many years a member of the Junta, and by Balcarcel and Castillo, Professors in the College. These witnesses also swear to the handwriting of the dispatch, and of the copies of the letters, to the genuineness of the signatures of Vicente Segura, President of the Junta, and of Gondra, the Chief Clerk of the Junta, who certifies to the copies which accompany the dispatch. On the margin of the dispatch is an acuerdo, or order signed "Tornel," directing it to be sent to the Junta Facultativa. The fact that Tornel was Director of the College, and the genuineness of his signature, are also proved by the same witnesses.

From the archives of the Junta Facultativa of the College, are produced traced copies of the minutes of a session of the Board of the 24th April, 1846 (erroneously dated 24th March). These minutes show a resolution of the Board that a report of what had been done, and the result of the assay made by Professor Herrera, be communicated to the Director of the College. From the same archives is produced a traced copy of the reply of the Director General Tornel to the report of the Board, in which he acknowledges the receipt of a letter from Moral, President of the Faculty, of the 24th April, communicating the result of the assay. The accuracy of these traced copies, and the existence, genuineness and handwriting of the originals, are proved by the Professors Castillo and Balcarcel, who were present at the meeting of the Faculty, and who not only swear to their personal recollection of the facts, but also testify that the cinnabar was received, an assay made, a meeting of the Faculty on the subject held, and the specimens deposited with appropriate labels in the Cabinet of the College, where they now remain.

From the archives of the Junta de Mineria, are produced traced copies of the office copy of the communication sent to the Director of the College, the original of which is found, as we have seen, in the archives of the latter; also, a traced copy of the communication received from the Director of the College, announcing the result of the assay—with a marginal note directing it to be transmitted to the Government, signed by Segura, President of the Junta de Fomento.

There is also produced, from the same archives, a traced copy of the borrador or office copy of the communication thereupon addressed by the Junta to the Minister of Justice, together with a traced copy of his reply, dated May 9th, 1846.

There is also produced from the same archives a traced copy of the borrador or draft of a second communication from the Junta to the same Minister, transmitting to him the petition of Castillero for aid, etc., and recommending it to the favorable consideration of the Government.

A traced copy of the reply to this communication by the Minister of Justice, is also produced from the same archives.

Appended to the espediente containing it, are certificates of Manuel Couto, Secretary of the Administration of the Mining Fund, and in charge of the archives of the Office of Mineria.

A certificate of Vicente Segura, certifying to the official character and handwriting of Couto.

A certificate of P. Almazan, Chief Clerk of the Ministry of Encouragement, Colonization, etc., certifying to the official character of Segura, Administrator of the Mining Fund, and to that of Couto, the Secretary, and that the archives of the office are in charge of the latter, and also to their signatures and seals.

A certificate of J. Miguel Arroyo, Chief Clerk of the Ministry of Exterior Relations, certifying to the official character of Almazan, and to his signature and the seal of his office.

And, finally, a certificate of John Black, U. S. Consul, certifying to the official character and the signature of Arroyo, and also that he is the person authorized by law to legalize Mexican documents to be used in foreign conntries, and that the seal of the Department affixed to the documents is the same used in the legalization of all documents by that officer.

The accuracy of the traced copies, the existence and handwriting of the originals, where those originals are borradors or drafts, and the existence and handwriting of the signatures to the originals, where they are the original communications reeeived by the Junta, are proved by Mr. Bassoco, by the Professors Balcarcel and Castillo, by Miranda and Yrisarri, who were employed in the Ministry of Justice, by Manuel Couto, who testifies that he copied Castillero's petition from his rough draft, and by Castillo Lanzas, the former Minister of Relations of Mexico.

From the archives of the Ministry of Relations, to which the archives of the Ministry of Justice have been transferred, are produced traced copies of the communication addressed by the Junta to the Minister of Justice, informing him of Castillero's discovery, and embodying the communication received by the Junta from the Director of the College, informing the Junta of the results of the assay.

A traced copy of the draft of this communication, as we have seen, is produced from the archives of the office from which it emanated. A traced copy of the borrador or draft of the Minister's reply to this communication, is also produced from the same archives, in all respects conforming to the original reply, a traced copy of which, as before stated, is produced from the archives of the office to which it was directed.

There is also produced, from the same archives, a traced copy of the communication of the Junta, inclosing and recommending Castillero's petition, corresponding with the borrador produced from the archives of the Junta, and a traced copy of the borrador of the reply of the Minister, in like manner corresponding with the original produced from the archives of the Junta.

On the margin of the communication of the Junta is the usual membrete or memorandum of its contents, and an "acuerdo" or order of the Minister in regard to it. The latter is signed with the rubric of the Minister, and the official dispatch transmitted to the Junta conforms entirely to the acuerdo or resolution taken on the subject, and noted in the margin of the communication. There is also produced from the same archives a certified copy of the draft of the communication addressed by the Minister of Justice to the Minister of Relations.

From the archives of the Ministry of Relations is produced a traced copy of this last communication of the Minister of Justice, the borrador of which is found in the office of the latter; and a borrador of the communication or dispatch addressed, in pursuance of the order of the Minister of Justice, by the Minister of Relations, Castillo Lanzas, to Pio Pico, Governor of California. And, finally, the claimants produce from their own custody, the original of the last mentioned dispatch, signed by Castillo Lanzas, and addressed to the Governor of California.

It may here be observed, that from the Archives of the Ministry of Relations is also produced a traced copy of the communication of Pio Pico, of February 13, 1846, addressed to the Minister of Relations, informing him of Castillero's discovery, and transmitting Castillero's letter of December 10th, 1845. There is also produced from the same Archives a traced copy of the borrador of the reply of the Minister, dated April 6th, 1846.

We have already seen that the borrador of Pico's communication, and the original of the Minister's reply, are found among the Archives of California in this city. Their genuineness is undisputed.

To the traced copies from the Archives of the Ministry of Justice are affixed the certificate of Arroyo and seal of his department, as also the certificate of Black, the United States Consul, in the same terms as those already mentioned.

The accuracy of the copies, the existence of the originals in the Archives of the offices to which they belong, their handwriting, and the genuineness of the signatures they bear, are sworn to by the escribientes, or clerks, by whom they were copied, some of whom are still connected with the Ministries by M. de Bassoco, and by the ex-Minister Castillo Lanzas himself.

There is also produced a traced copy of an extract from a book now existing among the records of the Ministry of Justice. It contains various entries or notes, purporting to have been made from May 11th, 1846, to May 20, 1846. Amongst those made on the 20th, is an entry of the membrete or memorandum of contents of the Junta's letter to the Minister of May 14th, and of the acuerdo or resolution taken by the Minister on the subject. On the top of the first page on which the communication of the Junta is written, are found these letters and figures: "L. g^{l} 15, S. f. 140 v^{ta}."

José M. Yrisarri, the Fifth *Oficial* of the Ministry of Justice, being interrogated as to the meaning of this inscription, testifies that it means "Libro General, vol. 15, reverse of page 140," and that the entry already mentioned is found in the volume and page referred to. He further states that this inscription, or reference, was made by himself, as was also the entry on the book to which it refers.

The original dispatch of Becerra to the Minister of Relations, is stated by Miranda and Yrisarri, to be in the handwriting of the former. The "acuerdo" on the margin is said by Castillo Lanzas to be in his own handwriting and signed with his genuine rubric; and the draft of the dispatch addressed by him to Pio Pico, to be in the handwriting of Mr. Quintanar, an employé of the Ministry of Relations in 1846. The original dispatch addressed to Pio Pico is proved to be in the handwriting of A. J. de Velasco, by Castillo Lanzas and by Velasco himself. The signature of Castillo Lanzas is proved by Lafragua, former Minister of Relations in Mexico, by Velasco, who wrote the dispatch, and by Castillo Lanzas himself.

It is also testified by Mr. Negrete, that this identical dispatch was handed to him in December, 1846, by Castillero; that a copy of it was inserted in the instrument by which Castillero ratified the McNamara contract, and the original sent by him (Negrete) to Alexander Forbes, of Tepic, on the 19th December, 1846.

In corroboration of this statement of Mr. Negrete, there is produced a series of letters written by him to Alexander Forbes, from December 5th, 1846, to February 6th, 1847. In these letters Mr. Negrete informs Mr. Forbes of the state of his then pending negotiation with Castillero; and in his letter of December 18th, advises him that he transmits "the document showing the grant which the Supreme Government made in favor of Don Andres Castillero for two leagues of land," etc.

These letters, which Mr. Negrete swears he saw for the first time since they were written, when produced to him by the claimants in this country, he testifies are in his own handwriting and that of his clerk, Oruña, and signed by himself. He also identifies three checks or orders drawn by himself on his banker, Don Donato Manterola: one in favor of Castillero for four thousand dollars, and receipted by the latter, and two in favor of Don Nazario Fuentes, the Notary, for one hundred and thirty-seven dollars and twenty-five cents, and twenty-nine dollars and seventy-five cents, respectively, both of which are receipted by Fuentes.

The testimonio or authenticated copy of Castillero's instrument of ratification, containing the Lanzas dispatch, is exhibited. It is signed by Nazario Fuentes, the Notary Public, whose signature and *signo* are attested by three Notaries Public in a certificate under the seal of the National College of Notaries of Mexico, dated December 19, 1846.

A second copy of the same instrument, issued from the office of the Notary Fuentes under his hand and seal, is also produced. It is dated February 6, 1847, and is certified by three Notaries Public, under the seal of the National College of Notaries. Among the three Notaries signing these certificates is Mr. Villalon. This gentleman has been examined as a witness. He testified that his signature and *signo* on each of the certificates are genuine, that they were affixed at their respective dates, that the signatures of the other Notaries are genuine, as is also the seal of the National College. It is also shown that this instrument of ratification was brought to California by Mr. Walkinshaw, in 1847, when he took charge of the mine; and the terms of James Alexander Forbes' ratification, in which the language of Castillero's Act is copied, show that the latter must have been before him when writing his own ratification of the McNamara contract—a conclusion rendered certain by the very distinct allusion in Forbes' letter of May 5th, 1847, "to the possession of two sitios ordered to be given by the dispatch of Señor Castillo Lanzas."

The claimants have also produced from the Archives of the Junta de Fomento, traced copies of the original borradores, or drafts, of the minutes of the Junta in April, May, September, November and December, 1846; also a traced copy of the clean copy made from those minutes, and authenticated by the rubrics of the members of the Junta who assisted at the sessions; and finally a copy of the entire Volume 3d of the Minutes of the Junta from April 2, 1846, to June 30, 1847.

In these Minutes, amongst a great number of other entries, we find a record of the action of the Junta from the reception of the specimens of cinnabar to the payment of the Notary Calapiz, when further proceedings were abandoned.

In the Actas of the session of April 23d, 1846, is an entry of the receipt of specimens of cinnabar from the Presidio of Santa Clara, in California, and a resolution that they be sent, with copies of Castillero's letters, to the Director for assay.

On the 4th May, the receipt of the letter of the Director inserting the report of the Junta Facultativa is noted, and it is resolved that it be transcribed to the Supreme Government, representing that a reply has been made to Señor Castillero, asking him what kind of protection or assistance he requires.

On the 6th May, the Actas show that Don Andres Castillero appeared and made a verbal report, etc., and the Junta resolved that Señor Castillero should present his indications in writing. On the 14th May, the receipt of the communication from the Minister of Justice dated May 9th is noted.

On the 25th May, is a like note of the receipt of Becerra's dispatch of the 20th, approving the agreement made with Castillero, etc., and a resolution of the Junta that the proper judicial agreement be drawn up immediately, etc.

On the 29th May, is a note of an order for the payment of twenty-five dollars to the notary Calapiz, for proceedings in the instrument of agreement which had been made with Castillero to assist his quicksilver enterprise, etc.

The accuracy of the traced copies of the Actas is testified to by Mr. Bassoco, who compared them with the originals; and he also proves the existence and authenticity of the originals in the Archives of the Junta, and the genuineness of his own rubrics and those of his colleagues affixed to them.

The claimants have also produced, and filed as an exhibit, an original r.port made by the Junta de Fomento to the Minister of Justice, relative to the matters confided to its care. This report is embodied in a report made to the National Congress by José M. Lafragua, Minister of Relations, and read before that body on the 14th, 15th and 16th December, 1846.

The original manuscript "Memoria," or report by the Junta, is stated by Mr. Bassoco, to have been procured by himself from Escalante, the agent of Lafragua. Its proper place of custody was the Ministry of Relations, but M. de Bassoco supposes that it had probably been taken by Mr. Lafragua to his own house, when the latter was preparing his report to the National Congress, and accidentally remained among his papers. He identifies the signatures and rubrics of Vicente Segura and Isidro R. Gondra, which are affixed to it, and states his conviction that it is the identical document sent in by the Junta to the Minister.

On referring to the "Actas" of the Junta, we find it noted on the 5th November, 1846, that a dispatch was received from the Minister of Relations, dated November 3d, calling for an account of the labors of the Junta, to be furnished within eight days.

It also appears, that on the 9th the reading of the report was commenced; that it was concluded at the session of the 16th, and a resolution adopted, that the report should be transmitted to the Government; and that on the 5th December, a communication from the Minister acknowledging its receipt was received by the Junta.

Two copies of the Report of Lafragua, in which the Memoria was embodied, are also offered in evidence by the claimants. It is a printed volume of considerable size.

Of these copies, one was originally produced by the claimants, and identified by Mr. Lafragua, Mr. Bassoco and others.

A second copy has recently been produced and identified by the Hon. J. P. Benjamin, one of the counsel in the cause, as having been received by him in 1849, from Don José Garay, the validity of whose grant he was then investigating. The volume remained in Mr. Benjamin's possession until about two years ago, when Mr. Rockwell, also of counsel for claimants, called on him to retain him in this cause. In the course of conversation Mr. R. alluded to an official report of Mr. Lafragua, which, in his opinion, contained conclusive proof of the genuineness of the title of the claimant. From his description Mr. Benjamin thought he recognized the volume in his possession as the one referred to, and immediately procured it from an adjoining room. On examination it was found to contain the passages relating to the discovery of the mine, etc., which are found in the copy previously produced by the claimants. Mr. Benjamin was not, until that time, aware that it contained anything in reference to the mine. At Mr. Rockwell's request, he allowed him to retain the volume, which he recognizes as the one now produced.

The claimants have also offered in evidence files of the "Diario," the "Republicano," and the "Monitor Republicano," newspapers, in which the reading of Lafragua's report on the 14th, 15th and 16th December, 1846, is noticed.

It is unnecessary to extract at length the passages in this report, in which reference is made to Castillero's discovery, and the action of the Government upon it.

They merely contain an account of the presentation of the specimens to the Junta by Señor Moral—the assay, the inquiry of Castillero as to the assistance he desired, his petition, and the Junta's agreement to it; the approval of the agreement by the Supreme Government, and the failure to carry it into effect owing to the order of the Supreme Government of May 10th, 1846, directing the suspension of all payments from the public Treasury.

In the foregoing statement of the documentary and other proofs on which claimants rely to show the action by the

Mexican authorities in reference to the important discovery of Castillero, much evidence as to various handwritings, signatures, etc., has been omitted.

Enough has been set forth to show the nature and the force of the proofs offered in support of the genuineness of the documents exhibited.

It will be seen that these proofs do not consist of any one set of papers derived from a single office, the archives of which might have been falsified and the officials corrupted.

Each document is found in two and some in three distinct repositories. The borradores are produced from the offices from which the communications emanated—the originals from the offices to which they were sent—and in some instances the communications are, according to the Mexican custom, inserted in dispatches from the office to which they were originally directed, and those dispatches are found in the archives of the Ministry to which they are addressed.

All the papers are so intimately connected and complicated with each other, that it is almost impossible to suppose any one to have been fabricated unless the whole series be spurious. They are written in various handwritings, with a multitude of signatures, rubrics, etc., of well known individuals attached to them. The same document contains, in some instances, no less than four different handwritings—viz., that of the clerk who drew it, of the official who signed it, of the clerk who wrote the membrete and acuerdo, and that of the Minister by whom the latter was signed.

The writing is sworn to be that of clerks attached for many years to the offices from which the papers emanated. Their handwriting must therefore be well known, and a forgery of it could readily be detected.

When we consider the long series of forgeries, and the almost innumerable perjuries which must have been committed if these documents are not genuine, the crimes imputed to the witnesses are as appalling as the extent and almost endless ramifications of the conspiracy to commit them are incredible.

We must suppose that Professors in a National College, forsaking their scientific pursuits, have carefully fabricated false minutes of the proceedings of the Faculty of which they were members; that they have made a tedious and dangerous journey to sustain, by carefully prepared perjuries, the forgeries they had committed; and that they have had ingenuity and depravity enough to give to their statements the appearance of truth, by inventing circumstantial details as to the reception of the specimens, the assay made of them, their deposit in the cabinet of the College, and even the purport of the tickets or labels upon them, which as they state can be seen by any visitor to the College.

With regard to the Junta de Fomento, the forgeries and perjuries imputed are still more complicated and improbable. Not only must the various dispatches alleged to have been addressed by them to the Director of the College, and to the Minister of Justice, with the signatures of the President of the Junta and of the Secretary, the handwriting of the clerk who drew the marginal notes upon them, and the rubric of the Minister appended, have either been forged, or falsely sworn to have been written at their dates, but a series of *actas* which record the proceedings and resolutions of the Junta must have been fabricated, or extensively interpolated, and the rubrics of the members forged. And, as if reveling in supererogatory crimes, they must also have fabricated the borradores, or rough drafts, from which the clean copies of the minutes were made out—the existence of which would hardly have been suspected, and which it would naturally be supposed had been destroyed.

They must also have prepared a voluminous report to the Minister, in which has been inserted an account of these proceedings, precisely such as, if they had taken place, we should expect to find. The manuscript of this report, which is claimed to have been accidentally left among the private papers of the Minister to whom it was addressed, must have been forged, or the interpolated passages inserted in it, in a manner to defy detection. And they must also, at least as early as 1848, or in the beginning of 1849, before this case was presented or a tribunal constituted to decide upon it, have procured the same interpolations to be made in the printed report of the Minister Lafragua, read to the National Congress in December, 1846, a copy of which was in the hands of Mr. Benjamin at least as early as the fall of 1849.

After procuring these various forgeries and interpolations to be made, the claimants must have induced numerous witnesses to elaborate and swear to a series of perjuries—minute, circumstantial and plausible—the invention of which displays nearly as much skill and ingenuity as the testimony in regard to them, if false, discloses moral turpitude.

They must have succeeded in suborning, not merely a few nameless and obscure individuals, but numerous persons in high official and social positions; and especially Mr. de Bassoco, a gentleman venerable for his years, and respectable for the singular intelligence and amenity with which he sustained the protracted, acute, and most searching cross-examination of the counsel for the United States.

They must also have procured two ex-Ministers of Relations to perjure themselves, not merely by false testimony before a Commissioner in Mexico, but in Court, within our jurisdiction and subject to our laws, with a full knowledge that the Government alleged the claim to be spurious, and that no efforts would be spared to detect and punish those who were concerned in the supposed conspiracy to defraud it.

Again, this vast conspiracy, from its nature, could not have been successfully carried out without the complicity or connivance, not only of nearly all the officials in the various offices at the alleged dates of the papers, from the lowest probationary clerk or "meritorio," up to the Minister of State himself, but also of those officers employed when the forged papers were afterwards placed in the Archives, as well as of all those who still more recently have certified to their genuineness; and yet, from all these persons concerned in or cognizant of the crime, no whisper has been heard betraying the important secret. Mr. Black, the United States Consul, continues to attach his certificate to the papers without suspecting that he might be lending his aid to a conspiracy to defraud his own Government; and Mr. Forsyth, the United States Minister to Mexico, and for some time resident at the Capital, examines the documents at the various Ministries, and states that "they are found in the several offices where they appropriately belong, were produced by the officers having custody of them, and that he saw nothing whatever to cause him to doubt their being genuine originals."

But the proofs of the genuineness, in part at least, of the documents obtained from Mexico, are obtained from another and an unquestionable source.

Among those documents was found, as has been mentioned, the dispatch of Pio Pico, with the original letter of Castillero of December 10th, 1845, conveying to the Supreme Government the first news of the discovery. There was also produced a traced copy of the borrador of the reply of the Minister.

It is stated by counsel, that the reception of these documents from Mexico first suggested to them the propriety of instituting a search for evidence of the correspondence in the Archives in the Surveyor-General's office.

The search was accordingly made, and there was found the draft of Pio Pico's letter to the Minister, the original of the Minister's reply, together with a letter from Castillero, clearly referring to a previous one of the 10th December.

I am not aware that the genuineness of these documents produced from the Archives in this city is questioned.

It thus appears that the Archives from Mexico are corroborated on the only points where, from their own nature, they were susceptible of corroboration by other records.

The existence of the documents now relied on to establish the title of the claimants, at least as early as the spring of 1847, and prior to the date of the supposed forgery, is also shown by testimony adduced by the United States.

We have already seen, that in James Alexander Forbes' letter of May 5th, 1847, he alludes to "the possession of two sitios ordered to be given by the dispatch of Señor Castillo Lanzas."

In his letter of July 14, 1847, he speaks of the "two leagues conceded to Castillero and socios," and throughout his correspondence frequent and unmistakable allusions occur to the Lanzas dispatch, with reiterated expressions of distrust of its validity.

The same objections made to the document in 1847, are repeated and enforced up to February, 1850, long after the date of the alleged forgeries, but without the slightest intimation that during that interval a second Lanzas dispatch had been fabricated. The document now exhibited is open to all the objections, and liable to every criticism originally made, and so constantly repeated, to the document received by Mr. Forbes in 1847. He complains, in 1850, that his suggestions relative "to the attainment of the important document," explained in his memorandum left at Tepic, in 1849, have not been acted upon. He expatiates upon the insufficiency and discrepancies of the Castillo Lanzas dispatch; but he nowhere breathes a word of reproach or complaint that an abortive and absurd forgery had been committed, the only result of which had been to leave the title as "imperfect and ambiguous" as before.

Had such been the case, we learn enough of Mr. Forbes' disposition from this correspondence to feel sure that reproaches would not have been spared.

There is one other consideration, and it is the last to which I shall advert, which naturally leads us to infer, independently of the proofs, that some proceedings similar to those alleged to have been had, must have taken place in Mexico.

So far back as the Ordinances of 1783, quicksilver had been the subject of distinct and special legislation. The fact that it was indispensably necessary to the extraction of the precious metals, gave to it an exceptional character, and an ample and cheap supply of it had been recognized as essential to the development of the mineral wealth of Spain and Mexico.

It is unnecessary to recapitulate the various decrees and laws of those countries designed to promote the discovery and production of this metal. It is sufficient to say, that out of the public revenues of Mexico a part had been devoted to the formation of a fund called the "Fondo de Azogues," to be used in searching for and developing mines of quicksilver. On every quintal produced a bounty was paid, and to those who should succeed in producing a specified quantity per year, a large sum of money was to be given.

The hope of discovering rich mines of quicksilver within the Republic had led the Junta to institute expensive explorations in various parts of the country, and on all sides it seems to have been considered a national object of primary importance to liberate the Republic from its almost entire dependence on the mines of Almaden, from which the chief supply was obtained.

When, therefore, Castillero discovered a mine of which the "ley" surpassed in richness any that had previously been known, and when shortly afterwards he proceeded to Mexico, it is not conceivable that he should have neglected to inform the Junta of his discovery, and requested of it the assistance in the prosecution of his enterprise which it was one of the most important objects of its institution to furnish. That he would have desired the ratification of his mining possession, and especially a grant of two leagues, we may infer from the fact that he had already solicited a similar grant from the Governor of California.

That the Junta would have received the announcement with the utmost satisfaction, and zealously co-operated with him by recommendations to the Supreme Government, and aiding him by all means in its power, we might conclude even without any proofs of the facts; and proceedings similar to those alleged to have occurred, would have been the natural and almost inevitable consequence.

These proceedings may, it is true, have been interrupted by the breaking out of war and the alarming condition of public affairs; nor do the considerations last suggested authorize us to assume that the dispatch of Becerra, or that of Lanzas, were in fact written at their dates; but they justify the conclusion that the proceedings were initiated, and that the records of them produced from Mexico are at least in part genuine.

Having thus given an imperfect summary of the proofs offered by the claimants, I proceed to consider some of the objections urged on the part of the United States.

It is contended that neither Lazaro Piña, who is alleged to have carried the specimens of cinnabar to Mexico, nor Castillero himself, could have arrived in that city at the time indicated by the documents produced. If this be true, and an *alibi* can be proved as to those persons, we may well regard with suspicion documents found to be false in so important a particular. But the proofs offered by the claimants on these points are too clear to admit of doubt.

We have already had occasion to notice the letters addressed by Castillero while in this country to Gen. M. G. Vallejo, at the christening of whose child he had assisted, and who thus became his compadre.

In a letter adressed to Vallejo, and dated February 21, 1846, Castillero says: "By the brigantine schooner which brought these communications" [referring to communications spoken of in the preceding sentence of the letter] "we have received information," etc. "*This vessel sails shortly, and will carry communications of what has occurred lately. Myself or Piña will leave in it, or both together.* I am only detained waiting the arrival of the division which may touch here in a day or two."

In another letter, dated March 11, 1846, to the same person, he says: "Piña embarked on the 4th of this month in Monterey, and was dispatched in perfect order. He will travel post to Mexico."

These letters are produced by General Vallejo. He swears that they were received shortly after they were written. The signature and handwriting of Castillero are not disputed. If antedated, they must have been written by Castillero in Mexico, and sent on to Vallejo to be produced and sworn to by him—a supposition extravagant in itself, and disproved by the intrinsic evidence of the letters themselves, which contain allusions to passing events, and are couched in a style impossible to invent after the lapse of years.

In corroboration of this statement, the consular books of Mr. Larkin, then United States Consul at Monterey, have been produced. They are identified by Mr. Swasey, the consular clerk at the time. From these books it appears that the brigantine schooner "Hannah" was noted as about to sail for Mazatlan on the 4th of March, the day on which Castillero supposed she had actually sailed. It also appears that the Consul, desirous of sending dispatches to the United States, detained her three days, and a note in his memorandum book shows that she in fact sailed on the 7th.

Shortly after the sailing of the Hannah, becoming alarmed for the safety of Colonel Frémont, who was then encamped on the peak of Gavilan, and expecting an attack, Mr. Larkin sent by a special courier dispatches to Santa Barbara, in the hope of intercepting the Hannah at that port, and of having them conveyed by him to Commodore Sloat at Mazatlan.

That the Hannah arrived at Mazatlan on the 1st April, we learn from various sources.

1st. The "Diario Oficial," a newspaper published in the City of Mexico, contains in the number published on the 22d April, under the head of "Marine news. Mazatlan,—arrivals of vessels," a notice of the arrival at Mazatlan, on the 1st April, of "the American brigantine schooner 'Hannah' of eighty-nine tons, Captain Benjamin F. Thusum, and a crew of ten men."

2d. From a letter of Mott, Talbot & Co., merchants of Mazatlan, addressed to Mr. Thomas O. Larkin, and found among his papers since his decease.

This letter is dated "U. S. S. Portsmouth, 1st April, 1846," and informs Mr. Larkin that his letters have this moment arrived "per 'Hannah.'"

3d. Mr. Larkin's letter to Capt. Gillespie, a copy of which is found in his consular book, which Mr. Swasey swears to have written himself.

In this letter Mr. Larkin says: "Capt. Montgomery, of the Portsmouth, being under sailing orders (the 1st or 2d instant), was waiting at Mazatlan for the Mexican mail when Commodore Sloat heard *per brig Hannah*, of the situation of Capt. Frémont near St. Johns, and immediately dispatched the ship; she was twenty-one days from Mazatlan to Monterey."

4th. The positive statement of Mr. Swasey, clerk to Mr. Larkin, that the latter sent dispatches by the brig Hannah, in March, 1846, in consequence of which the Portsmouth came to Monterey.

These proofs leave no room for doubt as to the sailing of the "Hannah" from Monterey, in the early part of March, with Piña on board as a passenger, unless, indeed, we adopt the theory of the Government, and assume that the letter of Castillero to Vallejo is forged, and that the latter has committed perjury; that the notes of entries and departures in Larkin's consular book are also forged; that the letter of Mott, Talbot & Co., is forged; that the letter of Larkin to Frémont, of March 8th, as also his letter to Capt. Gillespie of April 23d, are forged; that a number of the "Diario Oficial," purporting to be dated April 22d, has been prepared and procured to be printed, and a false entry of pretended marine intelligence from Mazatlan inserted in it; and, finally, that Mr. Swasey, and probably Mr. Larkin's son, have committed deliberate perjury in swearing to the genuineness of the books and papers of the deceased Consul. All this we must assume on the faith of a single statement made by Captain Paty, of the bark Don Quixote, to the effect that "Don Andres Castillero and his servant (*Lazaro Piña, I think, was his name*), were passengers" on board his vessel, on her voyage from Monterey, in April, 1846. But, even supposing that Capt. Paty's memory is accurate, and that Lazaro Piña did not sail in the Hannah, but remained to accompany Castillero in April, it only proves that the latter was mistaken when he wrote to Vallejo from Santa Clara that Piña had embarked on the 4th March from Monterey. It may have happened, that in the three days during which Larkin detained the Hannah, something occurred to induce Castillero to countermand his orders to Piña, and to send his letters and specimens by another hand; for, it must be borne in mind, that proofs of the precise mode in which a few letters and specimens of ore were sent to Mexico fourteen years ago, cannot reasonably be exacted of the claimants. It is surely enough if they show that a vessel sailed about the time supposed, in which Lazaro Piña, or any other messenger of Castillero, might have been a passenger. If the United States contend that the letters and specimens were not and could not have been received in Mexico at the time indicated in the documents produced from that city, and therefore that those documents are false, it is for them to establish the fact.

It is also suggested that Castillero was not in Mexico at the time at which he is alleged to have presented himself before the Junta de Fomento.

The evidence relied on by the United States to support this assertion, consists of a publication in an evening paper in Mexico, of the 6th May, 1846, of the receipt by the Government at the last moment before the paper went to press of important intelligence from California. As this intelligence was undoubtedly contained in the dispatches sent by the vessel which carried Castillero, it is inferred that he could not have arrived in time to be present on the 6th at a meeting of the Junta; and therefore that the actas are false.

That Castillero might have reached Mexico in the first days of May, is evident from the fact that he left Acapulco on the 24th April.

As to the precise time at which the dispatches of which he was the bearer, or which had been sent by the vessel which conveyed him to Acapulco, arrived in Mexico, we have no means of ascertaining, except from the publication referred to.

The paper purports to have been published at 3 o'clock, P. M. of the 6th; as the dispatches were addressed to the Government, and not to the newspaper, it may be assumed that they were first delivered at the appropriate Ministry. What the diligence or energy of Mexican journalists may be, in obtaining the latest news, and how long an interval would probably elapse before they would possess themselves of the contents of a Government dispatch, we are wholly uninformed. That the news was communicated to the newspapers shortly after 12 M. of the 6th, may be inferred from the fact that it was in print at 3 P. M. It is not surely unreasonable to suppose that the dispatch reached the Government on the previous evening, or early on the same morning. I see no reason why Castillero might not, after delivering his dispatches, have presented himself to the Junta on the same day. It appears from the minutes of the session of the 4th, that having learned the result of the assay, the Junta had made a reply to Castillero, asking him what aid he required, Castillero would naturally therefore have presented himself to the Junta immediately upon his arrival; for, besides the invitation of the Junta, and his other reasons for expediting the business, he had engaged the master of the "Don Quixote" to remain for him at Acapulco on her return voyage. This Captain Paty testified he did. But after waiting at Acapulco from the 21st April to the 18th May, he received news that Castillero would meet him at Mazatlan or San Blas. He touched at those places, but heard nothing of him. The circumstance, apparently unimportant, that Castillero determined to rejoin the vessel at Mazatlan, and not at Acapulco, as originally intended, is in precise accordance with the arrangement alleged to have been entered into by him with the Junta, viz : that he was to receive the sum of $5000 in the form of a draft on Mazatlan.

The importance of this incidental corroboration is perhaps not great. It seemed, however, worthy of mention.

But with regard to the inferences sought to be drawn against the genuineness of the actas, from conjectures as to the probable time of the arrival of the dispatches in Mexico, it seems to me obvious, on any hypothesis, that those dispatches must have been delivered, and Castillero have arrived in Mexico, in time for him to present himself before the Junta on the 6th, as their minutes show.

From the foregoing, it appears that the evidence on the part of the United States is insufficient, not only to disprove, but even to raise a doubt as to the fact of the reception of Castillero's letters and specimens, or of his own appearance before the Junta at the dates mentioned in the actas of that body.

But it is objected that the documents produced from Mexico are not admissible in evidence.

This objection is based on the ground that all muniments of title are incident to the land, and pass with it as if a part of it.

That, therefore, all Archives of Mexico relating to the disposition of public lands in California, were included in the treaty and passed to the United States with the cession of the soil. It thus became the duty of the political power to execute the treaty with reference to the muniments, as well as the land, and until that is done, and the political power obtains those muniments and presents them to the Courts, the latter cannot judicially recognize their existence.

No authority directly in point has been cited in support of this position, but a vivid picture has been drawn of the possible evils which might result if adverse claims to the lands of the United States were allowed to be set up, founded on alleged public records existing in a foreign country, and proved by the depositions and certificates of foreign officials.

It will be seen that this objection would apply, although the genuineness and the sufficiency of the documents to convey title were undisputed. If a public and formal grant of a certain tract had been made, to establish which and to show the proceedings which led to it, evidence from the Mexican Archives were necessary—if the Congress had by law conveyed a title to an individual, of which the only evidence existed in the reports of committees and the journals of that body, the principle contended for would require the Court to reject all such documentary evidence, no matter in what way proved or authenticated; for they are to be rejected, *not* because their genuineness is doubtful, but because they *are* Archives and muniments of title to land.

It is admitted that, as a general rule, the right to muniments of title passes with the land, and he who owns the latter is owner of the evidences of his title to it.

But in the cases submitted to this Court under the Act of 1851, the inquiry always is,—who is the owner, the United States or a private individual?

The United States, in consenting to be sued, and in submitting her rights to the determination of Courts, has abdicated, *pro tanto*, her prerogatives as a sovereign, and appears before the Court precisely as any individual who asserts an ownership in land.

To say, then, that all muniments of title belong to the United States as owners of the land, and cannot be noticed by the Courts until commended to them by the Political Department of the Government, is to assume the very point the suit was instituted to determine; for the question is—does the United States own the land? The claimant avers that she does not, and never did, and in support of his claim he produces muniments of title which, on the very principle contended for, belong to him and not to the United States, for they are the muniments of title to his own land.

I cannot perceive, therefore, that the familiar doctrine of the common law, which regards title deeds as incident to the land, and as passing with it, has any application.

In the argument submitted by the counsel for the United States, the distinction seems to have been lost sight of between the political rights of the United States as a sovereign and her purely proprietary rights as an alleged owner of land, which are alone passed upon in this class of cases.

In defining the boundaries of the Territorial Sovereignty of the United States; in determining whether a particular tract is within the limits of a territory, the sovereignty of which has been ceded by treaty to the United States, the Courts must always adopt the construction given to the treaty by the Political Department (Elam *vs.* Neilson, 11 Peters R. 282). But when the United States consent to appear merely as a suitor in the Courts, and to litigate her rights with an adverse private claimant, the rights of both must be determined by the application of the ordinary rules which prevail in actions between private individuals.

It is remarked by the counsel for the United States: "If the Judiciary were authorized to say what land was intended to be transferred, and what papers as muniments of title and incidental to the land, it might designate land and accredit papers which the political department did not, and thus conflict might arise within the Government itself."

But this is precisely what the highest authority of the Nation has, by the law of 1851, enjoined upon the Courts to do. The very object of that law was, that the Courts *should* ascertain what lands passed to the United States by the Treaty, and what lands were private and did not pass. The question, by that law, was converted from a political into a judicial one, and no conflict could possibly arise, for the political and all other departments are by law required to be governed by the decision of the Court, which determines what is public land belonging to the United States, and what is private land belonging to individuals.

There is something repugnant to reason and justice in the idea that the United States, after consenting to appear as an ordinary litigant before the Courts, and submit her proprietary rights to their determination, should suddenly, in the midst of the suit, throw off her character as a mere party to a suit respecting the ownership of land, or rather, without ceasing to be such, should resume and assert her sovereign rights, and announce to her antagonist that evidences of title he offers, though genuine and conclusive, shall not be admitted by the Court unless presented to it through and by herself; while, at the same time she refuses to obtain them from the foreign Government, or to receive them if offered, or to present them to the Court if received.

Compared with such manifest injustice, the evils which might result from possible impositions practiced on the Courts by means of forged archives, etc., are insignificant.

I think the general objection to the admissibility of the documents because they are Mexican archives, not recognized as such by the political department of the United States Government, cannot be maintained.

Assuming, then, the documents from the archives of Mexico to be genuine and admissible, I proceed to consider their legal effect:

1. As to the alleged ratification of acts of the Alcalde Pico. This ratification is supposed, by the claimants, to be contained in the dispatch of Becerra to the Junta, of May 20, 1846, and in the marginal "acuerdo" on the letter of Vicente Segura, signed with the rubric of the same Minister.

The dispatch of Becerra announces, as we have seen, to the Junta, that "His Excellency [*i. e.* the President] has been pleased to approve, in all its parts, the agreement made with that individual [viz. Castillero,] in order to commence the working of said mine."

The "acuerdo" on the margin of Segura's communication is as follows: "Granted, in the terms which are proposed, and with respect to the land, let the corresponding order issue to the Minister of Relations for the proper measures of his office, with the understanding that the Supreme Government accedes to the petition."

An "acuerdo," or order on the margin of a letter, petition

or communication of any kind, is merely an expression of the determination of the Minister or other functionary to whom it is addressed, in regard to its subject matter. Its chief use was to direct the clerks or other subordinates in the preparation of the reply, or in taking other action with regard to it. If the proceeding has been interrupted after the "acuerdo" is affixed, but before the dispatch is written or title issued as directed, it may be regarded not unreasonably as a species of equitable title, or as sufficient, coupled with other equitable circumstances, to justify the party in asking the completion of the proceeding so initiated. But when the title has issued, or the dispatch been written in pursuance of the "acuerdo," when the latter has been submitted to the Minister, and approved and signed by him, the dispatch so approved and signed is the highest and best evidence, not only of the action of the Government in the premises, but of the true intention of the "acuerdo;" for, surely, no argument is necessary to prove that an official reply, signed by a public officer, is better evidence of his resolution with regard to a particular application, than a direction to his subordinates as to the form in which the reply is to be drafted.

Dismissing, then, the "acuerdo," or rather treating it as intending precisely what the dispatch prepared in obedience to it expresses, let us consider the true import and effect of the latter.

It will be observed that the dispatch of Becerra does not, in terms, profess to ratify any mining possession or grant, either of lands or of pertenencias. Nor does it announce that the President has been pleased to make any such ratification. It merely informs the Junta that His Excellency has approved an agreement made by the Junta with Castillero.

It is not pretended that any such agreement was, at that time, or afterwards, formally entered into between the parties. The propositions of Castillero are dated May 12th. The communication of the Junta is dated May 14th, and Castillero himself, in the preamble to the statement of his propositions, expresses his persuasion that the Junta will accede to his request "so far as may be within its powers, and that it will send up to the Supreme Government with a recommendation that which may require the decision of the latter."

From the communication of the Junta it is evident, that the authorization of the Supreme Government was necessary to enable it to furnish Castillero with the iron retorts and flasks belonging to it, as also to make him the loan he solicited of five thousand dollars, payable in quicksilver at one hundred dollars per quintal, and without the five per cent. premium per annum which the law required it to exact.

Until the approval of the Supreme Government of this proposed arrangement could be had, no formal contract could be entered into. It was, therefore, not until May 25th, and after the receipt of Becerra's communication approving the proposed contract, that the Junta resolved "that the proper judicial agreement be drawn up immediately, and that application be made for the draft for the five thousand dollars on Mazatlan or Guadalajara," as appears by the Actas of that day.

That the agreement was never so drawn up and executed is admitted; and on the 29th May an order was made for the payment of the Notary Calapiz "for proceedings relative to it," its consummation having been prevented by the order suspending all payments out of the quicksilver fund.

The language of the dispatch is, therefore, evidently inaccurate in speaking of the approval of the agreement *made* or "*convenio celebrado*" with Castillero. Its evident intention was to signify the approval by the Government of the agreement proposed to be made, and which the Junta had expressed its willingness and even anxiety to enter into.

What the agreement was, which, after the approval of the Government had been obtained, the Junta and Castillero had fixed upon and nearly consummated by a formal act before a Notary, we learn from the Report of the Junta of Nov. 17th, 1846, produced by Señor de Bassoco, and embodied in Mr. Lafragua's report in December of the same year.

In this report, the Junta, after giving an account of the presentation of cinnabar ore, etc., by Señor del Moral, of its assay, and of their inquiry of Castillero what aid he required, proceeds as follows:

"The Señor presented his petition in due form, and it having been very attentively examined by the Junta, he made his propositions, to which this Junta agreed, to wit: That there should be delivered to him five thousand dollars in money, eight iron retorts of those which the Junta ordered to be made for the examinations previously made, and all the quicksilver flasks it had in the negotiation of Tasco; Señor Castillero obligating himself, on his part, to repay said advance in quicksilver at the rate of one hundred dollars per quintal, within six months from his leaving the port of Mazatlan. This agreement was approved by the Supreme Government on the 20th of the same month; but on account of the declaration of blockade made by the United States of the North, when he was about to receive the draft on Mazatlan, the Ministry issued the order of September 19th of this year, directing the suspension of all payments of the branch of quicksilver, except those for the support of the College and the expenses of the office."

In this account of the cause, and the date of the abandonment of the agreement with Castillero, the Junta are evidently inaccurate; for their own Actas show that the communication informing them of the blockade of Vera Cruz and Tampico, and directing the suspension of all payments for the extraction of quicksilver, was dated on the 27th May, and received by the Junta on the 28th; and on the 29th, the Notary Calapiz was paid for his proceedings in relation to the intended contract.

The communication of the 19th September ordered that the assets of the quicksilver fund should *continue to be used merely for the support of the College*, and it demanded a loan of twenty-five thousand dollars from the Dotal fund. This was strenuously opposed by some of the members, on the ground that the Dotal fund was the private property of the creditors of that fund. As these discussions occurred less than two months previous to the date of the report of the Junta, and were no doubt fresh in its recollection, and as the report was prepared in great haste, only eight days being allowed for the purpose, the Junta fell into the error of ascribing the breaking off of the negotiation with Castillero to the order of September 19th, instead of to that of the 28th May.

But with respect to the agreement made with Castillero, and

approved by the Supreme Government, the report is very explicit. It sets forth the terms of that contract with a clearness which leaves no room for doubt as to what it was that the Supreme Government approved.

The agreement thus entered into embraced all the subjects upon which the Junta had authority to act. Nor can it be said that the approval of the Supreme Government was only required as to those propositions of Castillero which related to the ratification of his mining possession and a grant of two leagues, for we learn from the letter of the Junta that that body had no authority, without the approval of the Government, either to sell the retorts and flasks desired by Castillero, or to lend him a large sum without interest, to be repaid in quicksilver. But with the granting of lands the Junta had nothing to do, and whatever might have been the resolution of the Government on Castillero's seventh and eighth propositions, it would never have been communicated to the Junta in the form of an approval of an agreement into which they were supposed to have already entered.

Some stress has been laid on the use of the word "concedido," or "granted," in the marginal "acuerdo" of Becerra.

Had this word appeared alone, and been written on the margin of Castillero's petition, it might, perhaps, have been considered evidence that the whole prayer of the petition had been granted. But it is written on the margin of the Junta's letter, and clearly imports that its request was granted, viz., that the proposed agreement was approved, as is unequivocally shown by the official dispatch written on the same day, and in pursuance of the marginal order of the Minister.

We shall presently see that the petition of Castillero was for land, and not for additional mining pertenencias.

The acuerdo therefore adds:—

> "With respect to the land, let the corresponding order issue to the Minister of Relations, for the proper measures of his office, with the understanding that the Supreme Government accedes to the petition."

The corresponding order did issue—we have it in Becerra's dispatch to the Minister of Justice. The proper measures were taken in his office—we have them in his dispatch to the Governor of California; and from it alone can we learn what was done by the Government "with respect to the land" petitioned for by Castillero.

It has already been stated that the Act of Possession of the Alcalde Pico embraced two distinct objects: first, the judicial possession of a mine, with the number of pertenencias allowed by law—but how many, both the Alcalde and the parties seem to have been uncertain; secondly, a grant of a tract of land extending three thousand varas in every direction, as a "*gracia*" or gift to Castillero. The distinction between these two acts of the Alcalde is not only admitted but strenuously insisted on by the counsel for claimant, and it is contended that the first was legal and valid, while the second is conceded to be utterly nugatory and void.

When, therefore, Castillero asked that the Junta would recommend the approval by the Supreme Government of the possession which had been *given him* of the *mine*, in the same terms as those in which he then held it, he must have intended to ask either for a ratification of the possession of the mine, or for an approval of the grant of three thousand varas of land, or for both.

That he did not ask for three thousand varas to be given him as additional pertenencias, is admitted by one of the able and eminent counsel who argued the cause for the claimant.

In the printed report of his argument, he is asked by Mr. Randolph, of counsel for United States:

"You argue, then, that the Junta, misunderstanding this document of Castillero's, supposed it to be for additional pertenencias, and as such recommended its confirmation?"

Mr. Benjamin.—"Certainly."

But this was not the only error into which the Junta fell; for they not only supposed that Castillero was seeking additional mining pertenencias, and not merely a tract of land for his hacienda, etc., as well as two square leagues to supply wood for his burnings, but they supposed the three thousand varas so desired, would only amount to fifteen pertenencias, whereas they would amount to nine hundred. The Junta evidently supposed that Castillero solicited a tract three thousand varas long, and of the width of one pertenencia. As a pertenencia is two hundred varas in length, a tract three thousand varas long would comprise exactly fifteen pertenencias. They overlooked the fact that the tract, was to be three thousand varas "*in every direction*," or six thousand varas square, making nine hundred pertenencias.

If, then, the Supreme Government had formally and unequivocally signified its assent to this recommendation of the Junta, and ratified the possession as represented by them, it may well be doubted whether in a Court of Equity it could be deemed to have ratified any more than a possession of fifteen pertenencias, which was all that Castillero, speaking through the Junta, demanded.

But the fact, that the Junta thought it necessary to devote so much time, and to suggest so many arguments, to induce the Supreme Government to ratify a supposed mining possession of fifteen pertenencias, justifies the supposition that had they known it to have comprised nine hundred pertenencias, they would probably have withheld their recommendation.

That both the Junta and the Supreme Government were willing to assist the enterprise of Castillero by every means in their power, is evident. Their object in so doing, was not to confer a favor on Castillero personally, but to promote the production of quicksilver in the largest quantities and at the cheapest rates possible.

The same policy would have forbidden them to give to a single miner nine hundred mines, of one pertenencia each, and thereby to exclude from so large a tract all miners who might otherwise have discovered and developed new mines in the vicinity, and increased the production and diminished the price of the metal the Government was so anxious to obtain.

It has appeared to me that the very considerations urged by the counsel of the claimant with regard to the policy and interest of Mexico in promoting the production of quicksilver, render it impossible that it could, consistently with that policy, have consented to a monopoly by a single miner of a mining tract of such enormous dimensions.

It is clear, therefore, that the Supreme Government could not have intended to ratify the possession of the three thousand varas as mining pertenencias.

But we have already seen that the approval by the Government of the agreement made by the Junta, is conclusively shown not to import a ratification or grant of any pertenencias or lands whatsoever; for the terms of the agreement so approved are disclosed to us by the Junta itself, and can be ascertained as exactly as if the formal instrument had been executed by the parties, the approval of the Government appended to it, and were now before us.

It does not appear that the Act of Possession of Pico was ever exhibited either to the Junta or the Supreme Government. If it had been, it would have disclosed the fact that the mine had been denounced and possession of it given, as on the lands of José Reyes Berreyesa. It would also have been seen that the Alcalde had assumed to grant a tract three thousand varas in every direction from the mouth of the mine, which must have included a large portion of the land of a private individual, even supposing that the mine itself might not have been within his limits. Adopting the obvious construction of the Act of Possession contended for by the claimants, and regarding that act, and the ratification asked for by Castillero, as referring to a tract of land and not to additional pertenencias, and assuming with the distinguished counsel for the claimants, that the Junta was mistaken in supposing that any number of additional pertenencias, whether fifteen or nine hundred, were asked for, we may well doubt whether the Supreme Government, if informed that this tract would in great part include private property, would have so readily made the grant. We have no reason to suppose that, as between individuals at least, rights of property are not as scrupulously respected and enforced by the Mexican as by other nations.

Again, it is contended that in addition to the grant of three thousand varas in every direction, made by the Alcalde and approved by the Supreme Government, there were also granted to Castillero two square leagues of land, to be measured in like manner from the mouth of the mine.

The mode adopted and the precautions observed by the Supreme Government in signifying its willingness that such a grant should be made, will hereafter be adverted to.

Our only concern with it at present is to observe, that on the claimant's theory, the Supreme Government first ratified a concession of six thousand varas square, or more than a league and a quarter in extent, and then issued orders for a further grant of identically the same land, with three-quarters of a league in addition. Its resolution with regard to the latter is formally and regularly communicated to the Governor of California, who was directed to take the proper steps to carry out the intention of the Government; while with regard to the former, its determination is supposed to be expressed in a declaration that it approves a contract, the terms of which we know, and which has no reference, nor could it have had, to grants of land; and the approval of which, if it could by possibility be construed to mean an approval of all Castillero's propositions, would import a grant of the two sitios, as clearly as it would import a ratification of the concession of the six thousand varas square.

But the dispatch of Becerra to the Minister of Relations informs us to what part of Castillero's petition the President thought proper to accede, in language too explicit to be misunderstood.

After transcribing his letter to the Junta, Becerra says: "And I have the honor to transcribe it to your Excellency, to the end that with respect to the petition of Señor Castillero, to which His Excellency the President *ad interim* has thought proper to accede, *that* there be granted to him two square leagues as a colonist," etc.

It is insisted by the counsel for the claimants, that the words "*that there be granted to him two square leagues as a colonist*," are descriptive of the petition of Castillero to which the President acceded. The observation is just. Such is no doubt the true construction of the dispatch, and it establishes beyond doubt, that in acceding to the petition the President meant only to accede to that part of it which asked for a grant of two leagues as a colonist, without expressing any resolution as to the application for a ratification of the concession of three thousand varas.

As, then, the supposed ratification is not contained in the approval of the contract of the Junta, nor in the acceding by the President to the petition for two leagues in colonization, it must be found, if at all, in the word "concedido" or "*granted*" in the acuerdo. But for the reasons given above, I am satisfied that no such signification can be attached to that word in the face of the dispatches written in pursuance of the acuerdo, and which embody and explain its meaning.

But if any doubt could remain as to the true intention and effect of the approval of the Junta's contract, it would be dissipated by the evidence afforded by the acts of the parties, of the construction placed upon it by themselves.

Throughout the whole negotiation for the purchase of barras or shares in the mine, conducted by Mr. Negrete on behalf of Mr. Forbes with Castillero in person, the latter, though urged to exhibit his documents of title, produces only the dispatch of Castillo Lanzas for two leagues. In the instrument of ratification the mine is spoken of as of three pertenencias in extent, and Castillero "cedes in favor of the contractors of supply (aviadores), and for the sixteen years of this contract, the two square leagues of land of which the Government has made him a concession, as shown by the official document which he presents, that it may be inserted at the end of the present instrument."

The Lanzas dispatch is accordingly copied in the instrument, but not the slightest allusion is made to any other grant of three thousand varas in every direction made by an Alcalde, and approved by the Supreme Government.

In all the transactions between the parties, the idea is but once suggested, that the approval of the Junta's contract with Castillero imported a ratification of the Alcalde's concession of three thousand varas.

It occurs in Alexander Forbes' letter to James Alexander Forbes of February 3, 1850. In that letter Mr. Forbes says:

"We think at present it may be the best plan to get an authenticated copy of the approval by the Mexican Government of the three thousand varas given by the Alcalde on giving possession of the mine. As a doubt may be started as to whether the Alcalde, acting as the Juez de Mineria, had a right to make this grant, yet, if approved by the Mexican Government before the possession of the country by the Americans, there could be no doubt on the subject. This takes in our hacienda, and *unless opposed by the Berreyesas*, would, I should think, settle the question. *Castillero says such an approval was given*, and that on his arrival in Mexico he will procure a judicial copy of it. *This is the plan we shall adopt*, if we hear nothing from you to alter this resolution.

"Since writing the foregoing, I have looked over your private letter to William Forbes, dated 18th October, in which you state the limits or boundaries as follows: 'The boundaries must be expressed as joining on the north and northwest by lands of the ranchos de San Vicente and de los Capitancillos; and the east, south and west, by Serrania or Tierras baldias.'

"Castillero is *not certain of accomplishing this latter plan*, but thinks the first, *that is the three thousand varas, the best.*"

It will be observed that this letter unmistakably discloses the "plan," which James Alexander Forbes had suggested and Alexander Forbes adopted, of obtaining fraudulent and antedated documents from Mexico expressing the boundaries of the two sitios, etc. No reliance can therefore be placed on the statement that Castillero said the approval was given.

But whatever he may have told Mr. Forbes, the approval and the agreement approved are before us, and we have already seen that they contain no allusion to any concession of land by an Alcalde.

The silence of Castillero during his negotiations with Mr. Negrete, is far more significant than any statement made four years afterward to Mr. Forbes; nor can it be said that at the time of those negotiations he was ignorant of the action of Government; for the Castillo Lanzas dispatch, then in his possession, and inserted at the end of the contract of ratification, recites the Becerra dispatch, which contains the approval of the contract. The parties were in Mexico; the public offices were accessible, and it would have been easy to ascertain what was the contract approved. That Castillero knew what that contract was cannot be doubted; and yet he and all the other parties, through a series of years, treat the dispatch as a concession of two leagues, but never suspect it to contain evidence of, or to be in itself, a ratification of the Alcalde's gracia, or gratification. Even so late as the date of Mr. Halleck's affidavit, James Alexander Forbes, with the notarial copy of the Lanzas dispatch in his possession, never seems to have imagined that the dispatch of Becerra inserted in it, and announcing the approval of the Junta's contract, constituted the ratification he so much desired of the Alcalde's grant of three thousand varas.

I think it clear, therefore, that the dispatch of Becerra cannot be construed to import a ratification of the action of the Alcalde, either in respect to the possession of the mine, or to the grant by him of three thousand varas.

As to the alleged grant of two leagues. It cannot be denied, that if the documents produced by the claimants be genuine, they show that Castillero presented a petition for two square leagues of land; that this petition was by the Junta de Fomento submitted to the Supreme Government; that the Junta was formally apprised by the Minister of Justice that the proper communication had been sent to the Minister of Relations, that suitable orders might be issued by him with respect to that part of Castillero's petition; that a communication was accordingly sent to that Minister, informing him that the President had acceded to Castillero's petition, and requiring him to issue the corresponding orders; and that the Minister of Relations, in pursuance of these instructions, transcribed the communication to the Governor of California, in order that in conformity with what the laws and dispositions on colonization provided, he might put Señor Castillero in possession of the said two leagues.

It is manifest that this dispatch of Castillo Lanzas does not by its terms grant the land solicited. It contains no words translative of title; it is not addressed to the supposed grantee, but to the Governor of California; it cannot have been intended to serve as a muniment of title to Castillero, for otherwise it would have contained formal words sufficient to vest the estate in him. It is merely an official communication addressed by one executive officer to another, which, for aught that appears, might as well have been sent to the Governor of California by a courier as by the hands of Castillero. It contains, however, an unequivocal official declaration that the President of the Republic had thought proper to accede to Castillero's petition, and an order to put him in possession conformably to the laws on Colonization.

The case thus resembles in some respects that of United States *vs.* Lecompte, 11 How. 124, where the claimant had obtained an order from the Lieutenant-Governor directing the Procurador del Comun to put him in possession, if in so doing no prejudice would result to third persons.

The petition solicited two leagues at the place called Llanacoco, to be located so as to include the entire prairie of that name. The Supreme Court held that the order of the Lieutenant-Governor could not be construed to signify an absolute unconditional grant of any specific land; and, as it was never presented to the Procurador, and the land had never been severed from the public domain by that officer, and no occupation was shown which could supply the deficiency by giving certainty and definiteness to the claim, it was rejected.

If the reasoning of the Supreme Court be attentively considered, it will be seen that they refused to treat the order of the Lieutenant-Governor as an absolute grant, for two reasons: 1st. Because it directed the petitioner to be put in possession, "if in so doing no prejudice could result to third persons," and the matter was referred to an officer, to whom the duty was confided of ascertaining the means, etc. of the petitioner and "judging of the propriety of the grant." And 2d. Because the land was not severed from the public domain by the description in the concession, or by an authorized survey, or by any definite occupation.

It may, I think, be inferred from the whole opinion, that if the concession had described any tract which could be identified, and the petitioner had occupied it, the claim would have been confirmed, notwithstanding the omission to present the order to the Procurador. In the case at bar, the Governor is ordered to put the petitioner in possession "in *conformity with what is prescribed in the laws and dispositions on colonization.*"

The 2d article of the law of 1824, declares that "those lands of the nation, which are not the property of any individual, corporation or town, are the subject of this law, and may be colonized."

The first step to be taken by the Governor, in execution of the order, would have been to ascertain whether the land was within the colonization law—that is, whether it was the land of the nation, or belonged to any private individual; precisely as the Procurador was to ascertain whether by putting the petitioner in possession, any injury would result to third parties.

To construe this dispatch as an absolute grant of a specific tract, we must suppose Castillero to have practised on the Government a gross fraud, either by concealing or misstating the facts. For, even admitting that he had reason to believe that the mine itself was not within the limits of Berreyesa, he must have known that a tract of two leagues measured in all directions from the mouth of the mine, would certainly have included a portion of his land.

The counsel for the claimants, feeling, no doubt, the force of this objection, suggested that it was intended that inquiry should be made by the Governor, and the two leagues were to be located in such a way as not to include private land. But this admission proves that the duties to be performed by the Governor, were exactly those asigned to the Procurador del Comun, in the case referred to; and if the order to put in possession was not a grant in that case, neither can it be so considered in this.

The case at bar is, in some respects, stronger than that reported; for here, the order was addressed to the Executive of a Department to whom, by the laws, in conformity to which the order was to be executed, it belonged to issue the formal title for the land, while the functions of the Procurador were merely to inquire into and report the circumstances of the petitioner, and to mark off and sever from the public domain the land granted.

Had the Governor and Departmental Assembly, in ignorance of the application of Castillero, and the action of the Supreme Government upon it, regularly granted the same land to another person, in strict conformity with the colonization laws, I cannot doubt that his title, though subsequent in date to the dispatch of Lanzas, would have prevailed; and on the reception of that dispatch the Governor would either have refrained from executing it at all, or would, more probably, have allowed Castillero to take his two leagues out of the nearest body of ungranted land.

In the very ingenious brief filed by the counsel for the United States, it is observed:

"The theory of government for the Mexican Territories or Departments was, that all the powers of government were exercised immediately by the local Political Chief or Governor, through whom the will of the Supreme Central authority at the City of Mexico was transmitted, and by the action of which functionary, and not otherwise, it became operative on persons and things. It was a government of governments—the plan which Spain established in the beginning for the government of the Indies, and which Mexico continued to this extent without a change. Its model was the organization of an army. It was the same whether the Sovereign's will was expressed in the form of a general law, or some particular disposition like a grant; all were alike instructions to an inferior officer. They were not binding on him until known, nor effective until obeyed—until he had done or suffered another to do that which was required.

"The local representative of the National Sovereignty was invested with all the active powers of Government in this Department. He, alone, could manifest the grantor's will, and from Mexico no more could come than the impulse which should move him to act. In these titles, nothing was done here, even the mining title being an original grant in Mexico. The local representative has never acted, and therefore there has been no expression of the grantor's will."

In this lucid exposition of the general theory on which the vast governmental machinery of the ancient Spanish monarchy, and measurably that of Mexico, were constructed, I entirely concur. But if it be argued that because the will of the Sovereign was ordinarily communicated to and executed by subordinate agents, he had no power himself directly to act upon persons and things without the intervention of the local authority, I cannot assent to the conclusion; for the will of an absolute sovereign can be manifested in whatever form he may choose to adopt; and had the King of Spain seen fit to make, under his own hand and seal, a grant to a subject in a remote province, I cannot suppose that the royal patent in the subject's hands would not have been respected by the subordinate authorities, notwithstanding that the title had been transferred, and the grant consummated without their intervention.

Nor is it certain that the order of the Sovereign to a subordinate to make a grant to a subject, or do any other act in the performance which he is interested, is a mere nullity, until known to and obeyed by the inferior officer.

If, as is admitted, the impulse which moves the subordinate to act, rightfully proceeds from the central authority—if the subordinate has no discretion in the premises, but must obey, and his duties are purely ministerial and executive, it would seem that the sovereign disposition which required him to act, cannot be regarded as a mere nullity, even though never in fact obeyed.

Such, I understand to have been the ruling of the Supreme Court in a recent case, where an order to the Governor that a particular island should be assigned to an individual was held itself "to adjudicate the title;" the formal issuing of the title papers being a merely ministerial act to be performed by the Governor.

It is true, that, in that case, the order was received and obeyed and the title issued. But, the language of the Supreme Court is explicit, that the dispatch "operated of itself *to adjudicate the title to the claimants.*" This case will be more fully considered hereafter.

But whether a notice of the *superior will* must have been given to the local representative of the Government, and that will must have been obeyed before it could affect the rights of persons, or the condition of things; or whether, as seems to be considered by the Supreme Court, the sovereign disposition, though unknown to, and unobeyed by, the local authority, had an immediate and independent operation, it is clear that by the uniform practice of the Spanish and Mexican Governments, the dispositions of the sovereign authority always contemplated the instrumentality of the local subordinate.

If, therefore, in this case the Government had known that the tract solicited was public land, and that no objection whatever existed to making the grant, it would have been a signal departure from its ancient and established practice to issue the grant directly to the applicant.

In treating, then, this dispatch as an order to make a grant, we suppose the Government to have conformed to its immemorial and traditional usages. While to consider the dispatch as itself conveying the title, and containing merely an order to the Governor to put the claimant in possession of land already granted to him by the direct act of the Supreme Government, is to suppose the latter to have adopted an exceptional mode of proceeding, inconsistent with that pursued in the Islands case, and with the theory on which the whole system of Government was organized.

For these reasons I cannot regard the dispatch of Lanzas as a direct grant of the two leagues referred to.

Taking this view of the dispatch, we can account for the very explicit declarations of the Mexican Commissioners, that no grants had been made of land in California subsequent to May 13, 1846.

For, even if their researches had extended to the official correspondence Ministers of Relations who had held office from the date of the declaration, and they had discovered the dispatch of Lanzas, they regarded it but as an order to the Governor to make a grant, which they knew could never have been acted on. The truth of their declarations, therefore, to the American Commissioner, is thus entirely consistent with the genuineness of the documents, while, if the dispatch be considered to import an absolute and present grant, we are driven to choose between two alternatives—one, that the Commissioners were guilty of a deliberate falsehood—the other, that the dispatch itself is a forgery.

But the dispatch, though not itself a grant, is, nevertheless, evidence that the Supreme Government acceded to Castillero's petition; and it is, at least, an order that a grant should be made to him by the Governor of California, in conformity with the colonization laws, *i. e.* if the land were vacant and no other insuperable objection existed.

Giving, then, this construction to the dispatch, let us consider its legal effect—and here we are fortunately not without authority to guide us.

In the case of Andres Castillero vs. The United States, for the Island of Santa Cruz (23 How. 464), the claimant relied on a dispatch of the Minister of Interior, in many respects resembling that of Castillo Lanzas.

In that dispatch the Minister informs the Governor, that in consideration of the services and merits of Castillero, the President directs him (the Minister) "to recommend Castillero very efficaciously to your Excellency and the Departmental Junta, in order that before proceeding to the distribution which should be made conformably to the laws, and as is directed in the order of this date, of the lands of the islands adjacent to that peninsula, there be assigned to that individual the one which he may select of those nearest to the place where he should reside with the troops under his orders."

It will be noticed that the terms of this dispatch are not in some respects so strong as those of the dispatch of Lanzas.

It is not said that the President has acceded to a petition for any particular island or tract of land. Castillero is merely "recommended very efficaciously to the Governor and the Departmental Junta," and this recommendation is made in order that there be "*assigned*" to him the island he may select, etc.; contemplating, evidently, the execution and delivery by the Governor of the formal title for the island so selected. In the Lanzas dispatch the President's assent to the petition is communicated to the Governor, "in order *that he may put Castillero in possession* of the land"—an expression which has afforded room for the construction that no further title paper was designed to be issued.

In the Santa Cruz Island case, the Supreme Court held that "the dispatch of the Government operated to adjudicate the title." Its language is:

"They [the Governor and the Junta] were accordingly

directed not to proceed to make adjudications under the previous order until the assignment of the title to this claimant was perfected, but they were *not required to make the assignment or to cause it to be made.*

"To accomplish that purpose, and to carry into effect the command of the President, two things only were necessary to be done; one was to be performed by the claimant, and the other was a Ministerial act. It was the claimant who was to make the selection, and if it was a proper one, near the place where he was stationed with his troops, nothing remained but to make the assignment as described in the dispatch. Emanating as the dispatch did from the supreme power of the nation, *it operated of itself to adjudicate the title to the claimant, leaving no discretion to be exercised by the authorities of the Department.* Neither the Governor nor the Assembly, nor both combined, could withhold the grant after a proper selection, without disobeying the express command of the Supreme Government. *Nothing, therefore, remained to be done, but to issue the title papers, and that was the proper duty of the Governor as the Executive organ of the Department.*"

This language would seem too clear for misconstruction. It seems to me an express decision, that an order of the Supreme Government directing the Governor of a Department to make a grant, operates to adjudicate the title to the land specified. It could not, however, have been meant that the order itself was an absolute grant; for it is evident that the formal grant was to be made by the Governor, and besides, it did not refer to any particular island, but to such island as Castillero might select; and it was only after the selection was made and found to be a proper one, that the title attached to any particular piece of land.

If such were the effect of the order of Pesada, I am unable to perceive why the Castillo Lanzas dispatch must not be considered to have had a like operation. The direction in that dispatch to the Governor to put Castillero in possession in conformity with the laws and dispositions on colonization, confided no more discretion to him than the duty of seeing that the island selected was a proper one, confided to the Governor in the case before the Supreme Court; and if the issuing of the title papers in that case was a "merely ministerial act," to be done in respect of lands, the title of which had already been adjudicated to the claimant, the same view must be taken of the action, which, construing the Lanzas dispatch least favorably for the claimants, Governor Pico was in this case ordered to take.

It is contended that the President of Mexico had no authority to make the order for a grant contained in the Lanzas dispatch.

But, 1st, it is apparent, both from the action of the Supreme Government in this case, as well as in that of the islands on the coast, that it exercised the power.

The presumption, therefore, arises, that it had the authority it exercised. " The public acts of public officers purporting to be exercised in an official capacity, and by public authority, are not to be presumed to be usurped, but a legitimate authority previously given or subsequently ratified, which is equivalent." —The United States vs. Arredondo, 6 Pet. 728.

2. The Colonization Law of 1824, while it enjoined upon the States of the Confederation the duty of making laws or regulations for colonizing within their respective limits, committed the whole subject of colonization in the territories to the Supreme Executive.

The 16th article of that law provides, that "the Executive shall proceed in conformity with the principles established to the colonization of the territories."

In pursuance of this authority the Supreme Executive in 1828, framed the regulations which prescribed the mode in which the colonization of the territories should be effected.

It was from the dispositions thus made by the Supreme Executive, that the Governor and Junta in the Territory of California derived all their powers with respect to the granting of land.

Had the President seen fit to confide the authority to grant, either to the Governor alone, to the Prefects of the Partidas, or to local Commissioners, he might have done so, or he might have retained it exclusively to himself.

The regulations, in fact, provided that concessions made by the Governor should not be definitively valid unless approved by the Departmental Assembly; and, in case its approval was not obtained, the Governor was to report to the Supreme Government for its decision.

Grants made to "empresarios" were, by the 7th regulation, not to be held as definitively valid until the approval of the Supreme Government was obtained.

It thus appears, not only that the authority of the Governors with respect to colonization was not immediately conferred by any law of Congress, and owed its existence to discretionary regulations of the Supreme Executive, but that by those very regulations the Executive reserved to itself an important part of the granting power. After the adoption of the Central System, and the division of the whole Republic into Departments, the right to dispose of all the lands belonging to the nation seems to have been confided to the Supreme Executive. The law of 1837 gave to the President authority to sell or pledge them, and by his decree of 11th March, 1842, other important and fundamental changes in the colonization laws with regard to foreigners were made by General Santa Anna.

It is said that by making a grant directly to an individual, or in directing the Governor of a Department to make one, the President violated an existing law, which even an absolute monarch cannot do; for he may abrogate or modify the law, yet while it remains unrepealed he cannot violate it.

The general principle is admitted, but its application to this case is not perceived.

That the President, by the law of 1824, could have reserved to himself the whole right of making grants in the territories, has already been shown. Such a disposition, though not in accordance with the ordinary policy of the Spanish and Mexican Governments, which intrusted the administration of local affairs to local subordinates by whom the orders of the Supreme Government were carried into effect, would, nevertheless, have been legal and within the limits of the discretion confided to the Executive by the law of 1824.

This power he still retained, notwithstanding that he had framed general regulations on the subject for the guidance of the Governments of the Territories; for those regulations were the mere creature of the President, and could not deprive him or his successors of the general powers given him by law, or of the right to act directly in special cases by making the grant himself, or by ordering the Governor to do so.

The Governor of a Department had no power to grant lands by virtue of his office, or conferred on him as such by law.

All his authority to grant was derived from the regulations of the Executive, of whom he was but the agent and the instrument. He was at all times subject to executive instructions, and the President might at his discretion withdraw any lands from colonization, prescribe new qualifications for grantees, or in any other manner modify the Governor's authority with respect to grants, or direct him as to its exercise.

That he did so interpose with regard to certain Mission lands which the Governor and Assembly were about to grant is well known, and this Court has decided grants in violation of the order of the Executive to be invalid. The islands cases and the case at bar furnish additional instances of the exercise of the same power.

I confess myself unable to understand how a grant by the President, still less an order to his local subordinate to make a grant, can be deemed such a violation of the law as no absolute monarch could commit, or indeed any violation of law whatsoever.

3d. The question is decided by the Supreme Court, in the case which has been referred to.

If the Supreme Government had power to direct the title papers to be issued, and the dispatch operated to adjudicate the title in that case, it must be deemed to have possessed the same authority, and a similar operation must be attributed to the dispatch in the case at bar.

But it is urged that a distinction should be drawn between the cases, on the ground that islands on the coast were not within the colonization law of 1824, and therefore might be granted directly by the Supreme Executive, but that he had no authority to act in relation to lands embraced within the provisions of that law, except in obedience to it, and in conformity with the regulations of 1828.

It has already been shown that under the Colonization Law the President had authority either directly to grant or to order the Governor to grant public lands in the territories. But his power to grant islands was also derived from the same law, and in making the grant of the island of Santa Cruz, the validity of which has been affirmed by the Supreme Court, he acted in strict obedience to it.

It will also be seen that the judgment of the Supreme Court is not based on the supposed existence of any authority in the Executive not derived from the law of 1824; and also that his right to repeal or modify at his will, in a particular case, his own general regulations which imposed rules on the subordinate local authorities, is impliedly recognized in the decisions referred to.

The 4th article of the law of 1824, provides that "the lands embraced within the twenty leagues bordering on any foreign nation, or within ten leagues of the seacoast, cannot be colonized without the previous approbation of the Supreme Executive power."

As by the previous section of the law, the Congresses of the various States were directed to enact laws and regulations for colonization within their respective territories, while by Art. 1.6 a similar duty was enjoined upon the Supreme Executive with respect to lands within the territories, it is obvious that the 4th article was intended chiefly to restrict the power of the States rather than that of the Executive, whose assent to the grant was all that was required.

In the case of the United States *vs.* Arguello (18th How. 548), it was held by the Supreme Court that the "colonization" spoken of in the 4th article must be construed to mean colonization by foreigners, and not the distribution of lands to individuals and families.

The power of the Governors of California to grant lands within the ten littoral leagues might perhaps have been sustained, even if the 4th article be construed to apply to grants to individuals, on the ground that the absence of any express prohibition in the regulations, and the constant exercise of the power with the full knowledge of the Supreme Government, authorize the presumption that the approval required by the 4th article was in fact given.

However this may be, it is clear that the Governors of California did not assume to grant the islands on the coast without the previous permission of the Supreme Government. Application for such permission was accordingly made, and it was finally communicated to the Departmental authorities in the dispatch of Pesada of July 20, 1838.

When, therefore, the President ordered a grant of an island to be made, which order the Governor obeyed by issuing the title papers, the grant was in strict conformity with the Colonization laws.

For that law confided to the Supreme Executive, as has been observed, the whole subject of colonization within the territories, nor did it impose any limits on the exercise of his discretion, except that the colonization was to be conducted according to the principles established by the law.

Those principles were of a general character, and fixed nothing as to the particular agencies or mode to be adopted in conferring the title upon the colonist.

In the case of lands within the ten littoral leagues, the law itself forbade their colonization without the previous approbation of the Supreme Executive power. The general regulations, therefore, by which the Supreme Executive authorized the Governors and Juntas of the Departments to grant public lands, were never construed to authorize them to grant the islands on the coast, and as observed by the Supreme Court, the power to make such grants was neither claimed nor exercised by the Departmental authorities prior to the 20th day of July, 1838, when the "previous approbation of the Supreme Executive" required by the law was communicated to them.

That approbation having been thus obtained, the Departmental authorities proceeded to grant; and in so doing, acted in precise conformity with the Colonization laws.

It had appeared to this Court, that the effect of that dispatch was simply to communicate to the Departmental authorities the assent of the Supreme Executive that the islands should be granted, and thus to bring them within the general regulations which prescribed the mode in which all grants should be made.

Those regulations required the concurrence of the Departmental Assembly to give definitive validity to the grant by the Governor; but inasmuch as the Supreme Court had decided that even without the concurrence the grant was valid, unless the grantee's rights had been forfeited by abandonment, it seemed to me that a similar rule should be observed with respect to the grants of islands which by the previous assent of the Supreme Executive had been brought within the general regulations.

This view, however, the Supreme Court decided to be erroneous, and held that the dispatch prescribed a new rule on the subject, and that the general regulations did not apply to it. The mode of granting indicated in the dispatch of Pesada was, therefore, to be strictly followed, and inasmuch as the Departmental Assembly had not concurred, the grant to Osio was adjudged to be void.

But the reversal of the decision of the District Court on this point, in no way shows that the Supreme Court did not consider grants of islands, when made in pursuance of either the general or special regulations of the Executive, as not made under the Colonization law.

On the contrary, it appears to me manifest, that neither in the case of Osio, nor in that of Castillero, do the Supreme Court base their decision on the idea that grants of islands were not within the Colonization law of 1824; but that they reject the first claim, because in their opinion the grant was not made in the manner prescribed by the Executive, to whom that law committed the whole power over the subject, and they confirm the second claim because the Executive instructions were followed. It is explicitly stated in the opinion that "it is immaterial whether or not the power to grant the islands on the coast was vested in the Governor" (*i. e.* by the General Executive regulations of 1828), for the effect of the dispatch "was to repeal the previous regulations on the subject, and to substitute a new one in their place."

As this power of making regulations with respect to colonization in the territories was conferred, in terms, on the Supreme Executive, by the Colonization law, and was precisely that which it exercised when the general Executive regulations of 1828 were framed, I confess myself unable to perceive how a grant of an island on the coast, made in obedience to Executive instructions, was not, in every respect a grant under the Colonization laws, nor can I discover any foundation for the distinction attempted to be drawn between the case at bar and that of the island of Santa Cruz. The lands, in both cases, were open to grant under the general law; and even, if the granting of the islands on the coast to individuals be considered to be embraced within the provisions of the 4th Article, and that grants of lands within the littoral leagues were not, the only distinction between the cases would be, that in one the previous assent of the Executive was necessary, while in the other it was not. Each, when regularly granted, must be held to have been granted under the Colonization laws. When, therefore, in the Santa Cruz Island case, the Supreme Court held that the dispatch of the Minister ordering one of the islands to be assigned to Castillero "operated to adjudicate the title," the same construction must be given to the dispatch in this case, which states that the President has acceded to a petition for two leagues, and orders the Governor to put the petitioner in possession.

As, then, the Lanzas dispatch "operated to adjudicate the title" to the claimant, he must be held to have acquired an inchoate title, which, if founded on such equitable considerations as would have bound the former Government to complete it by issuing the formal title papers, this Government is equally bound to respect.

Had the claimant been an ordinary colonist, and relying on the action of the Supreme Government on his petition, settled upon and occupied the land, building a house upon, and cultivating it, and had the United States found him in the enjoyment of an undisputed possession, it cannot, I think, be doubted that his possession would have been undisturbed and his title confirmed, even though he had neglected to obtain from the Governor the formal grant.

But no such possession was taken in this case, nor was the concession received in California or even known to have been made, until after the subversion of the Mexican authority.

The question therefore arises: Were there any antecedent equitable considerations on which the concession was founded, such as would have bound the conscience of the Mexican Government to perfect it?

That an antecedent consideration, such as the patriotic and public services of the grantee, is one which a Court of Equity cannot disregard, has been expressly decided by the Supreme Court.

In the case of Frémont *vs.* the United States the Court says:

> "The grant was not made merely to carry out the policy of the colonization laws, but in consideration of the previous public and patriotic services of the grantee; and although this cannot be regarded as a money consideration, making the transaction a purchase from the Government, yet it is the acknowledgment of a just and equitable claim, and when the grant was made on that consideration, the title in a court of equity ought to be as firm and valid as if it had been purchased with money on the same conditions."—17 How. 558.

If, then, antecedent considerations of this nature are to be looked to, in determining whether the former Government was under any equitable obligation to perfect the title of the claimant, it is perhaps not easy to imagine a case where the merits of the petitioner and the consideration rendered by him for a small tract of land in a remote department could be greater.

The immense value of the discovery he had made to the great mining interests of Mexico, need not be dwelt upon. Our own experience in California enables us at once to appreciate how indispensable is an ample and cheap supply of quicksilver to the development of mines of the precious metals. But to Mexico the discovery was, as justly observed by one of the counsel for the claimants, "the unsealing of a hidden fountain of wealth, as precious to her as the rains and dews and living streams are to the nations that live by tillage."

For years it had been the policy of Mexico to stimulate explorations for, and to encourage the working of, quicksilver mines. By the laws of February 20, 1822, and 7th October, 1823, which imposed duties on gold and silver, quicksilver was expressly exempted from contribution.

In 1842, the Junta de Fomento was established, and empowered "to fix the mode in which the working of quicksilver mines was to be supplied, rewarded, stimulated and protected."

By the decree of May 24th, 1843, rewards of $25,000, and

of $5 per quintal, were promised to successful miners; and by the decrees of July 5th and September 25th of the same year, the Junta was empowered to work, to supply and protect quicksilver mines, and to cause researches for them to be made throughout the Republic.

When, therefore, Castillero announced the discovery of a mine surpassing in richness that of Almaden in Spain, upon which Mexico had so long been dependent, and desired a grant of two leagues in a department where land was commonly distributed gratuitously in tracts five times as large, he had equitable claims upon the Government far surpassing "the public and patriotic services of Alvarado," which the Supreme Court declares to have been an equitable consideration, as strong as if the grant had been purchased with money.

Compared with the service rendered and about to be rendered to the Mexican nation by Castillero, the consideration on which the ordinary colonization grants were founded was insignificant; for that consideration merely consisted in building a house, cultivating a few acres of an immense tract, and suffering wild cattle to roam at will over the remainder.

The fact that he was working the mine showed that Castillero had already effected a settlement upon the land, and its further development insured an accession to the population of the country far greater than could have been obtained by any other disposition of the public domain.

The purpose for which he sought the land, apprised the Government that it was of a kind not usually fit for cultivation, for it was required to supply wood for his burnings. In thus assisting his enterprise, the nation had as great an interest as Castillero himself, for it was the attainment of an object to which their attention, their efforts, and no inconsiderable portion of their revenues had long been devoted.

It has appeared to me that all these circumstances constitute an equitable consideration for the inchoate title or concession obtained by Castillero, and that they were sufficient to create an equitable obligation on the former Government, and therefore, on this, to complete and make good the inceptive rights he had acquired.

It is urged by the counsel for the United States, that even if the Castillo Lanzas dispatch be considered a grant, it nevertheless is void, because no possession of the land was given before the 7th of July, 1846, when the Mexican authority in California was subverted, and the United States acquired the land by the adverse title of conquest.

It is not denied that, as maintained in the brief of the counsel for the United States, in questions of prize or no prize, the liability of the property captured to condemntion depends upon the fact whether the possession and actual control of it have passed from the hands of the enemy to those of a neutral.

Nor is it questioned that, where territory is ceded by one Sovereign to another, the nationality of the inhabitants of the ceded territory is not changed until the stipulations of the treaty are executed by a formal delivery given, and by possession taken.

It is also admitted that, by the Roman law, and by most systems of jurisprudence, the property in a thing cannot be transferred without a delivery of the possession of the thing, either actual, or feigned and constructive; and that ordinarily, he who first obtains possession shall hold the thing even as against a prior purchaser, to whom it has not been delivered. In the transfer of land the same principle prevailed at the common law, and a symbolical delivery of the land, or livery of seizin, was indispensable to render a feoffment operative.

This, however, is now unnecessary in conveyances under the statute of uses.

In the grants made by the Governors of California, we accordingly find that "the judicial delivery of possession by the corresponding judge" was always contemplated. This proceeding would seem to have been designed for a double purpose: 1st. To complete the transfer of the property, by a formal delivery or tradition of the thing, thus adding the *jus in re* to the *jus ad rem;* and, 2d. To designate and sever from the public domain the tract granted by measuring its extent and establishing its boundaries.

Whether, if the boundaries are distinctly designated in the grant, the judicial delivery of possession was in strictness necessary to complete the right of property in the grantee, may be doubted; for at common law the King's grant was held to import livery of seizin, and the same principle is said to prevail at the civil law.

But whether technically necessary or not, it is settled by the decisions of the Supreme Court, that the want of a judicial delivery of possession is no obstacle to the confirmation of a grant of lands in California.

The occupation and settlement which, in the Louisiana and Florida cases, were considered to constitute the true grounds of the claimant's equity, were required by the Supreme Court to be shown, not because the technical rule required a formal delivery of possession to complete the transfer of the right of property, but because the petitioner, by occupying and cultivating his land under an inchoate title, and an implied promise of a grant, had rendered to the former Government a consideration which bound its conscience and that of its successors to perfect the title.

The question then, in this and other cases, is not whether a formal and technical delivery of possession has been made, but whether a consideration has been given for the grant, either antecedent by public services, the payment of money and the like, or subsequent, by occupation, settlement, etc., which in equity required the former Government to convert the inchoate title actually obtained into a perfect title. If, at the acquisition of the country, the conscience of the former Government was bound by this obligation, it is equally binding upon us; and the claimant, whether a resident or a foreign Mexican, has a right of property which the United States have agreed by the Treaty to respect.

Whether the consideration rendered, and the equitable claims on the bounty of the Mexican Government possessed by Castillero, are sufficient to create such an obligation, is a question which, perhaps, depends rather on the spirit in which his claims are looked upon, than upon any definite rule of law. It has appeared to me, as before stated, that the consideration rendered by him to the Mexican Government, did not merely constitute a claim upon its bounty; but that when he had obtained the assent of the Supreme authority to a grant of a specific tract of land, when orders had been issued to make him a grant and to put him in possession, the execution of which was prevented solely by the outbreak of war, the inchoate title so obtained ought to be respected by the United States.

But if the fact of possession and occupation be insisted on as indispensable, it is to be remembered that the land solicited and ordered to be granted to Castillero, was two leagues "on the land of his mining possession."

It is not disputed that early in December, 1845, he had occupied and worked the mine.

Possession of it had been given to him by the Alcalde, in a loose and informal manner, it is true, but still sufficient to give an official sanction to his occupation, more than six months before the conquest of the country; and from December, 1845, he and his assigns have continued to hold it. As, then, the mine was within the two leagues solicited, and as he had already taken possession of, and was working it, he may, perhaps, be considered, after the order of Lanzas was issued, to have been in possession of the lands referred to in that order.

In no cases was any other possession taken by the Californian rancheros of the large tracts—sometimes eleven square leagues in extent—granted to them, than by building a rude house of adobe, cultivating a small portion of the land, and stocking the remainder with a greater or less number of wild cattle or horses.

I am aware that in this view of the claimant's equities, I have the misfortune to differ from the Circuit Judge.

But on the best consideration I have been able to give to the subject, it has appeared to me not only warranted by the decisions of the Supreme Court, but in accordance with the dictates of the enlarged, and, so to speak, generous justice which should animate a great and a conquering nation in dealing with the rights of the vanquished.

But it is said that if Castillero obtained from the Supreme Government a grant of two leagues on his mining possession, it proves one of two propositions,—either that he was guilty of a gross fraud in suppressing the fact that such a grant would include private land, or that the Supreme Government committed a violation of law equally gross, in attempting to grant the lands of private individuals.

Had the title set up been a formal and absolute grant of two leagues, to be measured in every direction from the mouth of the mine, the observation would have possessed much force. But such is not the import of the Lanzas dispatch; on the contrary, it directs the Governor to put the petitioner in possession of the land solicited "in conformity with the laws and dispositions on colonization," a direction which rendered it certain that in the location of the grant private rights would be respected. Had the Supreme Government known that a tract of two leagues, measured in every direction from the mouth of the mine, would include private land, the precautions used in framing the order to the Governor would have sufficed for the protection of the owner; and the dispatch is, in effect, but the expression of the President's assent to a grant of two leagues on land of the petitioner's mining possession, and an order to the Governor to execute the grant; provided, and so far as it could be done without injury to third persons.

It cannot, therefore, be inferred from this dispatch, either that Castillero practiced a fraud on the Government, or that the latter committed or intended to commit a violation of any private rights whatever.

The last objection to the validity of the Lanzas dispatch which I shall notice, is that contained in the seventh division of the printed argument filed by the counsel for the United States.

The same point had been raised and fully considered by the Court in the case of Palmer vs. The United States. As the decisions of this Court have not been reported, it has been thought most convenient to append that opinion to this, adding to it such further observations as may seem appropriate:

"Before proceeding to an examination of the merits of this case, a general objection to the validity of the grant must be considered. The grant purports to have been executed on the 25th June, 1856, subsequently to the declaration of war between the United States and Mexico.

It is contended, on the part of the United States, that on general principles of public law, grants made *flagrante bello* when conquest has been set on foot, and actual occupation is imminent and inevitable, have no validity against the subsequent conqueror. The question has not heretofore been presented to this Court. It has been discussed with much ingenuity and ability.

It is urged that in the conduct of war, and the determination of its objects, the political department is supreme; and that the judiciary are bound by the view taken by the political branch of the Government; that, although Congress has alone power to declare war, to the Executive is given the right of shaping it to its ends, or of declaring its objects.

To ascertain its objects, resort must, therefore, be had to executive acts, and as the executive acts in this case unequivocally indicate that a principal object of the war was to acquire California, that acquisition was thus brought within the scope of the war, and must be so regarded by the Courts.

To this point, the case of Harcourt vs. Gaillard (12 Wheat.) is cited. Such being the object or scope of the war, it is urged that the intended conquest of California embraced not only the establishment of sovereign rights in the Territory, but also the acquisition of the public property within it.

That the proprietary rights to be acquired by the conquest are as essential, though not as important a part of the fruits of conquest, as the political, the commercial, and other advantages proposed to be obtained, and that no part of these objects of the conquest is to be ignored.

The conquest of California, including the acquisition of the public domain, having been thus shown to have been the object, or brought within the scope of the war, it was urged that any grants of public land made after the conquest was projected, and when it was about to be effected, though before it actually occurred, must be deemed to be in fraud of the rights of the incoming conqueror, and invalid as against him.

The foregoing statement is believed to present the outline of the argument submitted on the part of the United States.

Both the premises and the conclusion must be examined.

If the conquest of California was the object of the war, it must be so considered because that object was avowed by competent authority when war was declared; or because it was made the object of the war, after its commencement, by the political branch of the Government.

It may be admitted that this Government had long regarded California, or the Bay of San Francisco, as an important and desirable acquisition. The instructions of the President to

Mr. Slidell indicate the wish of the Executive to obtain it by purchase and cession, as Louisiana and Florida had been acquired.

It by no means follows that the intention to obtain it by force of arms, or conquest, can be attributed to Congress, still less that such was its object or motive in declaring war.

The law by which war was declared, recognizes it as previously existing by the act of Mexico; and it is known that hostilities arose by the invasion by Mexico of a territory claimed by the United States to be within their limits. Such was not, therefore, the object for which war was declared, or its existence recognized—nor could it constitutionally have been.

It is observed by Mr. Ch. J. Taney, in Fleming v. Page (9 How. 614): 'The genius and character of our institutions are peaceful, and the power to declare war was not conferred upon Congress for the purpose of aggression or aggrandizement, but to enable the General Government to vindicate by arms, if it should become necessary, its own rights and the rights of its citizens. *A war, therefore, declared* by Congress, can *never be presumed* to be waged for the purpose of conquest or the acquisition of territory.'

As a limitation upon the power of Congress, this distinction may, practically, be unimportant. As every war in which the country may be engaged must be regarded by all branches of the Government, and even by neutrals, as a just war; and as nations can readily cloak a spirit of rapacity and aggression under professions of justice and moderation, it is at all times easy, should our country be actuated by such a spirit, to declare an aggressive war, to be undertaken in self-defense, and an intended conquest to be desired only as a compensation for past or security against future injuries.

But the distinction is important when a Court is asked to presume that conquest was the object of the war.

Under our Government, at least, such a presumption cannot be indulged.

The conquest of California being thus shown not to have been the object for which war was declared, we may next inquire whether, by the acts of the Executive under its power to conduct the war, it became such, or was brought within its scope, in the sense in which the phrase was used at the bar?

In his annual message to Congress, in December, 1846, the President distinctly states that the war originated in the attempt of Mexico to reconquer Texas to the Sabine. After adverting to the considerations which had induced the Executive to interpose no obstacles to the return of Santa Anna, the latter being more favorably disposed to peace than Paredes, who was then at the head of affairs, the President observed: 'The war has not been waged with a view to conquest, but having been commenced by Mexico, it has been carried into the enemy's country, and will be vigorously prosecuted there, with a view to obtain an honorable peace, and thereby secure ample indemnity for the expenses of the war, as well as our much injured citizens, who have large pecuniary demands against Mexico.'

Similar declarations are frequently and emphatically repeated by the President in various communications to Congress, and in the correspondence between the American Commissioner and the Mexican authorities.

The object of the war, therefore, as indicated by executive acts and declarations, was not conquest; or, if conquest, it was that of a safe and honorable peace.

It is true, that after the military occupation of California, and after our arms had been everywhere successful, and perhaps at the commencement of hostilities, the Executive and the nation may have confidently anticipated that by the treaty of peace we would acquire California. As Mexico was known to be impoverished, and distracted by civil dissensions, it was obvious that the only indemnity she could afford us for expenses of the war was the cession of a portion of her territory.

The instructions of the Secretary of State to Mr. Trist, show that the extension of the boundaries of the United States, over New Mexico and Upper California, for a sum not exceeding $20,000,000, was a condition *sine qua non* of any treaty.

The extraordinary successes of our arms, the fact that we already held possession of a great part of the territory of the enemy, and virtually of his capital, our great expenditures of blood and treasure, entitled us to retain a portion, at least, of our conquest as the only indemnity we could obtain. But we were willing to restore a considerable part of our possessions, and to pay for that retained by us a large amount of money.

But such views and intentions on the part of the Executive as to the condition on which the war should cease, are very different from waging it with a view to conquest. The war cannot, then, in any just sense, be deemed to have been declared by Congress, or conducted by the Executive, with a view to conquest.

The power of the President in the conduct of the war was that of commander-in-chief of the army and navy. He had authority to direct and control military operations. As part of the treaty-making power, he could determine when and on what conditions a treaty of peace should be made. But he had no power to impress upon the war a purpose different from that with which it was commenced, and which, as Mr. Ch. J. Taney declares, Congress could not constitutionally entertain. 'The law declaring war,' observes the same great authority in the case above cited, 'does not imply an authority to the President to enlarge the limits of the United States, by subjugating the enemy's country. The United States, it is true, may extend its boundaries by treaty or conquest, and may demand the cession of territory as the condition of peace, to indemnify its citizens for the injuries they suffered, or to reimburse the Government for the expenses of the war.

'But this can be done only by the treaty-making power, or the legislative authority, and it is not a part of the authority conferred upon the President by the declaration of war. His duty and his power are purely military. As commander-in-chief, he is authorized to direct the military and naval forces placed by law at his command, and to employ them in the manner he may deem most effectual to harrass and conquer and subdue the enemy. He may invade the hostile country, and subject it to the sovereignty and authority of the United States. But his conquests do not enlarge the boundaries of the United States, nor extend the operations of our institutions and laws beyond the limits before assigned them by the legislative power.'

It is true that in the case in which these observations are made, the point to be determined was, whether enemies' territory, which in the course of hostilities had come into our military possession, became a part of the United States, and subject to our general laws. But they are important to this case as defining the power of the President in war, to be merely that of the military commander-in-chief; that territory can be acquired only by the treaty-making and legislative authority, and, consequently, that the fact that hostilities are by the military authority directed against a particular portion of the enemy's territory, cannot be said to make the acquisition of that territory the object of the war.

It is therefore apparent that the war with Mexico cannot be regarded by the judicial department of this Government as commenced, or conducted, with the object of effecting the conquest of California.

The most that can be said is, that its military occupation was effected as a means of crippling and subduing the enemy, and with the expectation, on the part of the Executive, that we would retain and finally insist upon the cession of the territory so subjugated by our arms as an indemnity for our injuries and expenses.

The nature and amount of indemnity to be required, the

extent of territory to be ceded, depended upon the will of the Senate and the Executive as the treaty-making power, and until that will was expressed in the treaty, the intention to effect the permanent acquisition of all California cannot be attributed to the political power, any more than a similar intention with regard to those conquests which at the close of the war were restored.

If, then, it were a principle of public law that all alienations of public domain by a sovereign are invalid as against an enemy who has commenced or is prosecuting a war, with the object of conquering the territory within which the property is situated, or who has set on foot expeditions for the purpose, with sufficient power to attain the end, as proved by the event, the facts of this case would hardly admit of its application.

But assuming the facts as contended for by the United States, we proceed to inquire whether such a rule of law exists. The right of Mexico to dispose of her public domain in California before the war, is admitted. It is not denied that that right ceased, as against the United States, when the latter effected the conquest of the country, and subverted the Mexican authority.

If it ceased before the actual conquest and displacement of the Mexican authority, it must be because the determination of the United States to effect the conquest, and the making preparation to carry out its determination, gave to the latter some inchoate or inceptive right to the territory subsequently conquered, and the title consummated by the conquest relates back by a kind of fiction to the date of its inception.

We have been unable to discover any trace or intimation of such a doctrine in any writer on the laws of war.

The rights derived from conquest are derived from force alone. They are recognized because there is no one to dispute them, not because they are, in a moral sense, rightful and just. The conquest of an enemy's country, admitted to be his, is not, therefore, the assertion of an antecedent right.

It is the assertion of the will and the power to wrest it from him.

Even where a conquest is effected to obtain an indemnity justly due, it is not the assertion of any antecedent right to the particular territory conquered, but only of the general right to a compensation for injury.

The right of the conqueror is, therefore, derived from the conquest alone. It originates in the conquest, not in the intention to conquer, though coupled with the ability to effect his purpose, nor even in the *right* to conquer as means of obtaining satisfaction for injury.

It is the fact of conquest, not the intention or power to conquer, which clothes him with the rights of a conqueror.

The rights acquired by the conquest are temporary and precarious until the *jus post liminii* is extinguished; and if a reconquest is effected, the rights of the sovereign who has temporarily been displaced revive, and are deemed to have been uninterrupted.

The term 'title by conquest' expresses, therefore, a fact and not a right. Until the fact of conquest occurs, the conqueror can have no rights. To affirm that a title acquired by conquest relates back to a period anterior to the conquest, is almost, a contradiction in terms.

Until, then, the conquest is effected, the rights of the existing sovereign remain unimpaired. He can, therefore, dispose of the public property at his discretion, nor can that right be effected by the determination of an enemy to conquer the territory, and by his preparations for the purpose, though the event may demonstrate the conquest to have been practicable.

The case of Harcourt *vs.* Gaillard has been cited by the counsel of the United States in support of the doctrine contended for by them.

The distinction between that case and the case at bar is obvious.

In Harcourt *vs.* Gaillard the question was as to the validity of a grant by a British Governor, of land within a territory claimed to belong to the United States. As our Government had asserted and maintained by arms its title to the disputed tract, the judicial department were not at liberty to declare the claim to be wrongful, and to recognize the right of any other sovereign over the territory in question.

The title of the United States was in no sense acquired by conquest. Her title was antecedent to the war—it was merely maintained by arms and recognized by the treaty of peace.

The question presented was, in the language of the Court, 'one of disputed boundaries, within which the power that succeeds in war is not obliged to recognize as valid any acts of ownership exercised by his adversary.'

Had the claim been that of conquest alone, the case would have presented, says the Court, more difficulty. 'That ground would admit the original right of the Governor of Florida to grant, and if so his right to grant *might* have continued until the treaty of peace, and the grant to Harcourt might, in that case, have had extended to it the principles of public law which are applicable to territories acquired by conquest, whereas the right set up by South Carolina and Georgia denies all power in the grantor over the soil.'

The distinction is made still more apparent in a subsequent part of the opinion of the Court: 'War is a suit prosecuted by the sword; *and where the question* to be decided is one of *original claim to territory*, grants of soil made *flagrante bello* by the party that fails, can only derive validity from treaty stipulations. It is *not necessary* here to consider the *rights* of the *conqueror in case of actual conquest.*'—(p. 528.)

The latter is precisely the question to be considered in the case at bar.

The argument of the counsel for the United States can, therefore, derive no support from the case referred to.

It is proper, however, to observe that the case of Harcourt *vs.* Gaillard was not cited by the counsel as directly in point. It was thought to establish that all grants of territory brought within the scope of the war are invalid; that the case of disputed boundaries presents one illustration of the general principle, while the case at bar furnishes another.

It has seemed to me, however, that the principle of that decision relates exclusively to the case of disputed boundaries, and that the distinction is clearly drawn between that case and one like the present; that between them the obvious difference exists that the former is a case of 'original claim to territory,' while the other is 'one of actual conquest.'

It is said on the part of the United States, that if a belligerent can, after a declaration of war, grant any portion of his property, he can grant the whole, and thus might, by granting himself away, escape responsibility. The case supposed is an extreme one. It can rarely occur that a nation will seek safety by self-destruction.

But in such case the adversary might refuse to recognize such a voluntary suicide as affecting his rights. For the purpose of obtaining satisfaction he might justly treat the nationality sought to be extinguished as still existing. But in all Courts his rights would be enforced against the successor or grantee of the extinguished sovereign.

The question would then be purely political, for the new sovereign, whether to carry on the war or accede to the demands of the enemy of his grantor, and for the latter whether to prosecute the war against the new sovereign. Little aid, however, can be derived from the consideration of such extreme and improbable cases.

It is further urged, that the doctrine contended for on behalf of the United States is in the prize law.

It may, perhaps, be admitted that a theory of maritime prize formerly obtained, which assumed that a belligerent has a vested right by the declaration of war in all sea-borne private property of the other belligerent; that no such property can be the subject of lawful sale; that all contracts of sale touching belligerent property of any sort, though valid on land, are invalidated by the mere fact of such property being embarked on the ocean, and that if transferred to a neutral after the declaration of war it is a lawful prize to the other belligerent.

Such is not now the received law of nations. It is now admitted that the *bonâ fide* sale of the ships of belligerents to neutrals in time of war is lawful and valid unless made *in transitu*.

In the Johanna Emilia, 29th Eng. L. and Eq. R., p. 562, Dr. Lushington says: 'It is not denied that it is competent for neutrals to purchase the property of enemies in another country, whether consisting of ships or anything else. *They have a perfect right to do so, and no belligerent right can override it.*'

Such is the doctrine maintained by our Government. See opinion of Mr. Attorney General Cushing, October 8, 1855.

If a sale to a neutral of a ship *in transitu* is held invalid as against a belligerent, it is not by reason of any inchoate right or lien acquired by the latter by the mere declaration of war, or because the right of the enemy to dispose of his property is invalidated by the declaration of war, but because a sale of a ship *in transitu* is taken as proof of collusion and fraud, and as showing that no absolute transfer has, in fact, been made. The soundness of even this rule is doubted by the Attorney General in the opinion referred to.

A sale of a ship not *in transitu* by a belligerent to a neutral is valid as against a subsequent captor, no matter how imminent the danger of capture would have been had she remained enemy's property, and no matter what may be the number of hostile fleets fitted out to cruise against her and similar property of the belligerent.

It appears, then, that the law of nations, with regard to prize of war, does not recognize the principle contended for.

It is urged, however, that this principle lies at the foundation of the doctrine of *post liminii*.

It is argued that a state of war implies the reciprocal denial by each belligerent of all rights on the other.

That each relies upon force alone—force to retain or force to take.

They are thus in *aequali jure*.

The principle, therefore, by which, on a reconquest, the original title revives, and is deemed to have been uninterrupted, is founded on the presumption that the displaced sovereign intended a reconquest when he was displaced, and his title on a reconquest relates back to the time when he is presumed to have formed such intention. If, then, (it is argued,) the title by reconquest relates back to the time of the formation of the intention to reconquer, the title by conquest must relate back to a similar period; for a state of war implies the negation of all antecedent right on either side. The only difference between the cases being, that in the case of a reconquest the intention to reconquer is presumed until the *jus post liminii* is extinguished, while in the case of conquest that intention must be shown by the political acts and declarations of the conqueror.

The argument is ingenious, but the premises are, I think, erroneous.

It is assumed that a *new* title is acquired by a sovereign who recovers territories from which he has temporarily been driven.

On the contrary, he holds it by his original title, which could only have been displaced by a permanent conquest. But the fact that he recovers the territory, proves that what seemed a conquest was but a temporary dispossession. The invader, therefore, acquired no rights, nor did the original sovereign lose any. He continues to rule, not by a newly acquired title which relates back to any former period, but by his ancient title, which, in contemplation of law, has never been divested.

Nor is it true that war is the reciprocal denial of all rights by the belligerents, with respect to the territories of either.

A conqueror does not deny that the territory seized was at the time of the conquest the territory of his enemy, any more than the attaching creditor denies the property attached to be that of his debtor.

On the contrary, he asserts it to be his. He seizes it as the property of his enemy, and because it is his. He asserts no antecedent title in himself. He declares, not that the territory was his, but that he will make it his by conquest.

The title or right acquired by a conquest is not the same as that of the original possessor.

It is temporary and precarious, and ceases the moment the conqueror is expelled. If, indeed, a title by conquest can be said ever to have existed when the event has proved that the attempted conquest could not be maintained.

The title of the original owner is wholly unaffected by the temporary dispossession, and even during his dispossession it is treated as valid and subsisting until the *jus post liminii* has been extinguished.

The extinction of the *post liminii* is necessary to ripen the temporary and merely possessory right of the conqueror into such an ownership of the territory as neutrals can recognize.

If these views be correct, the case of a reconquest does not present the instance supposed of a title relating back to the period of the formation of the intention to reconquer.

But the further discussion of this subject would require more time and space than can be devoted to it.

It might, I think, be demonstrated, that a rule which supposes all rights of a sovereign, with respect to territory subsequently conquered, to cease as against the conqueror, *not* when war is declared, but when the war is prosecuted with the object of conquest, when expeditions are fitted out for the purpose, and when the conquest is 'imminent and inevitable,' is not susceptible of practical application as a rule of international law.

That those rights must continue until the date of actual conquest, or of the treaty of cession, or else must cease at the declaration of war, and that an attempt to estimate the "imminency" of the conquest at any intermediate period, or to try the validity of the exercise of sovereign rights, by calculating the chances of war at a particular moment, would be impracticable and illusory.

On the whole, I am of opinion that the right of Mexico to grant her public domain in California, continued until the conquest of the country by the United States.

It is further urged, on the part of the United States, that grants made after the 13th May are not protected by the treaty of peace, because such was not the intention of the parties.

That the Mexican Commissioners who negotiated the peace, and who represented the claimants as well as the Mexican Government, solemnly, and after special inquiry, declared that none such existed; and that the treaty was negotiated on the faith of this declaration.

It is admitted that such a declaration was made, and embodied in the *project* of the treaty submitted to the Senate.

Had this declaration been contained in the treaty as adopted and ratified, it might very possibly have been regarded as a covenant or stipulation that no such grants should be deemed valid by the United States.

But the clause containing it was struck out by the Senate, not by the general vote which struck out the whole of the 10th Article, of which this declaration formed a part, but by a distinct vote upon the question whether this particular clause should stand as a part of the treaty.

The Court cannot assume, therefore, that the treaty was assented to by the United States on the faith of this declaration by Mexico, else why strike it out? It may, not unreasonably, be supposed that the Senate refused to allow the declaration to remain, because they were willing that grants made after the 13th May, if any such there were, should be submitted to the courts, and rejected or confirmed, as might be just.

But assuming that the treaty was concluded on the faith of this declaration, the rights of an individual to his property cannot be affected by it.

The stipulation in the treaty by which the property of the inhabitants of the ceded territory was secured, conveyed to them no additional rights. 'An Article to secure this object, so deservedly held sacred in the view of policy as well as of justice and humanity, is always required and never refused.'—(12 Wheat. 536.)

'When such an article is submitted to the courts, the inquiry is, whether the land in controversy was the property of the claimant *before the treaty*.'—(United States *vs.* Arredondo, 6 Pet. 712.)

If, then, the land in controversy was the private property of the claimant when the country was acquired, it must have remained such, though no treaty had been made. The United States do not claim to have acquired the ownership of any other property than the public property of the enemy, nor could they justly have demanded that Mexico should assent by the treaty to the confiscation of any property the right to which was vested in private individuals.

If, then, the United States have been willfully or accidentally deceived, as to the amount of property held in private ownership in the ceded territory, they may have a right to demand a return of some portion of the pecuniary equivalent paid by them.

The fraud or mistake of the Mexican Commissioners can have no effect upon a private right held sacred by the laws and usages of all civilized nations, which was not derived from the treaty, and which, had it been known to exist, the United States would have been bound to respect.

These observations are made with reference to the general proposition maintained at the bar, viz., that the declaration by Mexico that no grants had been made subsequent to May 13, 1846, invalidated all such grants to the same extent as if a stipulation to that effect had been embodied in the treaty."

In the brief filed in the case at bar, the Court is invited to review the grounds of the foregoing opinion; and the question is discussed by the counsel of the United States with characteristic ingenuity and ability.

The authority chiefly relied on in support of the position taken by the counsel for the United States, is Bynkershoek.

"We make war," says that author, "because we think that our enemy, by the injury he has done us, has merited the destruction of himself and all his adherents; as this is the object of our warfare, it is immaterial what means we embrace to accomplish it." * * * "A nation which has injured another, is considered, with everything that belongs to it, as confiscated to the nation that has received the injury. To carry that confiscation into effect, may certainly be the object of the war, if the injured nation thinks proper."

The doctrine here maintained, that in war, poison and every species of fraud may rightfully be used, has received the general condemnation of mankind. It may be that the censure on Bynkershoek is not wholly deserved, inasmuch as he expresses no approval of those practices, but differs from other writers mainly in distinguishing between the absolute rights of war and those voluntary relinquishments of them which are dictated by humanity and generosity.

But if it be admitted that humanity, christianity, and the usages and rules observed by all civilized nations (which constitute public law), forbid even in war the use of certain means, the discussion whether such rights abstractly exist, would seem to be a disputation savoring rather of the subtlety of the schools than of that practical sense which seeks to discover and establish the actual rules by which nations in a state of war are governed.

That the rights of war, as deduced by Bynkershoek, from a consideration of its abstract nature, are mitigated by the laws of war as established by the general consent of nations, with respect to the effects of conquest, as well as to the mode of warfare, is proved by the general recognition of the principle that, on the conquest of an enemy's territory, private rights of property are to be protected.

But if "a nation which has injured another is to be considered as confiscated, with all that belongs to it, to the nation that has received the injury," this confiscation must extend to private as well as public property.

A declaration of war undoubtedly involves the assertion of the right to measure and forcibly to exact an indemnity for the wrong which has occasioned the war.

To seize, to conquer, or to destroy an enemy's goods, his territory or his armed adherents, are but the means of exacting this indemnity.

As a matter of theoretical speculation, we may consider the seizure, the conquest, or the destroying, as done by virtue of a previous fictitious or hypothethical confiscation of property, or forfeiture of life, incurred at the date of the declaration of war. But the necessity of such a theory is not very apparent. For the right to subdue the enemy being admitted, as a means of obtaining an indemnity for previous wrongs, the supposed constructive confiscation can add nothing to the rightfulness of those acts. It is for this reason said, in the opinion above cited, that "the conquest of an enemy's country, admitted to be his, is not the assertion of an antecedent right. It is the assertion of the will and power to wrest it from him." On which the counsel for the United States observes: "Then all Governments are highwaymen! Forcibly to take without antecedent right is a very good definition of robbery."

The inference is not just. Conquest is, undoubtedly, the assertion of a right, but it is the right to conquer which results from a state of war.

It is not the assertion of a previous *right or title* to the territories conquered.

Whether in so doing the belligerent is acting like a highwayman, depends upon the moral justification for the war, an inquiry into which neither neutrals nor the courts of the belligerent can enter.

The hypothesis of an antecedent confiscation, to enforce which the seizure is effected, in no way affects the question. The moral justification of the supposed confiscation has still to be considered—in other words, the justice and rightfulness of the war.

But whatever be the reasonableness or necessity of supposing this theoretic confiscation by belligerents, of everything belonging to the enemy, it is manifest that by the laws of nations the confiscation is waived where territory is conquered, so far as respects private property; and especially where the conqueror, by the terms of the treaty of cession, has bound himself to respect all rights of private property existing at the date of the conquest.

To repudiate that obligation with respect to any property held in private ownership on the ground that, though private property when the conquest was effected, it was public property ten or twenty or thirty years before, when the war commenced, and that a writer on public law has said, that the declaration of war is a confiscation of all the property of the enemy, and that the conquest was merely carrying into effect the confiscation, would seem an attempt to justify the breach of a plain and positive obligation, which needs but to be stated to be condemned.

The obvious and natural construction of the Treaty is, I think, manifestly the true one, viz., that all private property *bonâ fide* acquired, and held as such by a legal or equitable title obtained under the former Government, is to be respected by the belligerent, to whom by conquest and treaty the rights of sovereignty have been transferred.

I do not think it necessary further to discuss this question. It is enough to say that I have attentively considered all that is urged by way of argument or illustration in the brief filed by the counsel for the United States. I have found nothing to which the answer did not appear to me easy, or which has shaken my confidence in the justness of the views previously entertained by the Court.

The question might well have been dismissed without argument; for we have an authoritative decision of the Supreme Court on the point. In the case of The United States v. Pico (23 How. R. 326), the Court says: "In the Act of Congress of 1851, and the decisions of this Court that day [viz., July 7th, 1846, the date of the capture of Monterey, and, constructively, of the conquest of California], is referred to as the epoch at which the power of the Governor of California, under the authority of Mexico, to alienate the public domain, ceased."

As, however, the point then before the Court was the determination of the precise date of the subversion of the former Government, and to decide upon the validity of acts done under Mexican authority *after* that event—while the validity of acts done previous to it was not questioned, nor does the point raised in this case appear to have been presented to the Court—I have thought it not improper to examine at some length the acute and ingenious argument submitted by the counsel for the United States.

I have given to this case much and anxious consideration. The preparation of this opinion has required more labor than even its great length would indicate.

Voluminous as it is, I am nevertheless aware that it is in many respects incomplete.

To have treated at length every point in the case would have extended it far beyond all reasonable limits.

I cannot conclude my labors on this most important case, without acknowledging the great assistance which the Court has derived from the very able and eminent counsel engaged in it.

Their indefatigable and exhaustive industry has presented to the Court every argument, authority and illustration which profound and patient study, not only of the American and English, but of the Mexican and Spanish laws, could suggest; together with every view of the complicated facts in the case, and of their relations to each other, which could assist the Court in its study of the mass of depositions which have been taken.

To the Court has been left merely the duty of considering the suggestions, and collecting and combining the abundant materials contained in the briefs of counsel.

On the whole case my opinion is:

That the claimants are entitled to seven pertenencias, to be measured in the manner, of the form, and of the dimensions prescribed in the Ordenanzas de Mineria of 1783.

And, also, that they are entitled to two square leagues of land, to be located on the land of their mining possession, but in such a way as not to include any land granted in private ownership, by competent authority, previously to July 7th, 1846.

DECREE.

At a stated term of the District Court of the United States for the Northern District of California, held at the Court-room in the City of San Francisco, on Friday, the 18th day of January, A. D. 1861.

Present: Hon. M. HALL MCALLISTER, Circuit Judge.
Hon. OGDEN HOFFMAN, District Judge.

THE UNITED STATES *vs.* ANDRES CASTILLERO.	Dist. Court, No. 420. Land Com. No. 366. "New Almaden."

DECREE.

This cause came on to be heard on appeal from the final decision of the Board of Commissioners to ascertain and settle Private Land Claims in the State of California, under the Act of Congress approved March 3d, 1851, upon the transcript of the proceedings and decision of the said Board, and the papers and evidence on which the said decision was founded, and upon the papers and evidence filed in this Court; and it appearing to the Court that the said transcript was duly filed according to law, and counsel for the respective parties having been heard, and due deliberation had in the premises, it is by the Court hereby ordered, adjudged and decreed, that the said decision be, and the same is, reversed and set aside: And it is likewise ordered, adjudged and decreed, that the claim and title of the petitioner, Andres Castillero, to the mine known by the name of New Almaden, in Santa Clara county, Northern District of the State of California, is a good and valid claim and title, and that the said Andres Castillero and his assigns are the owners thereof, and of all the ores and minerals of whatsoever description therein, in fee simple: And it is further adjudged and decreed, that the said mine is a piece of land embracing a superficial area, measured on a horizontal plane, equivalent to seven pertenencias; each pertenencia being a solid of a rectangular base two hundred Castilian varas long, of the width established by the Ordenanzas de Mineria of 1783, and in depth extending from and including the surface, down to the centre of the earth; said pertenencias to be located in such manner as the said Andres Castillero or his assigns may select, subject to the following conditions: first, that the said pertenencias shall be contiguous, that is to say, in one body; and secondly, that within them shall be included the original mouth of the said mine known as "New Almaden."

And it is further ordered, adjudged and decreed, as to all other rights, property and interests set out and claimed in the petition filed in this case, that the same are not valid, and are therefore rejected.

M. HALL McALLISTER,
Judge Circuit Court U. S.

OGDEN HOFFMAN,
District Judge.

January 18, 1861.

Filed January 18, 1861,
W. H. CHEVERS, Clerk.

ORDER GRANTING APPEAL
IN BEHALF OF CLAIMANT.

At a stated term of the District Court of the United States of of America, for the Northern District of California, held at the Court-room in the City of San Francisco, on Saturday, the 19th day of January, in the year of our Lord one thousand eight hundred and sixty-one.

Present, the Honorable OGDEN HOFFMAN, District Judge.

THE UNITED STATES
vs.
ANDRES CASTILLERO.
} No. 420.

And now, at this day, on application of A. C. Peachy, in behalf of said claimant, made in open Court, it is ordered, that an appeal from the final decision of the Court rendered in said cause at the present term be, and the same is hereby granted; and that a certified transcript of the pleadings, evidence, depositions and proceedings in the said cause, be sent to the Supreme Court of the United States without delay.

Filed January 19, 1861,
W. H. CHEVERS, Clerk.

ORDER GRANTING APPEAL
IN BEHALF OF THE UNITED STATES.

At a stated term of the District Court of the United States of America, for the Northern District of California, held at the Court-room in the City of San Francisco, on Monday, the third day of December, in the year of our Lord one thousand eight hundred and sixty.

Present, the Honorable OGDEN HOFFMAN, District Judge.

THE UNITED STATES *vs.* ANDRES CASTILLERO.	No. 420. Order of Appeal for the United States.

In this case, on the application of Calhoun Benham, United States District Attorney, made in open Court, it is ordered by the Court, that an appeal in behalf of the United States from the final decision of this Court rendered in said cause at the present term be, and the same is hereby granted, and that a certified transcript of the pleadings, evidence, depositions and proceedings in the said cause, be sent to the Supreme Court of the United States without delay.

January 25, 1861.

Filed January 25, 1861,

W. H. CHEVERS, Clerk.

In the United States District Court,

NORTHERN DISTRICT OF CALIFORNIA.

THE UNITED STATES
vs. } No. 420.
ANDRES CASTILLERO

"NEW ALMADEN."

Proofs, Indexes & Authorities

IN

BEHALF OF CLAIMANTS.

SAN FRANCISCO:
COMMERCIAL STEAM BOOK AND JOB PRESSES.
1861.

CONTENTS.

PROOFS

OF THE

FACT OF REGISTRY, POSSESSION, AND WORKING

OF THE MINE.

There are two written statements in registry made by Castillero.

The first is dated 22d November, 1845, when Castillero thought that the ore contained gold and silver.

The second, dated 3d December, 1845, after he had discovered that the ore contained large quantities of quicksilver.

The juridical possession was given by Pico on the 30th Dec., 1845. The two assisting and subscribing witnesses to act of possession are Antonio Suñol and José Noriega.

For proof of these papers, see brief of testimony on the genuineness and custody of the Espediente of Registry and Possession in the Recorder's Office of Santa Clara County.

The fact of discovery and denouncement is proved by the following documents and witnesses:

I. The fact of possession and working is proved by Pico, *Trans.* 630; by Suñol, *Trans.* 25, 409; by Noriega, *Trans.* 22, 419; by José Fernandez, *Trans.* 604, etc.

The extent of the working is best proved by Mr. Wm. G. Chard, who was employed by Castillero to open the mine; to dig the *pozo de posesion*; to construct the first rude furnaces at the Hacienda. See his deposition generally, *Trans.* 1068.

II. Castillero's letters of the 2d and 21st December, 1845—to General Vallejo—*communicate discovery and working.*—*Trans.* 2603.

III. Castillero's letter to Governor Pico, of the 10th Dec., 1845, informs him,—" *The mine has been denounced by me, and between a few we have formed a Company,*" etc.; and asks for *vacant lands* to cut wood, etc.—*Trans.* 1805.

IV. Governor Pico's letter to the Minister of Relations —of the 13th Feb., 1846—inclosing the foregoing original letter, and saying: " *You will be informed of the fortunate discovery made in this department of a quicksilver mine.*" * * * I avail myself, etc., *to send to your Exlcy, etc., the quicksilver which, as a sample, was sent to me by Señor Castillero, and to which he refers in the above mentioned letter.*—*Trans.* 1805.

Copies of the foregoing letters are found in Exhibit Bassoco No. 9.—*Trans.* 1805.

The *borrador* of Governor Pico's letter was retained in the California archives, *Trans.* 2549. Proved by Pico, *Answer* 7, *Trans.* 2532; proved by Olvera to be in his own handwriting, *Answer* 5, *Trans.* 2610. Pico remembers that he received letter and quicksilver from Castillero, *Answer* 13, *Trans.* 2533. Hopkins' statement, 3068.

Governor Pico's letter, with its inclosed note from Castillero, was received by the Minister of Relations on the 6th April, 1846, as by endorsement.—*Trans.* 1805; Castillo Lanzas' deposition, *Answer* 23, *Trans.* 2232; *Answer* 30, *Trans.* 2233.

V. On the 6th April, 1846, the Minister of Relations answered Governor Pico's letter expressing the President's satisfaction on hearing of Castillero's discovery.—*Trans.* 1806. There is no proof that this letter was received.

VI. On the 15th Dec., 1845, Castillero addressed another letter to Governor Pico, referring to his previous one of the 10th same month, *stating discovery of the mine.*—*Trans.* 2552. Proved by Pio Pico, *Answer* 13, *Trans.* 2533. See Hopkins' statement 3068. This letter is in the California archives.

VII. On the 31st Dec., 1845, Manuel Castro, Prefect of the 2d District, officially announced to Governor Pico that Castillero "*has denounced, and is now working, a quicksilver mine,*" etc. —*Trans.* 2550. Proved by Pio Pico, *Answer* 9, *Trans.* 2532. In California archives. See Hopkins' statement, 3068.

VIII. The foregoing was answered on the 22d January, 1846. as appears by marginal note. The blotter of answer is in the California archives.—*Trans.* 2551. Proved by Pio Pico, *Answer* 10, *Trans.* 2533; Hopkins, 3068.

IX. About 22d January, 1846, Capt. John C. Fremont visited the mine, where he saw Castillero, the owner, who described to him the process of acquiring mines under the Mexican law, and said he had so acquired that one; Fremont wished to purchase it, with Leidesdorff and others.—Fremont's deposition, *Trans.* 2678.

X. On the 3d April, 1846, Thomas O. Larkin, U. S. Consul at Monterey, in a letter addressed to the American Minister in Mexico, informs him of Castillero's discovery.—Ex. Swasey No. 2, *Trans.* 2667–8.

XI. On the 2d May, 1846, Mr. Larkin gives same information to Capt. J. B. Montgomery, U. S. ship Portsmouth.—Ex. Swasey No. 4, *Trans.* 2660; and Ex. Swasey No. 9, *Trans.* 2676.

XII. On the 4th May, 1846, Mr. Larkin addressed a long letter to the State Department, on the mineral resources of California, in which he mentioned Castillero's discovery of this mine, and described the process of extracting the metal, etc.—Ex. Swasey No. 5, *Trans.* 2669.

This letter is in the State Department; certified extracts from it are in evidence.—*Trans.* 2657.

XIII. Another letter from Larkin to Mr. Buchanan, dated 28th March, 1848, mentions *avio* of the mine in 1846.—Certified copy of extract from State Department, *Trans.* 2658.

XIV. On the 24th June, 1846, Mr. Larkin addressed a letter to G. P. Judd, Esq., Minister of Finance, Honolulu, Sandwich Islands, *stating the discovery of this mine by a Mexican officer in* 1845, *its denouncement before nearest Alcalde*, etc.; and he also sent to Mr. Judd, we may infer, a copy of his letter to Mr. Buchanan, of the 4th May, 1846, for they are both published in the Polynesian, at Honolulu, on Saturday, July 25th, 1846.—See *Trans.* 2223.

The number of the Polynesian containing this letter is in evidence. Proved by Milo Calkin, *Trans.* 2645; and by C. E. Hitchcock, printer of that paper, *Trans.* 2647.

Mr. Judd's reply to Mr. Larkin, acknowledging receipt of foregoing letter and specimen of cinnabar.—*Trans.* 3020. Proved by F. H. Larkin, and C. E. Hitchcock.—*Trans.* 3018–3024.

XV. We were led to a knowledge of the contents of the Polynesian of 25th July, 1846, by seeing them noticed in the *Diario del Gobierno de la Republica Mexicana* of the 27th Dec., 1846.—*Vide* Castillo Lanzas' deposition, *Answer* 692, *Trans.* 2355.

XVI. In the spring of 1848 General Mason, the Governor of California, visited the Almaden mine, and gave an account of the operations to Adjutant-General Jones in a letter of the 17th August, 1848.—Cal. Message and Correspondence, 1850, Doc. No. 17, p. 534.

PROOF

OF THE

GENUINENESS AND CUSTODY OF THE ESPEDIENTE OF REGISTRY AND POSSESSION,

NOW IN THE RECORDER'S OFFICE OF SANTA CLARA COUNTY.

I. Testimony of José Noriega, assisting witness to act of possession, taken 19th March, 1858.—*Answers* 2–3; *Trans.* 22.

II. Of Antonio Suñol, the other assisting witness, taken on same day.—*Answers* 2–3; *Trans.* 25.

III. Of Antonio M. Pico, the Alcalde who gave possession, taken 11th Nov. 1857.—*Trans.* 630.

N. B.—Pico was first called in this case by the Government. —*Trans.* 169.

IV. Of Suñol and Noriega, on their recall by Government. *Trans.* 409–419.

V. Of José Fernandez, claimant's witness, taken 6th Nov., 1857.—*Trans* 599.

Was *Sindico del Juzgado* and *escribano* of the Court in 1845. —*Trans.* 600.

Was *Juez suplente* in 1846—*Trans.* 600.

Proves signatures to espediente.—*Trans.* 602.

Received the espediente in 1845, and deposited it in the archives.—*Answers* 19–20–21; *Trans.* 602–3.

Saw it again in 1849, when he was Second Alcalde.—*Answer* 28; *Trans.* 603.

VI. Of James Weeks, Government witness, who was called twice, 24th June and 14th August, 1857.—*Trans.* 165–305.

When first called, no questions asked about this espediente. —*Trans.* 165.

When next called, Government attempted to prove the non-existence of the espediente in archives, but failed.—*Answer* 13; *Trans.* 307.

Witness saw espediente in 1846–7 when Burton was Alcalde. —*Answer* 15; *Trans.* 307.

And again in 1848, when he was himself Alcalde.—*Answer* 15; *Trans.* 307.

Identifies this espediente.—*Answers* 17–35–42–43–44; *Trans.* 307–310.

VII. Manuel Castro, Prefect, on 31st December, 1845, in a letter to Governor Pico *states denouncement*, and incloses petition for two leagues.—*Trans.* 2546.

This letter is in the California archives.

VIII. *Borrador*, or rough draft of answer to foregoing, in the California archives.—*Trans.* 2551.

IX. Inventory of papers and effects in the Juzgado, when Pico went out of office on the 1st January, 1846; signed by Pico, and receipted by Chavolla, the incoming Alcalde, 2d January, 1846. Mentions in list of documents *Posecion de la Mina de Sta. Clara, à D. Andres Castillero.*

Produced by Chapman Yeates, Clerk of the City of San José, from archives of that place on the 30th January, 1858.—*Trans.* 769.

Proved by McCutchen.—*Trans.* 767.

X. Castro's petition to Alcalde Pacheco, dated 27th June, 1846. *Refers to espediente of denouncement in the archives, and prays to be included therein.* Offered by Government.—*Trans.* 622.

XI. Marginal decree of Pacheco on foregoing petition, *ordering the petition to be included in espediente, as prayed.* Offered by the Government.—*Trans.* 622.

XII. Certified copy of the whole espediente, Castro's petition included, signed by Chavolla on the 13th August, 1846. Proved by Salvio Pacheco, brother of Dolores Pacheco, the First Alcalde, to be in his own handwriting, and made at its date.—*Trans.* 664.

XIII. Record of suit between Grove C. Cook and the owners of the mine. Cook said they were working the mine on his land contrary to law, and prayed they might be removed. Measurement and survey ordered. Plaintiff put in for costs. Proceedings in March, 1847.—*Trans.* 815.

XIV. Existence of the espediente in archives in March, 1847, proved by Forbes' Consular letter to Burton, Alcalde, on the 14th August, 1847, complaining of recent trespasses by Cook *within the limits of the juridical possession of the mine of Santa Clara*, and referring the Alcalde to "*the documents which exist in your office, upon which was founded your conviction of the justice of your decision in relation to the claim of Mr. Cook in March last, etc.*"

This letter was produced by Chapman Yeates, Clerk of the City of San José, from the archives of that place on the 30th January, 1858.—*Trans.* 769.

And is proved by McCutchen.—*Trans.* 767.

XV. Not two months after James A. Forbes had seen this espediente in Alcalde Burton's office, as proved by the foregoing letter, to wit, on the 5th May, 1847, in a letter to Mr Alexander Forbes, he speaks of the "*juridical possession which was given of the mine by the local authority of this jurisdiction*," and also of "*the three thousand varas of land given in that possession as a gratification to the discoverers.*" *Vide* Exhibit O. H. X. No. 11.—*Trans.* 842.

XVI. Alexander Forbes' petition to Alcalde Weeks, on the 19th January, 1848, prays that the *former act of possession* may be amended. Offered by the Government.—*Trans.* 806.

XVII. Weeks' amended possession given 21st January, 1848, refers to the *original act of possession*, and declares that "*the right and title of the mine to the land granted as a reward in the original act of possession shall remain valid.*" See

Trans. 148 for the Spanish, and 802–805 for the translations which are erroneous. Offered by the Government.

This is the same Mr. Weeks, the Government witness, who proves the original espediente to have been in the Alcalde's office in Burton's time, 1846–7, and in his own, in 1848.

XVIII. In the possessory action between Walkinshaw and Forbes. Complaint filed 18th October, 1849; Walkinshaw claims "*one-eighth part of the mine by title derived under the original act of registry.*" Offered by the Government.—*Trans.* 272.

XIX. In Mr. Horace Hawes' denouncement of the mine on the 6th October, 1849, for abandonment and insufficient registry, he describes the mine as situated on Berreyesa's Rancho, and as known *in its original title of registry as the mine of Santa Clara.*

He says its last possessors were Andres Castillero, Alex. Forbes, James A. Forbes, Robert Walkinshaw, and the two Robles. Offered by Government.—*Trans.* 287.

Alcalde May's proclamation thereupon, 23d October, 1849, describes the mine as *known and designated in its original act of registry as that of Santa Clara, and now known by the name of New Almaden.* Offered by Government.—*Trans.* 297.

The Alcalde refusing to entertain jurisdiction of the denouncement, Mr. Hawes filed his protest on the 15th November, 1849, and said, that besides abandoning the mine, the former owners *by reason of the insufficient registry thereof, etc.*, never acquired any title to the mine, which he stands ready here in Court, with *witnesses, records and documents*, offering to prove, etc. Offered by Government.—*Trans.* 299.

XX. On the 18th January, 1851, Capt. H. W. Halleck, accompanied by Josiah Belden, Mayor of San José, removed this espediente from the office of the latter to that of the County Recorder of Santa Clara County, where it properly belonged, and at 3 o'clock, P. M. of that day, it was filed in the County Recorder's office, as appears by the indorsement of the Recorder.—*Trans.* 257, 794. *Halleck's deposition*, *Trans.* 332.

PROOF

OF THE

GENUINENESS AND CUSTODY OF THE "COPIA AUTORIZADA" OF THE ESPEDIENTE OF REGISTRY AND POSSESSION.

The *copia autorizada* of the registry and possession, is Exhibit Young No. 1.—*Trans.* 2688.

McAllister, Edwards and Rose, received it from Walkinshaw in December, 1852, or in January, 1853.—*Vide Hall McAllister's deposition*, *Trans.* 2698.

It remained in McAllister's office from that time until May, 1858, when he delivered it to Walkinshaw, with other papers, and took a receipt.—*McAllister's deposition*, *Trans.* 2698; *Reese's deposition*, *Trans.* 2712.

Walkinshaw, on receiving it from McAllister, handed it to John Young, his agent, and now his executor, by whom it was with the other papers enveloped and labeled, "*Papers relating to the disputed barra, between R. Walkinshaw,*" *etc.*

In July, 1860, the counsel of the New Almaden Company, having occasion to examine all the conveyances by Padre Real of his shares in the mine, caused search to be made among Walkinshaw's papers for such as might be there, which led to the discovery of the espediente in question.—*See Young's deposition*, *Trans.* 2684, *and Bell's deposition*, 2696.

The genuineness of the espediente is proved by the Alcaldes and all the assisting and subscribing witnesses, except José Suñol, whose death is proved.

See Pico's deposition, *Trans.* 2731; See Antonio Suñol's deposition, *Trans.* 2718; See José Noriega's deposition, *Trans.* 2803; See Pedro Chaboya's deposition, *Trans.* 2806; See Pedro Sansevain's deposition, *Trans.* 2809.

CONTENTS OF ESPEDIENTES FROM MEXICO.

No. 1.

Official communication, dated 21st April, 1846—Junta de Fomento to General Tornel, Director of National College of Mining—sending specimens for assay, with copies of Castillero's letters to Herrera and Moral.

Original in Exhibit Bassoco No. 10, *Trans.* 1809; *Office copy* in Exhibit Bassoco No. 7, *Trans.* 1783.

Original proved by Bassoco, *Ans.* 152, *Trans.* 1240; Balcarcel, *Ans.* 152, *Trans.* 1871; Castillo, *Ans.* 34, *Trans.* 1940.

Office copy by the same persons.

No. 2.

The minutes of the proceedings of the Junta Facultativa of the College, of the 24th April, 1846; dated by mistake the 24th March of that year; to be found in Exhibit Bassoco No. 11, *Trans.* 1813.

Proved by Professor Castillo, Secretary of the Faculty, *Ans.* 29–30, *Trans.* 1939; and by Professor Balcarcel, a member of the faculty, *Answers* 40–41–42–43–44. *Trans.* 1870–1–2–3.

Personal recollection of Castillo, *Trans.* 1943, *Answer* 39; of Balcarcel, *Answer* 45, *Trans.* 1872.

No. 3.

Official communication, dated 29th April, 1846—General Tornel, Director of the College, to Don Tomas Ramon del Moral, President of the Junta Facultativa of same—acknowledging receipt of a letter from Señor Moral, of 24th April, 1846, communicating result of assay; found in Exhibit Bassoco No. 11, *Trans.* 1813.

Proved by same witnesses as foregoing, and in same answers.

No. 4.

Official communication, dated 29th April, 1846—from General Tornel, Director of College, to the Junta de Fomento—communicating result of assay; found in Exhibit Bassoco No. 7, *Trans.* 1785.

Proved by Bassoco, *Answers* 139–141, *Trans.* 1237–8; Balcarcel, *Answers* 40–42, *Trans.* 1870–71; Castillo, *Answer* 36, *Trans.* 1942; *Answer* 37, *Trans.* 1943; Miranda, *Answers* 10–11, *Trans.* 2147–8; Castillo Lanzas, *Answers* 17–18, *Trans.* 2230–1.

No. 5.

Official communication, dated 5th May, 1846—from Junta de Fomento to the Minister of Justice—making known to the President Castillero's discovery, etc.

Original in Exhibit Bassoco No. 8, *Trans.* 1794–5–6.

Office copy in Exhibit Bassoco No. 7, *Trans.* 1786–7.

Original proved by Bassoco, *Answers* 144–145, *Trans.* 1239; Yrisarri, *Answers* 13–14, *Trans.* 2110; Miranda, *Answers* 13–14, *Trans.* 2149–50–51.

No. 6.

Official communication in reply to foregoing, dated 9th May, 1846.

Original in Exhibit Bassoco No. 7, *Trans.* 1787.

Office copy in Exhibit Bassoco No. 8, *Trans.* 1797.

Original proved by Bassoco, *Answers* 138–139–141, *Trans.* 1237–8; Balcarcel, *Answers* 40–42, *Trans.* 1780–1; Castillo, *Answers* 36–37, *Trans.* 1941–2; Yrisarri, *Answers* 9–10, *Trans.* 2108–9; Miranda, *Answers* 10–11, *Trans.* 2147–8; Castillo Lanzas, *Answers* 17–18–19, *Trans.* 2230–1.

No. 7.

Contract proposed by Castillero to the Junta de Fomento, dated 12th May, 1846.

Original in Exhibit Bassoco No. 8, *Trans.* 1800–1–2.

Office copy in Exhibit Bassoco No. 7, *Trans.* 1790–1–2.

Original proved by Couto, who copied it for Castillero from his rough draft; *Answers* 25–26–27–28, *Trans.* 1000.

No. 8.

Official communication, dated 14th May, 1846—from the

Junta de Fomento to the Minister of Justice—inclosing the original of foregoing proposed contract, and urgently recommending the President of the Republic to agree to it.

Original found in Exhibit Bassoco No. 8, *Trans.* 1798–9–1800.

Office copy in Exhibit Bassoco No. 7, *Trans.* 1787–8–9.

Proved by Bassoco, *Answers* 144–145, *Trans.* 1239; Yrisarri, *Answers* 13–14, *Trans.* 2110; Miranda, *Answers* 13–14, *Trans.* 2149–50–51.

Marginal decree of grant on foregoing proved by Bassoco, *Answer* 145, *Trans.* 1239; Yrisarri, *Answer* 14, *Trans.* 2110; Miranda, *Answer* 14, *Trans.* 2151.

Record of foregoing marginal decree in the 140th leaf of the 15th vol. of the Libro General; found in Exhibit Bassoco No. 12, *Trans.* 1816.

Proved by Yrisarri, who made the record, *Answers* 24–25, *et seq.*, *Trans.* 2113–14–15, etc.

No. 9.

Official communication, dated 20th May, 1846—from Minister of Justice to Junta de Fomento—ratifying and approving said contract in all its parts.

Original found in Exhibit Bassoco No. 7, *Trans.* 1792.

Office copy in Exhibit Bassoco No. 8, *Trans.* 1802.

Original proved by Bassoco, *Answers* 139–141, *Trans.* 1237; Miranda, *Answer* 11, *Trans.* 2149; Castillo Lanzas, *Answer* 18, *Trans.* 2231; Yrisarri, *Answer* 10, *Trans.* 2109.

No. 10.

Official communication, dated 20th May, 1846—from Minister of Justice to Minister of Relations—communicating the President's approval of Castillero's contract with the Junta de Fomento; the grant of two square leagues to Castillero on his mining possession; and directing Minister of Relations to issue the proper orders.

Original found in Exhibit Bassoco No. 9, *Trans.* 1807.

Office copy in Exhibit Bassoco No. 8, *Trans.* 1803.

Proved by Bassoco, *Answers* 147–8, *Trans.* 1239–40; Yrisarri, *Answers* 20–21, *Trans.* 2111–12; Miranda, *Answers* 16–17, *Trans.* 2152–3.

The body of this letter is in Miranda's handwriting; Castillo Lanzas, *Answers* 22–23, *Trans.* 2232.

The marginal decree on the foregoing letter in these words —*May* 23*d*, 1846. *Issue the orders referred to in this communication*—is in the handwriting of Quintanar, and bears the rubric of Castillo Lanzas; proved by Castillo Lanzas, *Answer* 23, *Trans.* 2232; and by Bassoco, Yrisarri and Miranda, as last cited.

No. 11.

Official communication, or decree, dated 23d May, 1846—addressed by the Minister of Relations to the Governor of the Department of Californias—setting forth the President's approval of Castillero's proposed contract; the grant of two square leagues to Castillero on his mining possession; and ordering the Governor to put him in possession of said land.

This is the two league grant. The original was delivered to Castillero, and by him filed in this claim, before the U. S. Land Commissioners.—*Trans.* 65.

An office copy was retained in the Ministry of Relations. —*Trans.* 1807.

Original proved by Lafragua, *Trans.* 17; by Velasco, who wrote the grant, *Answer* 6, *Trans.* 2200; and by Castillo y Lanzas, the Minister who signed it, *Answers* 26–27–28, *Trans.* 2233.

Mr. Negrete proves that Castillero handed to him this identical paper in December, 1846, in order that it might be copied in his ratification of the McNamara contract; and after having been there copied, that he, Negrete, sent it to Mr. Alexander Forbes, of Tepic, on the 19th December, 1846.—Negrete's deposition, *Answers* 32–36, *Trans.* 2401–2; and Exhibit Negrete No. 9, *Trans.* 2448–91.

No. 12.

Lafragua's Report. On the 14th, 15th, and 16th December, 1846, Don José Maria Lafragua, Minister of Interior and Exterior Relations, read a report to the Mexican Congress.

His report refers to Castillero's discovery of a quicksilver mine in Upper California—p. 65 of Report, of which see Extract in Exhibit J. P. Benjamin No. 1, *Trans.* 3039.

In his report is also incorporated the report which the Junta de Fomento sent in to him, as Minister of Relations, on the 17th December, 1846.

The identical manuscript report which the Junta de Fomento thus sent in, is in evidence. It is Exhibit Bassoco No. 4, and begins in *Transcript* on page 1712.

This report of the Junta made very full reference to Castillero's discovery, and set forth what the Junta had done to assist the development of the mine. Extracts of the important parts of it will be found in Exhibit J. P. Benjamin No. 1, *Trans.* 3039–40.

Mr. Lafragua's and the Junta's Reports are proved:

By Lafragua, *Answers* 7–8, *Trans.* 17–18. The extracts filed by Mr. Lafragua will be found on pages 71, 72, 73 of *Transcript.*

By Mr. Bassoco, *Answers* 114 *et seq.*, *Trans.* 1232 *et seq.* He also proves the Original Manuscript Report (Exhibit Bassoco No. 4), *Answers* 121 *et seq.*, *Trans.* 1233.

By Mr. Castillo y Lanzas, *Answers* 53–54 *et seq.*, *Trans.* 2237–8. The reading of this report is noticed in the *Diario del Gobierno de la Republica Mexicana* of 19th December, 1846, No. 135; of the 20th December, 1846, No. 136; and of 21st December, 1846, No. 137; and also in *Monitor Republicana* of the 24th, 25th, and 28th, same month and same year; and in the *Republicano* of the 18th, 20th, and 21st of same month and year. *Vide* Castillo Lanzas deposition, *Answers* 696–699, *Trans.* 2357–8.

By J. P. Benjamin, deposition *Trans.* 3026 *et seq.*

By the *Actas* of the Junta de Fomento of the 5th, 9th, and 16th Nov., 1846, and of the 5th December, 1846, *Trans.* 1702–3–4–5.

MEMORANDA

OF

SOME AUTHORITIES IN BEHALF OF CLAIMANTS.

Digest of Principles of Mining Law, as Established by Gamboa, under the Ordinances of 1584; with Note of the Changes made by the Ordinances of 1783.

[The references in (—) are to Heathfield's Translation, 2 vols.]

I. The title of the subject in his mine, is perfectly consistent with the King's right of *eminent domain* (*alto dominio*). The subject's title is a complete legal title. It is not only a title to the beneficial interest, but is also a legal title to the property itself. There passes to the subject the "dominio directo ó de propiedad" as well as the "dominio util."—*Ch. ii, secs. 24–5–6 (1, pp. 27–29), p. 19; Ch. xvii, sec. 18 (2, p. 92), p. 332; Ch. xxi, sec. 15 (2, p. 157), p. 377.

II. The sovereign's title is however retained to the one-fifth, which he reserves of the fruits, and therefore the title of the subject is only to *part* of the mine, but not to the *whole*, because the grant is of four-fifths only. The King's *alto dominio* or *eminent domain* is also reserved by him, by virtue of which he retains the right to make a new grant, if the first title be lost by reason of forfeiture for non-compliance with conditions subsequent.—Ch. ii, secs. 19, 20, 24, 25 (1, pp. 23–28), pp. 16–19.

III. The true nature and legal definition of the title of the subject is that of a "grant with conditions subsequent" (una donacion modal). The nature of such a title and the legal principles applicable to it fully defined.—Ch. ii, secs. 25–26 (1, pp. 28–29), pp. 19–20.

---

* The paging between parentheses refers to Heathfield's Translation of Gamboa.

IV. And the mine thus acquired is land. It is a "fundo," or land.—Ch. v, sec. 5 (1, p. 144), p. 103; Ch. xxi, sec. 15 (2, p. 157), p. 377. It is "bienes raices," or real estate; Ch. xvii, sec. 22 (2, p. 95), p. 335. It is "bienes immobles," or immovables; Ch. xxiii, sec. 6 (2, p. 258), p. 447. And all the rules of law relating to petitory and possessory actions for other lands, are also applicable to petitory and possessory suits for mines; Ch. xxiii, secs. 1, 6, 19 (2, pp. 255, 258, 265), pp. 445–7. And in these actions the claim may be for the *ores* only, or for the *property* itself.—Ch. xxiii, secs. 19, 21 (2, pp. 265, 266), p. 453.

V. And the title to the miner is of the surface land as well as of all under the surface, even to the antipodes or "el infierno." And the miner cannot go underground outside of the area of his surface measurement, but perpendicular lines are to be drawn from his surface lines toward the centre of the earth, and these lines inclose his land.—Ch. xii, sec. 38 (1, p. 339), p. 238; Ch. xiv, secs. 21, 22, 23, 25 (2, pp. 26–28), p. 290, *et seq.*; Ch. xxiii, sec. 7 (2, p. 259), p. 448.

ORDINANCES OF 1783.

The limitation of the miner's right to the boundaries of his own land, as measured out on the surface, is fully developed in the Ordinances of 1783.—T. 8, Arts. 8, 11, 14, 17; Halleck, pp. 237–239.

VI. And the title above described is acquired by the registry, which is the *fundamental title of the mine*, "el titulo fundamental de las minas."—Ch. v, secs. 3–10–17 (1, pp. 143, 147, 150), pp. 102 *et seq.*; Ch. vii, sec. 3 (1, p. 175), p. 123; Ch. xi, sec. 2 (1, p. 307), p. 216; Ch. xi, sec. 24 (1, p. 319), p. 223.

VII. And receiving this registry is a *judicial* act, to take place before *Justicias* or *Alcaldes Mayores.*—Ch. v, secs. 14–15–19–22 (1, pp. 148–9, 151, 153), p. 107; Ch. x, secs. 1–3–8 (1, pp. 297–8, 301), p. 211; Ch. xii, sec. 28 (1, p. 335), p. 234; Ch. xiii, secs. 4–10 (2, pp. 5, 9), p. 276–9; Ch. xx, sec. 4 (2, p. 136), p. 363; Ch. xxv, sec. 9–10 (2, pp. 289, 290), p. 469.

ORDINANCES OF 1783.

Registry is to be made before the deputation of mining of that "territorio," or the one nearest, if there should be none there.—T. vi, art. 4; Halleck, p. 224.

VIII. But a failure to register does not forfeit the rights of the discoverer, if he makes a registry before some one else denounces, even after the delays fixed by law.—Ch. v, sec. 17 (1, p. 150), p. 108.

IX. In case of his failure to register, or in case of abandonment, a third person may denounce the mine, and *after judicial proceedings pronouncing forfeiture*, may register it, and thus acquire the title forfeited by the discoverer.—Ch. v, sec. 22 (1, p. 153), p. 110; Ch. xi, sec. 6 (1, p. 310), p. 218; Ch. xiv, sec. 36 (2, p. 35), p. 295; Ch. xvi, secs. 1, 7, 8, 10, 11 (2, pp. 61–63–66), p. 312. These *judicial proceedings described.*—Ch. xviii, (2, p. 103 *et seq.*), p. 343. So of all other causes of forfeiture there must be an *adjudication.*—Ch. xi, sec. 10 (1, p. 311), p. 219.

ORDINANCES OF 1783.

Forfeitures are a penalty, which must be *lawfully established and proved.*—T. vi, art. 8, 9, 10, 11; Halleck, p. 226. T. ix, art. 14, 15; Halleck, p. 244–5.

X. And the titles acquired by registry and denouncement differ only in *form* not in *substance*, because even after the judgment of forfeiture the denouncer acquires no title, unless he also registers.—Ch. v, sec. 21 (1, 152), p. 108.

XI. Titles to mines are also acquired by *sale* from the King. Thus in Peru, one *pertenencia* adjoining that of the discoverer is marked out for the King. This is to be *sold* for him, or worked or leased.—Ch. ix, secs. 28–29 (1, pp. 294–295), p. 208.

XII. The ordinary miner can only take two mines by registry or denouncement, leaving between them a space equal to three mines.—Ch. viii, sec. 19 (1, p. 269), p. 190.

ORDINANCES OF 1783.

None but the discoverer could denounce two contiguous mines on the same vein.—T. vi, art. 17; Halleck, p. 229.

But in considerable enterprises, the miners might have allowed to them many pertenencias, by grant of the Viceroy on report of the Tribunal General.—*Ib.*

XIII. But *discoverers* are worthy of special rewards, and the discoverer may therefore take as *many mines as he please* in the vein which he has discovered, and *even the whole vein.*—Ch. viii, sec. 1–12–13 (1, pp. 259–266), p. 188.

ORDINANCES OF 1783.

Discoverers may take in an absolutely new mountain, in which no mines or pits are opened, *three pertenencias* continuous or interrupted in the principal vein, and one in each of the other veins. If discoverers of a new vein in a mountain worked in other parts, *two* pertenencias.—T. vi, art. 1; Halleck, p. 223.

XIV. And partnerships may take twice as many mines as there are partners in the partnership.—Ch. vii, sec. 21 to 34 (1, pp. 185–195), pp. 121–137.

ORDINANCES OF 1783.

Partnerships may take *four* new pertenencias, without prejudice to what they may have a right to as discoverers, if they be such.—T. xi, art. 2; Halleck, p. 252.

And shall also be encouraged by all the favors, assistance and exemptions which can be granted them, in the judgment and *discretion* of the Royal Tribunal of Mining.—T. xi, art. 1; Halleck, p. 252.

XV. And the forfeiture against one who registers or denounces in his own name a mine owned in partnership, is only in favor of his copartners, *and cannot be adjudged unless he has omitted the names fraudulently.*—Ch. vii, sec. 5 (1, p. 176), p. 125.

XVI. Possession is taken by any symbolical act, such as by receiving the title papers, throwing stones, or walking over any part of the land, or any fictitious act done with the purpose and intent of taking possession.—Ch. xvii, sec, 3 (2, p. 79), p. 325.

ORDINANCES OF 1783.

Possession is to be given to the discoverer *as soon as* (luego que) he has made a pit of the prescribed depth, for which purpose ninety days are allowed him, and the Judge shall *immediately* give him his authentic certificate of possession, (su fé de posecion), measuring out his pertenencia, and making him fix stakes, as afterwards provided. But any third person has a right until the end of the ninety days to appear and claim that he is the discoverer, whereupon a summary judicial investigation shall determine the fact, and award the right of registry to the true discoverer. After ninety days, third persons shall not be heard.—T. vi, arts. 4 and 5; Halleck, p. 224–5.

XVII. When possession is given, the miner ought to have his mine staked out; so marked on the surface as to define his possession.

But he was not *obliged* to do this, unless cited by some other miner near him who *asked for stakes*. When so cited, ten days were allowed the miner to choose the surface parallelogram that he wished to have staked out for himself. If he suffered this delay to pass, *the Judge was to stake out his mine for him.* Ch. x, secs. 1–3–8 (1, pp. 297–298–301), pp. 210–13; Ch. xii, sec. 8 (1, p. 325), p. 328; Ch. xii, sec. 12 (2, p. 10), p. 280.

ORDINANCES OF 1783.

There shall be measured out to the denouncer, when he takes possession, his pertenencia, and he shall be required to set up stakes.

But if he have no neighbors, he may at any time better his stakes (mejorar las estacas) and change his boundaries, with the intervention of the mining deputation of the district.—T. 8, art. 11; Halleck, p. 236.

XVIII. Of all the causes of forfeiture of right, the only one to which the law attached special importance was abandonment; but even this forfeiture being *penal* must be very strictly construed.—Ch. xvii, secs. 5–6–11 (2, pp. 80–86), pp. 326–330 —and may be excused by reason of any lawful impediment which hindered the working, such as war, riots, epidemics, famine, etc.—Ch. xxvii, secs. 16–17 (2, pp. 91–92), p. 333.

ORDINANCES OF 1783.

Besides the impediments to working resulting from pestilence, famine, war, etc., forfeiture shall not be pronounced if there is an equitable defense on the part of the owner, who may be engaged in soliciting an *avio*, or be in want of operatives or provisions, etc., etc.—Title ix, Art. 15; Halleck, p. 244.

XIX. All the foregoing principles apply as well to mines on private as on public lands, because mines are not included in the grant of land by the sovereign unless specially mentioned.—Ch. iv, secs. 2, 5, 11 (1, pp. 130–132–134), pp. 94–93. But private owners of the land were to be paid by the mine owner for the land which he took, not only for the actual mine itself, but for the land requisite for the houses, reduction-works, and all damages caused by the working of the mine.—Ch. iv, secs. 6–7 (1, pp. 133–134), p. 96.

ORDINANCES OF 1783.

The discoverer who denounces a vein on private lands *must pay for the surface land which he occupies*, besides the damage *immediately* resulting, as well as for the land and water required for his Hacienda, etc.—Title vi, Art. 14; Halleck, p. 228.

XX. Possessory actions for mines were to be tried very summarily, for mines must not remain closed, this being against the whole public policy and public interest.—Ch. xxiii, secs. 11–12–14 (2, pp. 261–262), p. 449; secs. 30 (2, p. 271), p. 456; secs. 36 (2. p. 274), p. 458.

XXI. Especially must the Judges avoid appointing *receivers* who receive the fruits and defraud the owners of them.—Ch. xxiii, secs. 23–24 (2, pp. 267–268), p. 454.

XXII. In New Spain the special mining judges do not exist, nor is the ordinance in force there on that subject. In their stead, the ordinary Judges must act, *Alcaldes* and *Corregidores*. Ch. xxv, secs. 1–5–6–10 (2, pp. 286–288–290), p. 467–9.

XXIII. FORFEITURE OF MINER'S TITLE.

The change of legislation between the ordinances of 1584 and those of 1783, is particularly striking on the subject of forfeitures.

In those of 1584 we count no less than fifteen cases of forfeiture of mines.

1. Art. 18. For failing to register anew mine already owned and worked.—Halleck, p. 77.
2. Art. 24. For registering a mine not really belonging to him who registers it.—p. 78.
3. Art. 21. For registry without declaring names of partners.—p. 78.
4. Art. 22. For failure to dig pits and fix stakes when pertenencias are defined.—p. 82.
5. Art. 32. For taking a mine for another when the person taking it has no power of attorney and is not the hired servant of that other.—p. 85.
6. Art. 35. For failing, within ninety days after registry, to sink a well three *estados* or seven varas deep (unless excused for causes stated in Art. 36).—p. 86.
7. Art. 37. For failing to work the mine for four consecutive months.—p. 87.
8. Art. 38. For failure on part of denouncer to deepen former works three *estados*.—p. 89.
9. Art. 42. For purchasing a mine which has not been sunk to the depth of three *estados*.—p. 92.
10. Art. 42. For not registering purchase, or other change of ownership.—p. 93.
11. Art. 53. Of mine and metal; for smelting in another's furnace without permission of the Administrador of Partido. —p. 100.
12. Art. 59. Of mine and silver; for reducing ores otherwise than with quicksilver, after obtaining license to reduce them in that mode.—p. 105.
13. Art. 67. Of mines; by certain officers who are forbid-

den to hold them, and by those who participate in the offense of these officers.—p. 112.

14. Art. 68. By certain salaried persons; for becoming owners of mines within two leagues of the Partidos where they are employed.—p. 112.

15. Art. 73. Of all property, and ten years' labor in the galleys for purchasing or dealing in unstamped metal.—p. 114.

ORDINANCES OF 1783.

But by Ordinances of 1783 there are no forfeitures for irregularities of any kind, and the only forfeitures pronounced are:

1. For abandonment.—Title 9, Art. 13; Halleck, p. 243.

2. For failure by Secular Ecclesiastics to sell mines belonging to them, after a certain delay.—Title 7, Art. 2, p. 231.

3. For removing or impairing the strength of pillars and arches in a mine.—Title 9, Art. 7, p. 241.

4. For repeated violations of rules relative to the security and preservation of mines.—Title 9, Art. 10, p. 242.

5. In cases of partnership, for omitting the partner's name in the denouncement, or refusing to contribute for four months towards expenses, in which cases the forfeiture inures to benefit of copartners.—Title 7, Art. 6, p. 233; Title 11, Art. 8 p. 254.

SPECIAL QUICKSILVER LEGISLATION.

Ordinances 1783; Title 6, Art. 22; Halleck, 230.—Quicksilver mines excepted from the general law, and the Viceroy to determine whether the subject may work them and deliver all the quicksilver to the King at a fixed price, or whether the King will take the mines and give the party some equitable reward.

26th January, 1811—Decree of Cortes of Spain; Provisions of Ordinances of 1783 repealed; Grants to discoverers of quicksilver absolute and perpetual ownership, (*la seguridad del dominio absoluto y perpetuo del minero*); exempts it from all payments to the crown, *even of the fifth or the part which the miner has to pay;* makes the sale of it free, and liberates the miner from all conditions, except that he shall *manage and work his mine in* accordance with the general law.—Halleck, p. 383; Ordenanzas de Mineria, p. 81.

26th January, 1811.—The above decree forwarded by Council of Regency to Royal Mining Tribunal, advising it to offer a large pecuniary reward for the discovery of a rich quicksilver mine, and promising that the Council of Regency will reward, also, and ennoble by honorable distinctions those who invest capital in it.—Halleck, p. 382; Ordenanzas, p. 79.

2d February, 1811.—Council of Regency again informs miners that the Cortes have provided for giving rewards to discoverers of quicksilver, and the greatest reward to whoever shall discover the richest and most productive.—Halleck, p. 386; Ordenanzas, p. 83.

19th December, 1818.—Quicksilver exempted from all duties whatever by Royal order to Viceroy of New Spain.—Halleck, p. 393.

20th February, 1822.—Mexico, on acquiring independence, decrees a revenue in favor of the Empire on gold and silver, but (Art. 13) pure quicksilver remains *absolutely free* from duty. —Halleck, p. 400.

7th October, 1823.—Law of mines modified except as to *quicksilver*, which is expressly exempted *from all contribution.*—Halleck, p. 404.

11th March, 1842.—Foreigners allowed to acquire in property mines (including quicksilver) which they may discover.—Halleck, p. 427.

2d December, 1842.—Junta de Fomento established, and its special attention directed to making regulations, fixing "mode in which working of quicksilver mines is to be supplied, rewarded and stimulated and protected."—Halleck, p. 437.

24th May, 1843.—Rewards of $25,000, and $5 per quintal to quicksilver miners.—Halleck, pp. 452-3.

5th July, 1843.—Junta empowered to work, to supply, and to protect quicksilver mines; to search for such mines, to examine those discovered, and to take partial measures for efficaciously encouraging production of quicksilver.—Halleck, p. 455.

25th September, 1843.—Further decree to assist and encourage the working of quicksilver; researches ordered to be made through the whole Republic.—Halleck, p. 463.

Definition of the Civil Law terms, "dominium eminens," "dominium plenum," "dominium directum," "dominium utile."

I. *Dominium eminens*, or *altum*, as *Gamboa* calls it.

"All separate interests of individuals in property are held of the Government under this tacit agreement or implied reservation. Notwithstanding grants to individuals, the *eminent domain,—the highest and most exact idea of property,*—remains in the Government, or in the aggregate body of the people in their sovereign capacity: and they have the right to resume the possession of the property in the manner directed by the Constitution and laws of the State."

Chancellor Walworth, in Beekman *v.* Saratoga and Schenectady R. R. Co., 3d Paige Ch. R., 72. Bowyer calls it "a part of the sovereign authority, and one of the *jura majestatis*."—Bowyer's Universal Pub. Law, pp. 227, 372.

Vattel, "the right which belongs to society or the sovereign of disposing in case of necessity and for the public safety of all the wealth contained in the State."—Book 1, Ch. 20, Sec. 244.

See also Cooper's Justinian, p. 456.

II. *Dominium plenum*, termed *allodium* in the Common Law writers, is subordinate only to the *dominium eminens;* it is the highest right of property in land that an individual can have. Its two parts are the "*dominium directum*" and the "*dominium utile.*" These two parts are united in the allodial proprietor.

"Allodial lands are enjoyed by the owner, independent of any superior or without any feudal homage. * * * No lands are allodial, unless the grantor who makes them over expressly exempt the grantee from the obligation of homage and fealty, and renounce the rights of superiority which were formerly competent to him against the vassal in the subject of the grant. By the usage of Scotland, no lands are allodial except first, those of the King's own property: second, *the superiorities which the sovereign, as the fountain of feudal rights, reserves to himself in the property-lands of his subjects:* and third, churches, church-yards, etc."

Erskine's Institute, B. II, T. 3, No. 8.

III. But in the feud or fee, these two parts are separate. The *dominium directum* is retained by the lord, and only the *dominium utile* is conveyed to the vassal, and the vassal's estate takes the name of feud or fee.

"*Dominio directo.*"—The right of superiority in any real estate, without the right of the beneficial ownership (propriedad útil). Such is the ownership of the proprietor of land alienated in fee (feudo).—Escriche.

"*Dominio útil.*"—The right of receiving all the fruits of a thing under service, or tribute paid to him who reserves the "dominio directo;" such is that of the vassal in the land which he holds in fee.—Escriche.

"*Dominio pleno.*"—The right which one has in a thing, to alienate it *without dependence on another person*, to enjoy *all* its fruits, and to exclude others from its use.—Escriche.

See also Escriche—*Verbo* "*Feudo,*"—a reciprocal convention between the lord and the vassal, by which the former transfers to the latter the "*dominio util*" of anything, and the latter acknowledges the former as the direct owner (*como dueño directo*) and promises him fidelity, military or other personal service, and sometimes the payment of some tribute (*derecho*).

Erskine, in the Institute of the Law of Scotland, says:

"The grantor of the feudal right is called the *superior*, because he stands in a higher rank than the grantee, who is styled the *vassal.* * * * But our writers, particularly Craig, have, after the example of the old feudists, expressed the interest retained by the superior by "*dominium directum,*" or, as we translate it, the superiority; and that conferred on the vassal by "*dominium util,*" or the property. The word *fee* is often used promiscuously for both.—Book II, Title 3, No. 10. See also Title 6, No. 1.

Dumoulin, as quoted by Pothier, defines the "Fief" as being "*benevola, libera et perpetua concessio rei immobilis vel œquipollentis, cum translatione utilis dominii, proprietate retenta sub fidelitate et exhibitione servitiorum,*" and as translated by Pothier, "a gratuitous grant made in perpetuity to another of a thing immovable or reputed immovable, on condition of the grantee's rendering faith and homage and military service, and with reservation of the direct lordship (*seigneurie directe*).—Pothier; Coutumes d'Orléans, sec. 1.

The *dominium directum* is purely a title of honor, thus described by Pothier: "The direct lordship (*la seigneurie directe*) which the lord retains in the land held by the vassal in fee, is purely one of honor, consisting in the right of calling himself the lord, and of having himself recognized as such by the vassal who owns the property.—Pothier, sec. 2, No. 8.

GENERAL INDEX.

INDEX TO MEXICAN ARCHIVES.

* This session is dated 24th *March*, by a clerical error, which is explained in the testimony at pp. 1872–3, 1941.

May 9.	Minister of Justice acknowledges receipt.....	1787
May 6.	Castillero makes a verbal report to Junta, and is required to make his statements in writing	1760
May 12.	Castillero makes written proposals, stating his denouncement, and possession, and grant of three thousand varas....................	1790
May 14.	Junta transmits proposals with urgent recommendations...........................	1787
May 20.	*Acuerdo*, granting Castillero's demands......	1798
May 20.	*Acuerdo* is recorded in the *Libro General*.....	1816
May 20.	Minister of Justice replies to Junta reporting that President approved of the arrangement *in all its parts*..........................	1792
May 20.	Minister of Justice communicates the President's orders to Minister of Relations about two league grant.........................	1803
May 23.	*Acuerdo* of Minister of Relations, directing order to issue.........................	1807
May 23.	Minister of Relations sends order to Governor of California to put Castillero in possession.	1807
	Indorsement on the espediente in Ministry of Relations, "Concession of two square leagues to Señor Castillero"....................	1808
June 12.	Castro's power to MacNamara to bargain with some English company to work the mine..	2469
Nov. 28.	Contract of *avio* made under foregoing power	2471
Dec. 17.	Castillero produces the two-league grant in City of Mexico, and a copy is annexed to an authentic act of ratification of foregoing *avio*................................	2468
Dec. 19.	A copy is made by the Notary, who delivers it to Negrete..........................	2474
Dec. 19.	It is sent by Negrete to Alexander Forbes...	2449
Dec. 29.	Alexander Forbes receives the copy in Tepic.	2452
1847.		
Feb. 3.	Castillero and Negrete present a petition asking for a second copy, and Judge Juan Hierra orders it............................	2466

INDEX OF WITNESSES WHO PROVE MEXICAN ARCHIVES.

INDEX OF PERSONS ENGAGED IN THE BUSINESS,

WHO HAVE DIED, AND PAGES WHERE DEATH IS PROVEN.

"DIARIO OFICIAL," AND OTHER MEXICAN NEWS-PAPERS.

INDEX TO CALIFORNIA GENERAL ARCHIVES.

CALIFORNIA LOCAL ARCHIVES.

ARCHIVES AT WASHINGTON.

1846.

May 4. Letter of Consul Larkin to James Buchanan, Secretary of State—details, discovery, and *working of the mine*;—extract............ 2657
(See entire Letter at page 2669.)

1848.

March 28. Letter of Consul Larkin to James Buchanan, Secretary of State—announces *avio* of 1846, working and produce of mine, etc.,—extract. —(Entire Letter at p. 2673)............. 2658

Aug. 17. Report of Col. Mason, Governor of California, to Adjutant-Gen. Jones—Ex. Doc. 17, 1st Ses. of 31st Congress, p. 534. (Not copied in Record.)

ARCHIVES OF U. S. CONSULATE AT MONTEREY.

1846.

Feb. 12. Consul notes in his memorandum-book arrival of "Hannah" from Mazatlan............ 3018

March 7. Consul notes "Hannah" for Mazatlan....... 3018

March 8. Consul advises Colonel Fremont about brig "Hannah"......................... 2667

April 1. Mott, Talbot & Co., write Consul, announcing arrival of "Hannah" in Mazatlan........ 3020

April 3. Consul writes to United States Minister in Mexico; announces departure of Castillero on Dn. Quijote, and his discovery of the quicksilver........................ 2667

April 23. Consul writes to A. H. Gillespie that "Hannah" carried news to Commodore Sloat of Fremont's situation; that Castillero is expected back on Dn. Quijote by 1st July.......... 2674

May 2. Consul writes to Capt. Montgomery, of U. S. ship Portsmouth, details discovery and working of the mine which he has twice seen... 2676

May 4. Consul writes to James Buchanan, Secretary of State; gives same details as preceding.. 2669

June 24. Consul writes to G. P. Judd, Minister of Finance, Honolulu; announces discovery, *denouncement before nearest Alcalde*, and the working of the mine by a Mexican who is the *owner*, and a priest.................. 2224

July 2. Consul writes to ; states same facts as above.......................... 2670

July 20. Letter from G. P. Judd, acknowledging receipt of Consul's letter of 24th June, announcing that he had had it published in "Polynesian"......................... 3020

1848.

March 25. Consul writes to James Buchanan, Secretary of State; announces *avio* of 1846; working 200 lbs. per diem; shipment of $20,000 of quicksilver; recalls his having sent specimens of the ore to Buchanan, Benton, *et al*; in January, 1846....................... 2673

ESPEDIENTE OF THE MINE.

CONTENTS OF ESPEDIENTE.

WITNESSES WHO PROVE CALIFORNIA ARCHIVES.

Names.	By whom offered.	Date.	Page.
1. José Noriega...	Claimants.	1855, March 19.	23–2803
2. Antonio Suñol..	"	"	25–2718
3. Frank Lewis....	"	"	27
4. S. O. Houghton.	"	1857, Oct. 23.	321
5. H. W. Halleck..	"	" " 26.	332
6. José Fernandez.	"	" Nov. 6.	599
7. A. M. Pico.....	"	" " 11.	628–2731
8. Salvio Pacheco..	"	" " 19.	664
9. Jno. M. Murphy	"	" Dec. 23.	743–752
10. Raphael Sanchez	"	1858, Jan. 30.	765
11. Wm. M'Cutcheon	"	" " "	766
12. Chapman Yates.	"	" " "	769
13. Wm. G. Chard..	"	" Dec. 15.	1069
14. Peter Davidson.	"	1859, Jan. 22.	1084
15. Pio Pico.......	"	" Oct. 19.	2351
16. J. M. Covarrubias	"	" Nov. 23.	2606
17. Augustin Olvera.	"	" " 30.	2609
18. José Castro.....	"	" " 28.	2612
19. Jno. Bidwell....	"	" Dec. 31.	2642
20. Jno. C. Fremont.	"	1860, June 26.	2678
21. Jno. Young.....	"	" July 17.	2684
22. Thos. Bell......	"	" " "	2696
23. Hall McAllister.	"	" " "	2698
24. Wm. S. Reese..	"	" " 20.	2715
25. Pedro Chaboya.	"	" " 27.	2806
26. Pedro Sainsevain	"	" Aug. 7.	2809
27. Geo. M. Yöell....	United States,	1857, June 24.	144–252
28. J. W. Weekes..	"	" " "	165–305
29. A. M. Pico.....	"	" " "	169
30. H. C. Malone...	"	" " "	260
31. Walkinshaw's petition	"	1849, Oct. 18.	272
32. H. Hawes' denouncem't	"	" " 6.	287
33. Judge May's proclam.	"	" " 23.	297
34. Horace Hawes' protest	"	" Nov. 15.	299

35.	Frank Lewis...	United States,	1857, Aug. 15.	316
36.	Antonio Suñol.	"	" Dec. 12.	409
37.	José Noriega..	"	" " "	418
38.	R. C. Hopkins..	Claimant.		

PROOFS OF POSSESSION AND WORKING OF THE MINE.

		Introduced by.	Date of working.	Page.
1.	Wm. G. Chard..................	claimant	Nov. or Dec. 1845	1069
2.	Peter Davidson................	"	" " " "	1084
3.	Jacob P. Leese................	"	" " " "	1102
4.	Mariano G. Vallejo............	"	" " " "	2597
5.	Berreyesas in their bill in chancery,	United States	" " " "	228
6.	A. M. Pico....................	claimant	" " " "	633
7.	John C. Fremont...............	"	January, 1846	2678
8.	Larkin's letter to Capt. Montgomery	"	2 May, 1846	2676
9.	Do. do. James Buchanan.	"	4 " "	2669
10.	Do. do. G. P. Judd......	"	24 June, 1846	2224
11.	G. C. Cook—suit v. Castillero....	United States,	17 March, 1847	815
12.	Fernando Alden................	claimant,	March and April, 1847	9
13.	Chester Lyman.................	"	February, 1848	12
14.	Walkinshaw's petition..........	United States,	April, 1847	272–279
15.	J. A. Forbes' answer under oath..	"	" " " "	275
16.	Report Col. Mason to Adj.-General................ Ex. Doc. p. 534.		Spring of 1848.	

SALE OF BARRAS.

1846.	Dec. 17.	Castillero to Alex. Forbes,	5 barras for	....	$4000	1203
1847.	March 1.	Castro " " "	1 " "	....	800	151
	April 12.	" " " "	1 " "	....	1000	154
	" 16.	" " " "	1 " "	....	1000	157
	May 10.	" " " "	1 " "	....	1000	160
	Sept. 14.	Robles to Jas. A. Forbes,	2 " "	....	3860	172
	Dec. 27.	Padre Real to A. Forbes,	3 " "	....	4500	149
1848.	April 14.	Robles to Walkinshaw	2 " "	....	7000	173
1850.	Aug. 9.	J. A. Forbes to Parrott,	1 " "	....	24000	179
1855.	March 30.	" to W. E. Barron	1 " "	....	30000	189
1856.	Dec. 7.	Jecker, Torre & Co. Barron, F. & Co.—all interest.				192
1856.	Oct. 2.	Castillero to Billings......................................				519
"	Dec. 17.	Billings to W. E. Barron..................................				520

www.ingramcontent.com/pod-product-compliance
Lightning Source LLC
LaVergne TN
LVHW021308110826
845150LV00003B/525

* 9 7 8 1 4 2 5 5 5 9 0 5 2 *